D1165460

Control Engineering

Series Editor

William S. Levine
Department of Electrical and Computer Engineering
University of Maryland
College Park, MD 20742-3285
USA

Editorial Advisory Board

Warren E. Dixon
Aman Behal
Darren M. Dawson
Siddharth P. Nagarkatti

Nonlinear Control
of Engineering Systems
A Lyapunov-Based Approach

Birkhäuser
Boston • Basel • Berlin

Warren E. Dixon
Oak Ridge National Laboratory
Engineering Science and
  Technology Division
Oak Ridge, TN 37831
USA

Aman Behal
Clemson University
Department of Electrical and
  Computer Engineering
Clemson, SC 29634-0915
USA

Darren M. Dawson
Clemson University
Department of Electrical and
  Computer Engineering
Clemson, SC 29634-0915
USA

Siddharth P. Nagarkatti
MKS Instruments
Advanced Technology Group
Methuen, MA 01844
USA

Library of Congress Cataloging-in-Publication Data

Nonlinear Control of engineering systems : a Lyapunov-based approach / Warren E. Dixon
... [et al.].
    p. cm. – (Control engineering)
    Includes bibliographical references and index.
    ISBN 0-8176-4265-X (alk. paper) – ISBN 3-7643-4265-X (alk. paper)
      1. Automatic control. 2. Nonlinear control theory. 3. Lyapunov functions. I. Dixon,
Warren E., 1972- II. Control engineering (Birkhäuser)

    TJ213.N568 2003
    629.8'36–dc21

                                     2003045389
                                             CIP

Printed on acid-free paper.
©2003 Birkhäuser Boston

Birkhäuser ®

ISBN 0-8176-4265-X   SPIN 10852085
ISBN 3-7643-4265-X

Typeset by the authors.
Printed in the United States of America.

9 8 7 6 5 4 3 2 1

Birkhäuser Boston • Basel • Berlin
A member of BertelsmannSpringer Science+Business Media GmbH

To my beautiful son, Ethan Noel Dixon
—Warren

To my parents, Swaran and Ashwani K. Behal
—Aman

To my faithful wife, Dr. Kim Dawson
—Darren

To my loving wife, Meghana S. Nagarkatti
—Siddharth

Contents

Preface

The properties of linear systems provide very powerful design and analysis tools but the behavior of a nonlinear system may be much more complex. By neglecting nonlinear behaviors, resulting control designs can have unpredictable stability and degraded performance results. In Chapter 1, several simple examples are provided to highlight some of the pitfalls experienced by linear systems theory because of their inability to encapsulate and compensate for these effects, including the limitations of linearization, loss of tracking performance, and peaking phenomena.

Given the strong impetus to study nonlinear systems, many texts have been written on this subject. Some texts target the theoretical mathematical issues with little treatment of physical applications, while other texts target specific application areas. The authors of this text were motivated by the desire to produce a reference book that provides a balance between mathematically rigorous theoretical development and the application of various Lyapunov-based techniques to a broad class of practical engineering systems. Many applications could be addressed by this methodology but this text is not intended (nor could be) a comprehensive exposition of all engineering systems. Rather, fundamental development is provided to target some classic problems as well as emerging engineering applications. As shown by the manner in which the methods are applied and explained, the authors intend this reference book to provide insight into control designs for additional engineering applications.

In Chapter 2, control applications for mechanical systems are examined. In the first section of the chapter, several adaptive controllers are developed and experimentally demonstrated for an autobalancing tracking application, which is motivated by the desire to develop an alternative means to mitigate the vibrational effects of rotating systems. Unmanned surface vessels are becoming an increasingly important tool in marine applications (e.g., offshore oil industry), and so several adaptive automatic ship control systems are developed in the second section. Based on the development of the full-state feedback adaptive controllers, design techniques are also developed that illustrate how an output feedback controller can be constructed to improve reliability and/or reduce the noise, inherent in velocity signals. In the third section of Chapter 2, design techniques based on a unit quaternion formulation are presented for the tracking control problem for a general class of Euler-Lagrange systems. As an example, these techniques are applied to an unmanned underwater vehicle application.

In Chapter 3, controllers are developed for electric machines, described by electrical subsystem dynamics, a torque coupling that represents the electrical to mechanical energy conversion, and mechanical subsystem dynamics. In the first section of this chapter, a brief review of standard field-oriented control schemes for induction motors is provided. Through systematic alterations of the legacy control methodology, nonlinear techniques are presented that illustrate how the injection of nonlinear control elements can be used to yield improved stability results and performance characteristics. An adaptive extension is also provided to address parametric uncertainty associated with the mechanical load. To address applications that require a high-precision machine, the second section of this chapter uses a general nonlinear model of a switched reluctance motor (SRM) to develop an adaptive controller and an associated commutation strategy. This strategy is based on the full-order model (i.e., the electrical dynamics are not neglected), compensates for uncertainty in electromechanical model, uses a flux linkage model that includes magnetic saturation effects, and eliminates all control singularities (i.e., the voltage control input remains bounded for all operating conditions). The performance of this strategy is demonstrated on an SRM testbed. Motivated by the increasing number of commercial high-performance applications in the domain of rotating machinery, the third section of this chapter targets the development of a nonlinear tracking controller with an associated commutation strategy for a six-degree-of-freedom active magnetic bearing system.

In Chapter 4, several different robotic applications are examined. In the first section, a repetitive learning control strategy using Lyapunov-based illustrates how additional system information can be exploited in a control design. As an example, the repetitive nature of some robotic tasks is exploited to develop a controller to mitigate disturbance effects and improve link position tracking performance. Since the learning-based controller estimate is generated from a Lyapunov-based stability analysis, development is also presented to illustrate how additional control terms can be integrated to compensate for nonperiodic components associated with the unknown robot dynamics. Experimental results are provided to illustrate these concepts. In many industrial/manufacturing applications, a robot manipulator is required to make contact with the environment. To address these applications, an adaptive tracking controller is developed, which ensures asymptotic position/force tracking performance despite parametric uncertainty for robot manipulators under contact force constraints. An extension is provided that also describes how the full-state feedback controller can be modified to eliminate the need for velocity measurements. Experimental results of the output feedback controller are presented. Advances in control and sensor technologies have spawned new robotic applications where the robot is required to operate in unstructured environments. Camera systems can provide a passive noncontact sense of perception in unstructured environments, so the third section of this chapter targets the development of fixed camera visual servoing control designs in which visual information is embedded directly in the feedback-loop of the control algorithm. An extension is provided that incorporates redundant robot manipulators, and an adaptive controller is also developed for the camera-in-hand problem for applications not well suited to the fixed camera configuration.

Chapter 5 focuses on various aerospace applications. Based on the desire to control the attitude of an object without singularities, a full-state feedback, quaternion-based attitude tracking controller is developed in the first section. This controller compensates for the nonlinear dynamics of a rigid spacecraft with parametric uncertainty in the inertia matrix. An output feedback extension is also provided to eliminate additional sensor payload. Typically, spacecraft use separate devices to provide energy storage and attitude control. In the second section of this chapter, an adaptive integrated power and attitude control system is developed for a nonlinear spacecraft model where the body torque is produced by flywheels operated in a reaction wheel mode. The adaptive quaternion-based controller forces a spacecraft to track a desired attitude trajectory while simultaneously providing exponential energy/power tracking. Another key technological concept in aerospace systems is the distributed functionality of a large spacecraft among smaller, less-expensive, cooperative spacecraft. Flying two or more spacecraft in precise formation is typically referred to as multiple satellite formation flying (MSFF) and presents a number of complex challenges. In the final section of this chapter, the full nonlinear dynamics describing the relative positioning of MSFF is used to develop a Lyapunov-based nonlinear adaptive control law that guarantees asymptotic convergence of the position tracking error in the presence of unknown, constant, or slow-varying spacecraft masses, disturbance forces, and gravity forces.

The engineering systems described in the previous chapters are fully actuated (i.e., the number of control inputs (actuators) equal the number of degrees of freedom). However, because of actuator failures or various construction constraints some applications are underactuated (i.e., the degrees of freedom exceed the number of control inputs). In Chapter 6, various design strategies are examined that exploit some coupling between the unactuated states and the actuated states to achieve the control objective. The first underactuated application examined is an overhead crane system. Motivated by the desire to achieve fast and precise payload positioning while mitigating performance and safety concerns, several controllers are developed in the first section of this chapter that exploit the coupling between the payload and gantry dynamics. In the second section, a dynamic oscillator-based control strategy is developed for the position and attitude tracking and regulation problems for vertical take-off and landing (VTOL) aircraft modeled by underactuated dynamics that are nonlinear, nonminimum-phase, and subject to nonholonomic (nonintegrable) constraints. As an extension to the design approach used to address the VTOL problem, an automotive steering problem and an underactuated surface vessel problem are also examined. In the third section of this chapter, the kinematics of an underactuated satellite are formulated in terms of the constrained unit quaternion, and a similar (quaternion-based) dynamic oscillator control structure is developed to achieve tracking and regulation. As an extension to the design, an integrator backstepping technique is used to incorporate the dynamic model of an axisymmetric satellite.

The material in this text is intended for readers with a background in undergraduate systems theory and is written to be beneficial for students and practicing research and development engineers in the area of controls. The material in this book (unless otherwise specified) is based on the authors' control systems research directed at engineering applications.

To produce this text, significant assistance has been provided by various individuals. Specifically, the authors would like to express our most sincere appreciation to Marcio de Quieroz of Louisiana State University and Erkan Zergeroglu of the Gebze Institute of Technology for their contributions and support of this work. We also acknowledge Vikram Kapila of Polytechnic University for his contributions and technical insight regarding aerospace systems, Oak Ridge National Laboratory research staff members John Jansen, Lonnie Love, and Francois Pin, and the support by the following past and present graduate students of the Department of Electrical Engineering at Clemson University, whose efforts helped to realize this book: Bret Costic, Youngchun Fang, Matthew Feemster, Jian Chen, Vilas Chitrakaran, Michael McIntyre, Pradeep Setlur, and Bin Xian.

Knoxville, Tennessee   Warren E. Dixon
Clemson, South Carolina   Aman Behal
Clemson, South Carolina   Darren M. Dawson
Methuen, Massachusetts   Siddharth P. Nagarkatti

Nonlinear Control of Engineering Systems

A Lyapunov-Based Approach

1
Introduction

1.1 Pitfalls of Linear Control

A common engineering practice is to assume that a system can be described
by a set of linear differential equations for some operating range of interest
as follows

$$\dot{x} = Ax + Bu \tag{1.1}$$

where $x(t)$ denotes the states of the system, A, B denote time invariant matrices, and $u(t)$ denotes the control input. Based on the assumption that
(1.1) accurately describes the system behavior, the control practitioner can
then exploit various properties from linear control theory. As stated in [10],
in the absence of the input signal (i.e., the unforced system), these properties include (1) a unique equilibrium point if the A matrix is nonsingular,
(2) the equilibrium point is stable if the eigenvalues of A have negative real
roots, and (3) the linear differential equations can be solved analytically,
allowing the transient response to be explicitly determined. When a control input $u(t)$ is present, linear time invariant systems exhibit properties
including (1) superposition, (2) asymptotic stability of the unforced system
ensures bounded-input bounded-output stability, and (3) a sinusoidal input
leads to a sinusoidal output of the same frequency.

While the properties of linear systems provide powerful design and analysis tools, the behavior of a nonlinear system may be much more complex
than a linear model can encapsulate. By neglecting nonlinear behaviors, re-

sulting control designs can have unpredictable stability and degraded performance results. In the following sections, several examples are provided to highlight some of the pitfalls experienced by linear systems theory due to the inability to encapsulate and compensate for these effects. Specifically, examples are provided to illustrate the limitations of linearization and the potential for counterintuitive destabilizing effects.

1.1.1   Limitations of Linearization

As described previously, linear control design and the associated properties of linear theory are based on the underlying assumption that the system under consideration can be effectively described by a set of linear differential equations. Unfortunately, the resulting approximate linear equations are typically only valid in a neighborhood of an operating point, and hence, only describe the behavior of the actual system in that vicinity. In addition, some system phenomena cannot be captured by a set of linear equations. For example, some natural nonlinear effects are discontinuous and the response from these effects cannot be approximated by a linear function. These discontinuous nonlinear effects (e.g., Coulomb friction, backlash, hysteresis) are often called "hard nonlinearities." Although hard nonlinearities cannot be linearly approximated, physical systems that exhibit these effects are often approximated by neglecting these effects. Additional nonlinear phenomena include finite escape time, limit cycles, and chaos (a more complete description of these and other phenomena are provided in [5, 10]). As previously stated, the stability and performance of linear control systems may be compromised by neglecting such effects.

To illustrate the impact of the loss of system information through linearization, consider the following first-order scalar system [10]

$$\dot{x} = -x + x^2 \tag{1.2}$$

where $x(t = 0) = x_0$. After linearizing (1.2) about the stable equilibrium point $x(t) = 0$, the resulting dynamics and respective solution can be determined as follows

$$\dot{x}(t) = -x(t) \qquad x(t) = x_0 \exp(-t) \ . \tag{1.3}$$

Hence, (1.3) indicates that for any initial condition x_0 the system will exponentially converge to the stable equilibrium point. However, by setting the left side of (1.2) to 0, it is clear that two equilibrium points exist (at $x(t) = 0$ and $x(t) = 1$) for the nonlinear system. To illustrate the impact of neglecting the second equilibrium point, the solution to the nonlinear system can be determined by integrating (1.2) as follows

$$x(t) = \frac{x_0 \exp(-t)}{1 - x_0 + x_0 \exp(-t)} . \tag{1.4}$$

From (1.4), it is clear that if the initial condition is $x_0 < 1$, then the system will converge to the stable equilibrium point as determined by the linear system model. However, if the initial condition is $x_0 > 1$, the system will rapidly escape towards infinity within a finite time (i.e., finite escape time). To illustrate this effect, Figure 1.1 depicts the response of $x(t)$ for various initial conditions. From Figure 1.1, the rapid escape towards infinity can be observed for initial conditions $x_0 > 1$.

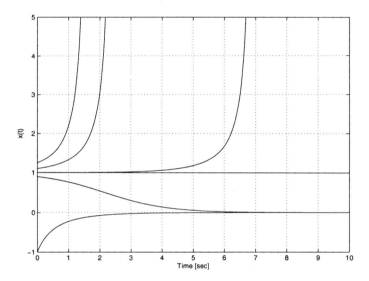

FIGURE 1.1. Nonlinear system response for various initial conditions.

In addition to the potential for unpredictable (from the linearized equations) responses as illustrated by the previous example, linearization may alter the structure of the system so that it becomes intractable. For example, consider the following nonlinear system

$$\begin{bmatrix} \dot{x}_1 \\ \dot{x}_2 \\ \dot{x}_3 \end{bmatrix} = \begin{bmatrix} \cos x_3 & 0 \\ \sin x_3 & 0 \\ 0 & 1 \end{bmatrix} \begin{bmatrix} u_1 \\ u_2 \end{bmatrix} . \tag{1.5}$$

The system introduced in (1.5) is a classic set of equations that are used to describe the motion of systems such as a unicycle or wheeled mobile robots (see [1] for examples of other physical systems). Clearly, these physical systems are controllable (see [2] for a mathematical examination of the controllability); however, if the system dynamics are linearized about the point $x_3(t) = 0$ as follows

$$
\begin{bmatrix} \dot{x}_1 \\ \dot{x}_2 \\ \dot{x}_3 \end{bmatrix} = \begin{bmatrix} 1 & 0 \\ 0 & 0 \\ 0 & 1 \end{bmatrix} \begin{bmatrix} u_1 \\ u_2 \end{bmatrix} ,
\tag{1.6}
$$

then the linearized dynamics for $x_2(t)$ become uncontrollable.

Another limitation is the loss of tracking performance caused by the requirement for the controller to be formulated based on a linear model of the system. To illustrate the potential for degraded performance, consider the following second-order scalar system

$$
m(1 + q^2)\ddot{q} + bq\dot{q} = u
\tag{1.7}
$$

where it is assumed that $q(t)$ and $\dot{q}(t)$ are measurable states, and $m = 1.5$ and $b = 2$ are known positive constants. The objective of a tracking control problem is to eliminate the mismatch between $q(t)$ and a desired time-varying trajectory, denoted by $q_d(t)$. To quantify the mismatch, a tracking error denoted by $e(t)$ is defined as follows

$$
e = q_d - q .
\tag{1.8}
$$

By neglecting the nonlinear terms, the system in (1.7) can be reduced to the following simple linear system

$$
m\ddot{q} = u .
\tag{1.9}
$$

Based on (1.9), a typical linear controller could be designed as follows

$$
u = \underbrace{m\ddot{q}_d}_{feedforward} + \underbrace{k_v\dot{e} + k_pe}_{feedback}
\tag{1.10}
$$

to yield the following stable closed-loop dynamics for the linear system

$$
m\ddot{e} + k_v\dot{e} + k_pe = 0
\tag{1.11}
$$

where $\ddot{q}_d(t)$ denotes the second time derivative of the desired trajectory, and k_v and k_p are positive constant control gains. By selecting $k_v = k_p = 1000$, the tracking performance of (1.10) for the linear model in (1.9) with $q_d(t) = 5 + 2\sin(t)$ is illustrated in Figure 1.2 with the associated control effort in Figure 1.3.

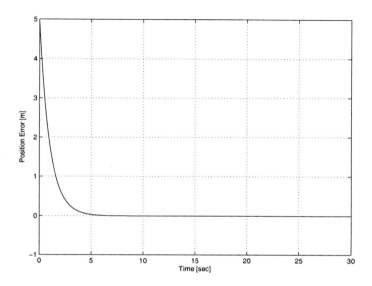

FIGURE 1.2. Position tracking performance for a linear controller applied to a linear model.

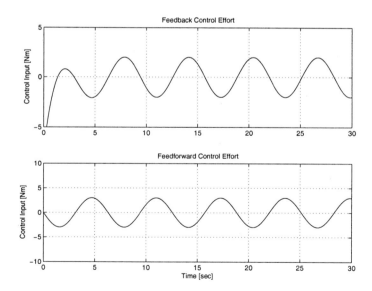

FIGURE 1.3. Feedback and feedforward control effort.

As illustrated in Figure 1.2, the linear controller can be applied to the linear model in (1.9) to yield perfect tracking response. However, when the same controller is applied to the actual nonlinear system given in (1.7), the tracking performance degrades as illustrated in Figure 1.4 with the associated control effort in Figure 1.5. The degraded tracking performance illustrated in Figure 1.4 is due to the fact that the nonlinear effects have been neglected and act as disturbances on the system. While the control gains introduced in (1.10) could have been adjusted to achieve different performance, perfect tracking of the reference input cannot be achieved (notice the steady state oscillations about zero depicted in Figure 1.4) because of the absence of a nonlinear feedforward term that can exactly compensate for the nonlinear components of the system. As a means to develop controllers that do not simply neglect the nonlinear terms, modern linear control practitioners employ various robust linear control approaches. The goal of these approaches is not to neglect the nonlinear terms; rather, the goal is to bound the effects of the nonlinear components. Unfortunately, these approaches are based on worst-case scenarios that require high-gain or high-frequency feedback, and the resulting performance is still degraded by the requirement to ultimately reduce the system model to a linear system to enable the use of linear control design tools. The subsequent sections of this chapter illustrate how controllers can be developed for this example problem from a nonlinear Lyapunov-based methodology to achieve perfect tracking.

1.1.2 Dangers of Destabilization

In some cases the use of linear controllers may result in counterintuitive destabilizing effects. For example, a peaking phenomenon occurs in linear systems that can have a destabilizing effect (potentially resulting in finite escape time) on the nonlinear subsystem dynamics. To illustrate this concept, consider the coupled partially linear system depicted in Figure 1.6 that is described by the following dynamics [11, 12]

$$\dot{x} = f(x, y) \tag{1.12}$$

$$\dot{y} = Ay + Bu \tag{1.13}$$

where $x(t) \in \mathbb{R}^n$, $y(t) \in \mathbb{R}^m$, and $u(t) \in \mathbb{R}^p$. The following assumptions are made for the system introduced in (1.12) and (1.13).

Assumption 1.1: The pair (A, B) is assumed to be controllable.

Assumption 1.2: The nonlinear function $f(\cdot)$ is assumed to be first-order differentiable with respect to time.

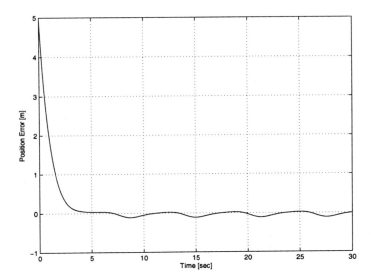

FIGURE 1.4. Position tracking performance for a linear controller applied to a nonlinear model.

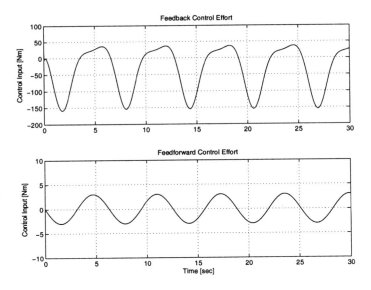

FIGURE 1.5. Feedback and feedforward control effort.

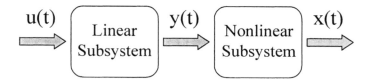

FIGURE 1.6. Partially linear system.

Assumption 1.3: The zero dynamics (i.e., $\dot{x} = f(x, 0)$) have 0 as a globally asymptotically stable equilibrium.

Based on (1.12) and (1.13) and Assumptions 1.1-1.3, it seems intuitive that a linear controller can be designed to exponentially drive the dynamics for $y(t)$ to zero, resulting in global asymptotically stable zero dynamics for the nonlinear subsystem. However, this strategy can lead to destabilization and even finite escape time of the nonlinear dynamics. To illustrate this point, consider the following simple example [12]

$$\dot{x} = \frac{-(1 + y_2)}{2} x^3 \tag{1.14}$$

$$\dot{y}_1 = y_2 \qquad \dot{y}_2 = u . \tag{1.15}$$

From (1.14), it is clear that Assumption 1.3 is valid. By designing a linear control input as follows

$$u = -a^2 y_1 - 2a y_2 , \tag{1.16}$$

repeated eigenvalues of the closed-loop linear subsystem will result at $-a$. Linear analysis methods can be used to determine the exact solution for $y_2(t)$ as follows

$$y_2 = -a^2 t e^{-at} . \tag{1.17}$$

From the solution given in (1.17), it is clear that the dynamics for $|y_2(t)|$ rise to a peak and then exponentially decay to zero. By setting the time derivative of $y_2(t)$ to zero, it can be determined that the peak occurs at $t = \frac{1}{a}$. Given (1.17), it is clear that larger values of a will drive $y_2(t)$ to zero faster. Hence, from Assumption 1.3 it seems that larger values of a will allow the nonlinear system to be stabilized faster. However, as stated in [6] this is simply a "high-gain mirage." To clarify this mirage, we can substitute (1.17) into (1.14) and then integrate the resulting expression as follows

$$x^2(t) = \frac{x_0^2}{1 + x_0^2 \left(t + (1 + at) \exp(-at) - 1\right)} \tag{1.18}$$

where $x(t = 0) = x_0$. The destabilizing effects of the peaking phenomena can now be understood by substituting values for x_0, a, and t into (1.18). For example, for $a = 10$ and $x_0^2 = 2.176$, the response for $x^2(t \cong 0.5)$ becomes unbounded (i.e., finite escape time). Interestingly, by approximating e^{-at} by $(1 - at)$ it is easy to verify that increasing the feedback gain a causes the system to escape at a faster rate! Additional examples and a discussion of these phenomena are provided in [6, 11, 12]. In Section 13.2 of [5], nonlinear (Lyapunov-based) control examples are provided to illustrate how the integrator backstepping technique can be used to achieve global stabilization of triangular systems such as (1.12) and (1.13).

1.2   Lyapunov-Based Control

In the previous section, various pitfalls of linear control theory were illustrated through several examples. These examples highlight the intuitive idea that control strategies that neglect some system effects have the potential for unexpected stability results and performance degradation. While linear control strategies may provide acceptable results when the effects of the neglected phenomena are not dominant (either naturally or due to imposed constraints on the operating range of the system), it seems apparent that control strategies that incorporate known system effects can provide significant advantages for a broader class of systems. Based on this concept, researchers have been motivated to investigate various methodologies for nonlinear control design and analysis. Although this investigation has led to several important methodologies (e.g., singular perturbation, describing functions, phase plane analysis), Lyapunov-based design[1] and analysis approaches have emerged as some of the most flexible, intuitive, and powerful tools.

1.2.1   Exact Model Knowledge Example

If the nonlinear effects can be exactly modeled (which is rarely possible from a practical standpoint), a potential approach is to exactly cancel the nonlinear dynamic effects with a model-based feedforward compensation term as follows

$$u = \underbrace{f_f(q, \dot{q})}_{feedforward} + \underbrace{f_b(q, \dot{q})}_{feedback}$$

(1.19)

---

[1] Lyapunov theory and its derivatives are named after the Russian mathematician and engineer Aleksandr Mikhailovich Lyapunov (1857–1918).

where $f_f(q, \dot{q})$ denotes the nonlinear feedforward term, and $f_b(q, \dot{q})$ denotes feedback terms. The structure of the feedforward and feedback components is motivated by the physics of the nonlinear system, the resulting closed-loop systems, and by the subsequent analysis. Lyapunov methods are based on a very simple result from calculus that if a differentiable function is positive for all time, and its time derivative is negative or zero, then the function will decrease towards zero or some positive constant. This result provides the basis for the iterative design philosophy that is used to construct (1.19) to shape the resulting closed-loop error systems. For example, a possible control design for the system given in (1.7) could be developed as follows

$$u = \underbrace{bq\dot{q} + m(1 + q^2)\ddot{q}_d}_{feedforward} + \underbrace{(1 + q^2)(kr + \alpha\dot{e})}_{feedback} .$$ (1.20)

In (1.20), α, k denote positive control gains, and $r(t)$ is defined as follows

$$r = \dot{e} + \alpha e .$$ (1.21)

Insight into the motivation for the design in (1.20) can be obtained through the following theorem and Lyapunov-based analysis.

Theorem 1.1 *The controller introduced in (1.20) provides global exponential tracking in the sense that*

$$e(t) = e(0)\exp(-\alpha t) + \frac{\dot{e}(0) + \alpha e(0)}{\alpha - \frac{k}{m}}\left(\exp\left(-\frac{k}{m}t\right) - \exp(-\alpha t)\right) .$$ (1.22)

Proof: To prove the result in Theorem 1.1, consider the nonnegative function $V(t) \in \mathbb{R}$ defined as follows

$$V = \frac{1}{2}mr^2 .$$ (1.23)

After taking the time derivative of (1.23) the following expression can be obtained

$$\dot{V} = r\left(\ddot{q}_d - \frac{u - bq\dot{q}}{(1 + q^2)} + \alpha\dot{e}\right)$$ (1.24)

where (1.7), (1.8), and the time derivative of (1.21) were used. After substituting (1.20) into (1.24), the following expression is obtained

$$\dot{V} = -\frac{2}{m}kV$$ (1.25)

where (1.23) was utilized. By integrating (1.25), the following result can be obtained

$$r(t) = r(0) \exp\left(-\frac{k}{m}t\right) \tag{1.26}$$

where (1.23) was utilized. Based on the definition of $r(t)$ introduced in (1.21) and the result in (1.26), linear methods can be used to prove the result in (1.22) (see also Lemma A.14 of Appendix A). □

Remark 1.1 *For the particular case when $\alpha = \frac{k}{m}$, a repeated root is obtained, and the expression in (1.22) becomes*

$$e(t) = [e(0) + (\dot{e}(0) + \alpha e(0))\,t]\exp(-\alpha t) \text{ for } \alpha = \frac{k}{m} . \tag{1.27}$$

To verify this result, the expression in (1.26) can be substituted into (1.21) and then integrated.

Remark 1.2 *Linear analysis arguments could have also been utilized to examine the stability of the tracking error. For example, by substituting (1.20) into (1.7), the following expression can be obtained*

$$\ddot{e} + \frac{(k+\alpha)}{m}\dot{e} + \frac{k\alpha}{m}e = 0 . \tag{1.28}$$

The expression in (1.28) can be integrated to yield the same expression as in (1.22).

1.2.2 Simulation Results

After numerically simulating the controller designed in (1.20) with $\alpha = k = 1$ for the system in (1.7) with the same desired trajectory as in the previous linear control example, the position tracking error and control effort depicted in Figure 1.7 and Figure 1.8, respectively, are obtained. From a comparison of the results illustrated in Figure 1.4 and Figure 1.5 with the results illustrated in Figure 1.7 and Figure 1.8, it is clear that the steady state error present in Figure 1.4 can be eliminated (theoretically) at the expense of increased control effort due to the nonlinear feedforward component.

Clearly, the performance of the approach in (1.19) is inherently tied to the ability of the model to accurately represent the physical system. However, developing a set of coupled differential equations that accurately represents the dynamic model is a challenging task. The challenge arises because the parameters of the model may not be known (or known precisely) and they may change over time (these model uncertainties are often referred to as

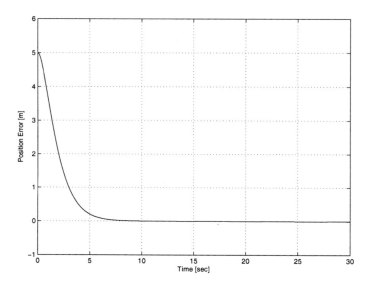

FIGURE 1.7. Position tracking error for a nonlinear controller.

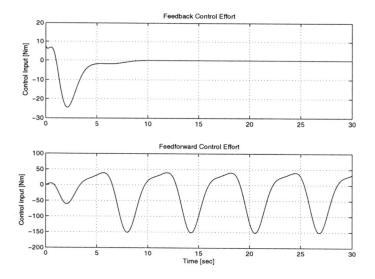

FIGURE 1.8. Feedback and feedforward control effort.

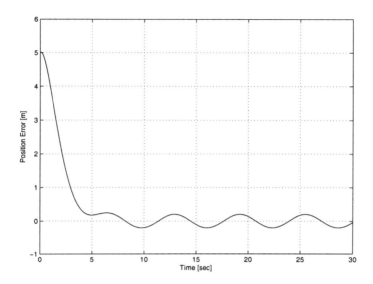

FIGURE 1.9. Position tracking error for a nonlinear controller with parametric uncertainty.

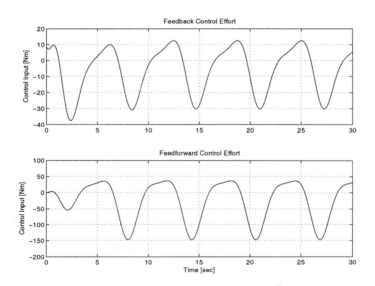

FIGURE 1.10. Feedback and feedforward control effort.

parametric uncertainty or structured uncertainty [3]). In addition to parametric uncertainty, the effects of some phenomena may be very difficult or impossible to accurately model, and hence, the effects may be neglected from the model (these disturbances are often referred to as unstructured uncertainty [3]). To illustrate the effect of model uncertainty, a numerical simulation was performed for the controller given in (1.20), where the constant parameters m and b of (1.7) are increased by 20% (however, the controller is based on the original values). Using the same control gains, the resulting tracking error and control effort are depicted in Figure 1.9 and Figure 1.10, respectively. From Figure 1.9, it is clear that the parameter mismatch results in degraded tracking performance.

1.2.3   Adaptive Example

To account for parameter mismatch, adaptive control designs can be constructed as follows

$$u = \underbrace{\hat{f}_f(q,\dot{q})}_{feedforward} - \underbrace{f_b(q,\dot{q})}_{feedback} \tag{1.29}$$

where $\hat{f}_f(q,\dot{q})$ denotes a partial knowledge feedforward term. Based on this approach, the controller introduced in (1.20) can be redesigned as follows

$$u = \underbrace{\hat{b}q\dot{q} + \hat{m}(1+q^2)\ddot{q}_d}_{feedforward} + \underbrace{(1+q^2)(kr+\alpha\dot{e})}_{feedback} \tag{1.30}$$

where the parameter estimates $\hat{b}(t)$, $\hat{m}(t) \in \mathbb{R}$ are generated by the following differential equations

$$\dot{\hat{b}}(t) = \frac{\Gamma_1 q\dot{q}r}{1+q^2} \qquad \dot{\hat{m}}(t) = \Gamma_2 \ddot{q}_d r \tag{1.31}$$

where Γ_1, $\Gamma_2 \in \mathbb{R}$ denote positive adaptive update gains. The stability of the tracking controller in (1.30) and (1.31) can be proven through a Lyapunov-based analysis as in the following theorem.

Theorem 1.2 *The controller introduced in (1.20) provides global asymptotic tracking in the sense that*

$$\lim_{t \to \infty} e(t) = 0 . \tag{1.32}$$

Proof: To prove the result in Theorem 1.2, consider the nonnegative function $V(t) \in \mathbb{R}$ defined as follows

$$V = \frac{1}{2}mr^2 + \frac{1}{2\Gamma_1}\left(b-\hat{b}\right)^2 + \frac{1}{2\Gamma_2}(m-\hat{m})^2 . \tag{1.33}$$

After taking the time derivative of (1.23) the following expression can be obtained

$$\dot{V} = -kr^2 + r\left((m-\hat{m})\ddot{q}_d + \left(b-\hat{b}\right)q\dot{q}\right) \qquad (1.34)$$

$$-\frac{1}{\Gamma_1}\left(b-\hat{b}\right)\dot{\hat{b}} - \frac{1}{\Gamma_2}(m-\hat{m})\dot{\hat{m}}$$

where (1.7), (1.8), the time derivative of (1.21), and (1.30) were used. After substituting (1.31) into (1.33), the following simplified expression can be obtained

$$\dot{V} = -kr^2 . \qquad (1.35)$$

Based on the expressions in (1.33) and (1.35), basic results from calculus can be used to conclude that the signals in $V(t)$ remain bounded (i.e., $r(t)$, $\hat{b}(t)$, $\hat{m}(t)$). From (1.7), (1.8), (1.21), (1.30), and (1.31), analysis can be performed to prove that $\dot{r}(t)$, $\ddot{q}(t)$, $u(t)$, $\dot{\hat{b}}(t)$, $\dot{\hat{m}}(t)$ are bounded for all time, and hence, $r(t)$ is uniformly continuous. Since $r(t)$ is uniformly continuous, $\dot{V}(t)$ is also uniformly continuous; therefore, since $V(t)$ is bounded, $\dot{V}(t) \to 0$ as $t \to \infty$. From (1.35), it is clear that $\dot{V}(t) \to 0$ only when $r(t) \to 0$; hence, from (1.21) it can be determined that $\dot{V}(t) \to 0$ only when $\dot{e}(t)$, $e(t) \to 0$ (see Lemma A.15 of Appendix A). Therefore, both global stability and convergence of the tracking error as indicated in (1.32) can be concluded.   □

Remark 1.3 *While $r(t)$, $\dot{e}(t)$, $e(t) \to 0$ as $t \to \infty$, the adaptive algorithm in (1.30) and (1.31) does not ensure that the parameter estimates converge to the actual values. A constraint that the desired trajectory be persistently exciting (i.e., sufficiently rich) may potentially be placed on the system to ensure that $(m - \hat{m}(t))$, $\left(b - \hat{b}(t)\right) \to 0$; hence, as stated in [9], an adaptive controller may be modified so as to guarantee exponential tracking.*

Remark 1.4 *The simplified analysis in Theorem 1.2 is provided to demonstrate the advantages of Lyapunov-based control designs in the presence of parametric uncertainty. In the subsequent chapters, various Lyapunov-based analysis tools will be more formally constructed. For an overview of nonlinear approaches that can be applied to compensate for unstructured uncertainty, see [8].*

1.2.4   Simulation Results

A numerical simulation was performed for the adaptive tracking controller designed in (1.20) for the system in (1.7) with the same desired trajectory as in the previous linear control example where

$$\alpha = 0.5, \quad k = 10, \quad \Gamma_1 = 2.0, \quad \Gamma_2 = 3.0 . \tag{1.36}$$

The resulting position tracking error and control effort are depicted in Figure 1.11 and Figure 1.12, respectively. The parameter estimates $\hat{b}(t)$ and $\hat{m}(t)$ are depicted in Figure 1.13. From a comparison of the results illustrated in Figure 1.9 and Figure 1.10 with the results illustrated in Figure 1.11 and Figure 1.12, it is clear that the steady state error present in Figure 1.9 can be eliminated because the adaptive parameter estimates compensate for the mismatch between the actual parameters and the estimated parameters. In comparison with the results in Figure 1.7, it is clear that the adaptive controller exhibits a longer transient. This observation can be attributed to the fact that the performance is linked to the speed at which the parameter estimates seem to converge to values that best accommodate for the parameter mismatch. Also note that the stability result for the adaptive controller becomes asymptotic as compared with the exponential results obtained in (1.22).

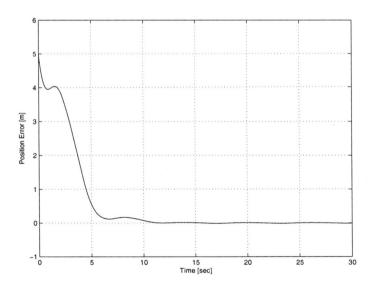

FIGURE 1.11. Position tracking error for an adaptive nonlinear controller.

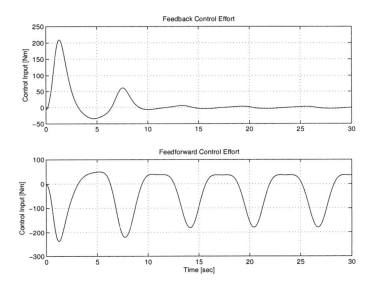

FIGURE 1.12. Feedforward and feedback control effort.

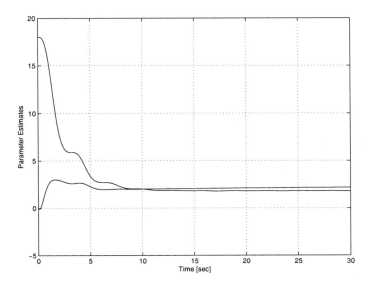

FIGURE 1.13. Parameter estimates.

1.3 Summary

In this chapter, motivation is provided for the use of a nonlinear Lyapunov-based approach to control design. To this end, several potential pitfalls associated with linear systems theory were demonstrated through various examples. Several Lyapunov-based designs were then presented to illustrate how improved performance can be achieved by using a feedforward control term to compensate for the nonlinear effects. From the analysis provided for Theorem 1.1 and Theorem 1.2, it can be determined that the control design and analysis are formulated through an iterative process. This process is facilitated by creative foresight into the interplay between the control input (including possible adaptive update signals) and the resulting error system embedded in the analysis. In subsequent chapters, a variety of Lyapunov-based design and analysis tools are developed and applied to different engineering systems.

References

[1] A. Bloch, M. Reyhanoglu, and N. McClamroch, "Control and Stabilization of Nonholonomic Dynamic Systems," *IEEE Transactions on Automatic Control*, Vol. 37, No. 11, Nov. 1992, pp. 1746–1757.

[2] C. Canudas de Wit, K. Khennouf, C. Samson, and O. J. Sordalen, "Nonlinear Control for Mobile Robots," in *Recent Trends in Mobile Robots*, ed. Y. Zheng, River Edge, NJ: World Scientific, 1993.

[3] J. J. Craig, *Adaptive Control of Mechanical Manipulators*, Reading, MA: Addison-Wesley Publishing Co., 1988.

[4] P. C. Hughes, *Spacecraft Attitude Dynamics*, New York, NY: John Wiley, 1986.

[5] H. K. Khalil, *Nonlinear Systems*, Upper Saddle River, NJ: Prentice-Hall, 1996.

[6] P. Kokotovic, "The Joy of Feedback: Nonlinear and Adaptive," *IEEE Control Systems Magazine*, Vol. 12, June 1992, pp. 7–17.

[7] M. A. Lyapunov, "Problème général de la stabilité du mouvement," *Ann. Fac. Sci. Toulouse*, Vol. 9, 1907, pp. 203–474. (Translation of the original paper published in 1892 in *Comm. Soc. Math. Kharkow* and reprinted as Vol. 17 in *Ann. Math Studies*, Princeton, NJ: Princeton University Press, 1949.)

[8] Z. Qu, *Robust Control of Nonlinear Uncertain Systems*, New York, NY: John Wiley, 1998.

[9] J. Slotine, "Putting Physics in Control – The Example of Robotics," *IEEE Control Systems Magazine*, Vol. 8, Dec. 1998, pp. 12–17.

[10] J. -J. E. Slotine and L. Wi, *Applied Nonlinear Control*, Englewood Cliffs, NJ: Prentice-Hall, 1991.

[11] H. J. Sussmann, "Limitations on the Stabilizability of Globally Minimum Phase Systems," *IEEE Transactions on Automatic Control*, Vol. 35, No. 1, Jan. 1990, pp. 117–119.

[12] H. J. Sussmann and P. V. Kokotovic, "The Peaking Phenomenon and the Global Stabilization of Nonlinear Systems," *IEEE Transactions on Automatic Control*, Vol. 36, No. 4, Apr. 1991, pp. 424–440.

2
Mechanical Systems

2.1 Introduction

In this chapter, several different control applications for mechanical systems are examined. The first system discussed is an autobalancing application. A perfectly balanced rotating object (i.e., the center of geometry and center of mass are coincident) will usually not undergo any vibration. However, due to the errors associated with geometric dimensions and the nonhomogeneity of the raw material, the construction of a perfectly balanced object is difficult to achieve using a standard manufacturing process. Due to the difficulty and/or expense required to construct a perfectly balanced object, some amount of vibration can be expected as an object rotates. This vibration can lead to performance degradation and/or failure of the mechanical system. These undesirable vibrational effects are often amplified during high-speed rotation. A simple solution to the imbalance problem is to introduce passive damping via selective placement of ball bearings. However, the use of a passive bearing often leads to an increase in friction, resulting in further degradation of the system performance. An alternative means to mitigate the vibrational effects of rotating systems is to produce frictionless forces (e.g., magnetic forces) that act on the rotating body. These forces can provide an autobalancing capability for the case of high-speed rotation-based systems (e.g., precision grinding, turbines, aircraft propellers, flywheels). However, since the slightest imbalance can induce very large and potentially destabilizing vibrations, an active control system that can generate the desired forces very precisely is needed. Moreover, an active control strategy that would not only be capable of stabilizing these vibrations, but also be able to identify the imbalance-related parameters of the system, would be extremely beneficial. By developing such a controller, the user would have the information necessary to make decisions based upon the imbalance parameters (i.e., shut the system down if it surpasses a predetermined maximum safety threshold). In the first section of this chapter, two adaptive control strategies are developed that address the autobalancing control objective of regulating the center of geometry of a high-speed rotating object that is tracking a desired angular velocity profile. Specifically, the first controller uses a gradient adaptive update law to compensate for the parametric uncertainty in the system, whereas the second controller uses a composite prediction error driven adaptive update law to identify and compensate for the uncertain imbalance parameters, provided a mild persistency of excitation and additional control gain conditions are satisfied. The performance of the second autobalance controller is demonstrated through experimental results.

Since unmanned surface vessels are becoming an increasingly important tool in marine applications (e.g., offshore oil industry), the development of automatic ship control systems has been a topic of considerable interest over the past decade. In the second section of this chapter, two adaptive controllers are developed for dynamically positioned ships (i.e., a ship system where the surge, sway, and yaw are controlled via thrusters and propellers of the ship [29]). The first controller utilizes full-state feedback (i.e., ship position and velocity measurements are available) to achieve global asymptotic tracking despite parametric uncertainty associated with the nonlinear ship dynamics. Motivated by the desire to eliminate the requirement for ship velocity measurements (e.g., due to the desire to improve reliability and/or reduce the noise that is inherent in velocity signals from the control system), an output feedback controller (i.e., the controller only requires the measurement of the ship position) is also developed. The adaptive output feedback controller is developed in tandem with a filter-based velocity estimator to achieve global asymptotic tracking. The performance of the output feedback controller is demonstrated through a simulation study.

In many mechanical applications the desired trajectory of the system is defined in terms of a constant inertial frame. For these applications, motivation exists to perform the control design with respect to the inertial frame (in the so-called task space of the system). However, precise tracking control of the orientation of a mechanical system with respect to an inertial frame (e.g., the orientation of a robot manipulator end-effector with respect to the base) is not straightforward. For example, several parameterizations exist to describe the orientation angles, including three-parameter representations (e.g., Euler angles, Rodrigues parameters) and the four-parameter representation given by the unit quaternion. Whereas the three-parameter representations always exhibit singular orientations (i.e., the orientation Jacobian matrix in the kinematic equation is singular for some orientations), the unit quaternion can be used to represent the orientation of a mechanical system without singularities. Thus, despite significantly complicating the control design, the unit quaternion may proffer some specific advantages when formulating orientation tracking control problems.

Given the motivation for a unit quaternion formulation, a unit quaternion-based control approach is applied to solve the tracking control problem for a general class of Euler-Lagrange systems. Specifically, a full-state feedback unit quaternion-based controller is formulated with respect to the inertial frame (as typically done for robotic systems). As an example application, the dynamics for an unmanned underwater vehicle (UUV) are cast into a similar form as the general control problem. However, for this application a controller is developed with respect to a body-fixed reference frame (as typically done for aerospace and marine applications). Simulation results are provided for the unit quaternion-based control development for the UUV.

2.2 Autobalancing Systems

In this section, two controllers are developed for the autobalancing problem of a rotating unbalanced disk (see Figure 2.1). Specifically, the design of a gradient-based adaptive control law is first presented that achieves global asymptotic tracking of a desired angular velocity profile while regulating the center of geometry of the unbalanced disk. A second controller is then developed that utilizes a desired compensation adaptation law (DCAL) [36] and a gain adjusted forgetting factor (GAFF) [38] to automatically identify the unknown imbalance-related parameters provided a mild persistency of excitation (PE) condition is satisfied. Provided this PE condition is satisfied, the second control strategy achieves exponential stability; whereas, if the PE condition is not satisfied, global asymptotic tracking/regulation is still achieved.

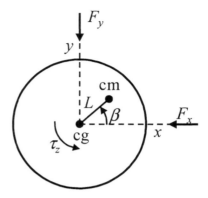

FIGURE 2.1. Diagram of the rotating unbalanced mass.

2.2.1   System Model

The mechanical system considered in this section consists of a rotating unbalanced[1] disk that vibrates in a plane perpendicular to its axis of rotation (see Figure 2.1). A control torque is applied to rotate the disk while a pair of perpendicular control forces are applied to regulate planar vibration. As typically done for various mechanical systems, Lagrange's method can be used to develop the dynamic equations. Specifically, the rotating unbalanced disk is a Lagrangian system that can be characterized by the following second-order differential equation

$$\frac{d}{dt}\left(\frac{\partial \mathcal{L}}{\partial \dot{q}}\right) - \frac{\partial \mathcal{L}}{\partial q} = u \qquad (2.1)$$

where $\mathcal{L}(t)$ denotes the Lagrangian defined as the difference between the kinetic energy, denoted by $T(t)$, and the potential energy, denoted by $PE(t)$, as follows

$$\mathcal{L} = T - PE. \qquad (2.2)$$

The position vector $q(t) \in \mathbb{R}^3$ given in (2.1) is defined as follows

$$q = \begin{bmatrix} x & y & \theta \end{bmatrix}^T \qquad (2.3)$$

where $x(t)$, $y(t) \in \mathbb{R}$ denote the linear position of the disk center of geometry along the x and y axes, respectively, and $\theta(t) \in \mathbb{R}$ denotes the angular

---

[1] The disk is unbalanced in the sense that the center of geometry does not coincide with the center of mass.

position of the disk. The position of the disk center of mass, denoted by $x_{cm}(t)$, $y_{cm}(t) \in \mathbb{R}$, is given by the following expressions

$$x_{cm} = x + L\cos(\theta + \beta) \qquad\qquad y_{cm} = y + L\sin(\theta + \beta) \qquad (2.4)$$

where $\beta \in \mathbb{R}$ represents the initial angle subtended by L with the positive x axis. The force/torque control input $u(t) \in \mathbb{R}^3$ given in (2.1) is defined as follows

$$u = \begin{bmatrix} F_x & F_y & \tau_z \end{bmatrix}^T \qquad (2.5)$$

where $F_x(t)$, $F_y(t) \in \mathbb{R}$ denote the perpendicular and planar control force inputs, and $\tau_z(t) \in \mathbb{R}$ denotes the torque applied to the disk. The kinetic energy for the rotating unbalanced disk can be computed as follows

$$T = \frac{1}{2}m\left(\dot{x}_{cm}^2 + \dot{y}_{cm}^2\right) + \frac{1}{2}I_{cg}\dot{\theta}^2 \qquad (2.6)$$

where $m \in \mathbb{R}$ denotes the mass of the disk, $\dot{x}_{cm}(t)$, $\dot{y}_{cm}(t) \in \mathbb{R}$ denote the velocities of the center of mass of the disk along the x and y axes, respectively, $\dot{\theta}(t) \in \mathbb{R}$ denotes the disk angular velocity, and $I_{cg} \in \mathbb{R}$ is the moment of inertia about the center of geometry given by the following expression

$$I_{cg} = I_{cm} + mL^2 \qquad (2.7)$$

where $I_{cm} \in \mathbb{R}$ is the moment of inertia about the center of mass, and $L \in \mathbb{R}$ denotes the distance between the center of geometry and the center of mass. After substituting (2.7) and the time derivative of (2.4) into (2.6) and then using some basic trigonometric properties, the kinetic energy can be expressed as follows

$$T = \tfrac{1}{2}m\left[\dot{x}^2 - 2L\sin(\theta + \beta)\,\dot{x}\dot{\theta} + \dot{y}^2 + 2L\cos(\theta + \beta)\,\dot{y}\dot{\theta} + L^2\dot{\theta}^2\right]$$

$$+ \tfrac{1}{2}\left(I_{cm} + mL^2\right)\dot{\theta}^2. \qquad (2.8)$$

Since no work is performed on the system by conservative forces, the potential energy of the system is given by the following expression

$$PE = 0. \qquad (2.9)$$

After substituting the kinetic and potential energy expressions given in (2.8) and (2.9) into (2.2), the dynamic model of the rotating unbalanced disk can be determined from (2.1) as follows

$$M\ddot{q} + V_m\dot{q} = u \qquad (2.10)$$

where $\dot{q}(t)$, $\ddot{q}(t) \in \mathbb{R}^3$ denote the disk velocity and acceleration vectors, respectively, and the inertia and Coriolis matrices, denoted by $M(q) \in \mathbb{R}^{3\times3}$ and $V_m(q, \dot{q}) \in \mathbb{R}^{3\times3}$, respectively, are defined as follows

$$M(q) = \begin{bmatrix} m & 0 & -mL\sin(\theta+\beta) \\ 0 & m & mL\cos(\theta+\beta) \\ -mL\sin(\theta+\beta) & mL\cos(\theta+\beta) & I_{cm}+2mL^2 \end{bmatrix} \tag{2.11}$$

$$V_m(q, \dot{q}) = \begin{bmatrix} 0 & 0 & -mL\cos(\theta+\beta)\dot{\theta} \\ 0 & 0 & -mL\sin(\theta+\beta)\dot{\theta} \\ 0 & 0 & 0 \end{bmatrix}. \tag{2.12}$$

The dynamic system given in (2.10) exhibits the following properties that are utilized in the subsequent control development and stability analysis.

Property 2.1: Symmetric and Positive-Definite Inertia Matrix

The symmetric and positive-definite inertia matrix $M(q)$ satisfies the following inequalities

$$m_1 \|\xi\|^2 \leq \xi^T M\xi \leq m_2 \|\xi\|^2 \quad \forall \xi \in \mathbb{R}^3 \tag{2.13}$$

where m_1, $m_2 \in \mathbb{R}$ are known positive constants, and $\|\cdot\|$ denotes the standard Euclidean norm.

Property 2.2: Skew-Symmetry

The inertia and Coriolis matrices given in (2.11) and (2.12) satisfy the following skew-symmetric relationship

$$\xi^T \left(\frac{1}{2}\dot{M} - V_m\right)\xi = 0 \quad \forall \xi \in \mathbb{R}^3 \tag{2.14}$$

where $\dot{M}(q)$ denotes the time derivative of the inertia matrix.

Property 2.3: Linearity in the Parameters

The left-hand side of (2.10) can be linearly parameterized as follows

$$M\ddot{q} + V_m\dot{q} = Y\phi \tag{2.15}$$

where the known regression matrix $Y(q, \dot{q}, \ddot{q}) \in \mathbb{R}^{3 \times 4}$ and the unknown constant parameter vector $\phi \in \mathbb{R}^4$ are defined as follows

$$Y(\cdot) = \begin{bmatrix} \ddot{x} & -\sin(\theta)\ddot{\theta} - \cos(\theta)\dot{\theta}^2 & -\cos(\theta)\ddot{\theta} + \sin(\theta)\dot{\theta}^2 & 0 \\[2mm] \ddot{y} & \cos(\theta)\ddot{\theta} - \sin(\theta)\dot{\theta}^2 & -\sin(\theta)\ddot{\theta} - \cos(\theta)\dot{\theta}^2 & 0 \\[2mm] 0 & -\sin(\theta)\ddot{x} + \cos(\theta)\ddot{y} & -\cos(\theta)\ddot{x} - \sin(\theta)\ddot{y} & \ddot{\theta} \end{bmatrix}$$

$$\tag{2.16}$$

$$\phi = \begin{bmatrix} m & mL\cos\beta & mL\sin\beta & I_{cm} + 2mL^2 \end{bmatrix}^T. \tag{2.17}$$

2.2.2 Control Objective

The objective of this section is to regulate the center of geometry to the position $(x = 0, y = 0)$ while ensuring that the angular velocity of the disk tracks a desired trajectory despite the parametric uncertainty associated with (2.17). The control objective is made under the assumption that $q(t)$ and $\dot{q}(t)$ are measurable. To quantify the velocity tracking control objective, we define the tracking error $e(t) \in \mathbb{R}$ as follows

$$e = \dot{\theta}_d - \dot{\theta} \tag{2.18}$$

where $\dot{\theta}_d(t) \in \mathbb{R}$ denotes the desired angular velocity of the disk where it is assumed that the desired trajectory is selected so that $\dot{\theta}_d(t)$, $\ddot{\theta}_d(t)$, $\dddot{\theta}_d(t) \in \mathcal{L}_\infty$. To facilitate the representation of the second-order dynamic model of (2.10) as a first-order differential equation, a filtered error signal $r(t) = \begin{bmatrix} r_1(t) & r_2(t) & r_3(t) \end{bmatrix}^T \in \mathbb{R}^3$ is defined as follows

$$r = \begin{bmatrix} -\dot{x} - \alpha_1 x \\ -\dot{y} - \alpha_2 y \\ e \end{bmatrix} \tag{2.19}$$

where $\alpha_1, \alpha_2 \in \mathbb{R}$ are positive control gains. As typically done for velocity tracking control applications, (2.19) has been constructed so that the signals $\theta_d(t)$ and $\theta(t)$ are not used in the feedback portion of the control, since they may become unbounded.

Remark 2.1 *The control objective given in this section is centered on the autobalance problem of regulating the center of geometry while ensuring angular velocity tracking. In the following development, it is assumed that a magnetic bearing system can produce the perpendicular and planar control force inputs given in (2.5). That is, only the mechanical dynamics of* *the magnetic bearing/rotating disk assembly are investigated. In Chapter 3, a controller is developed for a magnetic bearing system where both the mechanical and electrical dynamics are incorporated in the design.*

2.2.3 Adaptive Control

Control Formulation

To develop the open-loop error system for $r(t)$, we take the time derivative of (2.19), premultiply the resulting expression by $M(q)$, and then substitute (2.10) for $M(q)\ddot{q}(t)$ to obtain the following expression

$$M\dot{r} = -V_m r + M \begin{bmatrix} -\alpha_1 \dot{x} \\ -\alpha_2 \dot{y} \\ \ddot{\theta}_d \end{bmatrix} + V_m \begin{bmatrix} -\alpha_1 x \\ -\alpha_2 y \\ \dot{\theta}_d \end{bmatrix} - u \qquad (2.20)$$

where $V_m(q,\dot{q}) r(t)$ has been added and subtracted to (2.20). After utilizing (2.15), the expression given in (2.20) can be rewritten as follows

$$M\dot{r} = -V_m r + Y_1 \phi_1 - u \qquad (2.21)$$

where $Y_1(q, \dot{q}, t) \in \mathbb{R}^{3 \times p}$ denotes a measurable regression matrix, and $\phi_1 \in \mathbb{R}^p$ denotes a constant unknown parameter vector such that

$$Y_1 \phi_1 = M \begin{bmatrix} -\alpha_1 \dot{x} \\ -\alpha_2 \dot{y} \\ \ddot{\theta}_d \end{bmatrix} + V_m \begin{bmatrix} -\alpha_1 x \\ -\alpha_2 y \\ \dot{\theta}_d \end{bmatrix}. \qquad (2.22)$$

Based on the structure of the open-loop error dynamics given in (2.21) and the subsequent stability analysis, we design the control input $u(t)$ as follows

$$u = Y_1 \hat{\phi}_1 + K_s r - \begin{bmatrix} k_{p1} x & k_{p2} y & 0 \end{bmatrix}^T \qquad (2.23)$$

where $Y_1(\cdot)$ is given in (2.21), $\hat{\phi}_1(q, \dot{q}, r, t) \in \mathbb{R}^p$ denotes a parameter estimate vector for ϕ_1, and $k_{p1}, k_{p2} \in \mathbb{R}$ are positive control constants. Based on the subsequent stability analysis, the parameter estimate vector $\hat{\phi}_1(\cdot)$ is generated according to the following gradient update law

$$\dot{\hat{\phi}}_1 = \Gamma_1 Y_1^T r. \qquad (2.24)$$

The positive-definite control gain matrices $\Gamma_1 \in \mathbb{R}^{3 \times 3}$ and $K_s \in \mathbb{R}^{3 \times 3}$ given in (2.23) and (2.24) are defined as follows

$$\Gamma_1 = \text{diag}\{\Gamma_{11}, \Gamma_{12}, \Gamma_{13}\} \qquad K_s = \text{diag}\{k_{s1}, k_{s2}, k_{s3}\} \qquad (2.25)$$

where the notation diag$\{\cdot\}$ indicates a diagonal matrix with the diagonal elements given in $\{\cdot\}$. After substituting (2.23) into (2.21), the following closed-loop error system is obtained

$$M\dot{r} = -V_m r + Y_1 \tilde{\phi}_1 - K_s r + \begin{bmatrix} k_{p1}x & k_{p2}y & 0 \end{bmatrix}^T \qquad (2.26)$$

where the parameter estimation error $\tilde{\phi}_1(t) \in \mathbb{R}^p$ is defined as follows

$$\tilde{\phi}_1 = \phi_1 - \hat{\phi}_1. \qquad (2.27)$$

Stability Analysis

The stability of the adaptive tracking controller given in (2.23) and (2.24) can now be examined through the following theorem.

Theorem 2.1 *The controller given by (2.23) along with the adaptive update law given by (2.24) ensure global asymptotic regulation of the position of the geometric center and global asymptotic angular velocity tracking in the sense that*

$$\lim_{t \to \infty} x(t), y(t), e(t) = 0 \qquad (2.28)$$

where $e(t)$ was defined in (2.18) and $x(t)$, $y(t)$ were defined in (2.3).

Proof: To prove Theorem 2.1, we define a nonnegative function $V(t) \in \mathbb{R}$ as follows

$$V = \frac{1}{2}r^T M r + \frac{1}{2}k_{p1}x^2 + \frac{1}{2}k_{p2}y^2 + \frac{1}{2}\tilde{\phi}_1^T \Gamma^{-1} \tilde{\phi}_1. \qquad (2.29)$$

After taking the time derivative of (2.29), substituting (2.26) into the resulting expression for $M(q)\dot{r}(t)$, substituting the time derivative of (2.27) for $\dot{\tilde{\phi}}_1(t)$, and then cancelling common terms, the following expression is obtained

$$\dot{V} = -r^T K_s r - \alpha_1 k_{p1}x^2 - \alpha_2 k_{p2}y^2 \qquad (2.30)$$

where (2.14) and (2.19) have been used. The development given in (2.29) and (2.30) can be used to prove that $r(t)$, $x(t)$, $y(t)$, $\tilde{\phi}_1(t) \in \mathcal{L}_\infty$ and that $r(t)$, $x(t)$, $y(t) \in \mathcal{L}_2$ (see Lemma A.11 of Appendix A). Since $r(t) \in \mathcal{L}_\infty$, Lemma A.13 of Appendix A can be used along with (2.18) and (2.19) to prove that $\dot{q}(t) \in \mathcal{L}_\infty$; furthermore, (2.27) can be used to determine that $\hat{\phi}_1(t) \in \mathcal{L}_\infty$. Since the regression matrix $Y_1(\cdot)$ is made up of bounded arguments (where $q(t)$ is only present as an argument of bounded trigonometric expressions), $Y_1(\cdot) \in \mathcal{L}_\infty$; therefore, (2.23) can be used to prove that the control input $u(t) \in \mathcal{L}_\infty$. It follows from (2.26) that $\dot{r}(t) \in \mathcal{L}_\infty$; hence,

$\ddot{q}(t) \in \mathcal{L}_\infty$. Based on the facts that $r(t)$, $\dot{r}(t) \in \mathcal{L}_\infty$ and $r(t) \in \mathcal{L}_2$, Barbalat's Lemma (see Lemma A.16 of Appendix A) can be invoked to prove that

$$\lim_{t \to \infty} r(t) = 0. \tag{2.31}$$

The condition (2.31) and Lemma A.15 of Appendix A can now be utilized to prove the result given in (2.28). Alternatively, the expression in (2.31), the results $x(t)$, $y(t)$, $\dot{q}(t) \in \mathcal{L}_\infty$ and $x(t)$, $y(t) \in \mathcal{L}_2$, and Barbalat's Lemma can be used to prove the result in (2.28). $\square$

2.2.4 DCAL-Based Adaptive Control

In the previous section, a gradient-based adaptive controller was used to prove global asymptotic tracking of the angular disk velocity and global regulation of the center of geometry despite parametric uncertainty. As stated previously, a solution that could automatically identify the unknown imbalance-related parameters of the system would be beneficial. Therefore, building off of the previous result, a composite[2] adaptive, prediction error driven, DCAL-based controller is developed that forces the imbalance-related parameter estimate vector to converge to the actual values and also yields an exponential envelope for the tracking, regulation, and the parameter estimate error signals, provided a mild PE condition is satisfied.

Filtered Torque

To facilitate the construction of a composite adaptive controller that uses a prediction error-based update law, a filtered control input signal can be formulated as follows [26, 38]

$$u_f = f * u \tag{2.32}$$

where $u_f(t) \in \mathbb{R}$ denotes a measurable filtered control signal, $*$ denotes the standard convolution operation, $u(t)$ was defined in (2.10), and the filter function $f(t) \in \mathbb{R}$ is defined as follows

$$f = \gamma \exp(-\gamma t) \tag{2.33}$$

where $\gamma \in \mathbb{R}$ is a positive filter control gain constant. After substituting (2.10) into (2.32) for $u(t)$, standard convolution properties (see Lemma B.1

---

[2] The term composite [38] was coined because the adaptive update law is a combination of the gradient and least-squares update laws.

of Appendix B) can be used in conjunction with (2.15) to rewrite (2.32) in terms of the following linear parameterization

$$u_f = Y_f\left(q, \dot{q}\right)\phi. \tag{2.34}$$

In the linear parameterization given in (2.34), ϕ denotes the same unknown parameter vector defined in (2.17), and $Y_f\left(q, \dot{q}, t\right) \in \mathbb{R}^{3\times 4}$ denotes a known filtered regression matrix that is independent of acceleration measurements such that

$$Y_f\phi = \quad \dot{f}\left(t\right) * M\left(q\right)\dot{q}\left(t\right) + f\left(0\right)M\left(q\right)\dot{q}\left(t\right) - f\left(t\right)M\left(q\left(0\right)\right)\dot{q}\left(0\right)$$

$$+ f\left(t\right) * \left(-\dot{M}\left(q\right)\dot{q}\left(t\right) + V_m\left(q, \dot{q}\right)\dot{q}\left(t\right)\right)$$
$$\tag{2.35}$$

where $\dot{f}\left(t\right) \in \mathbb{R}$ can be determined as follows

$$\dot{f} = -\gamma^2 \exp\left(-\gamma t\right). \tag{2.36}$$

To foster the development of a controller that depends on a desired angular velocity regression matrix formulation, the following additional linear parameterization is designed

$$u_{df} = Y_{df}\phi_2 = f * Y_d\phi_2. \tag{2.37}$$

In the linear parameterization given in (2.37), the desired regression matrix $Y_d\left(q, t\right) \in \mathbb{R}^{3\times 3}$, and the constant parameter vector, denoted by $\phi_2 \in \mathbb{R}^3$, are defined as follows

$$Y_d\left(q, t\right) = \begin{bmatrix} -\sin\left(\theta\right)\ddot{\theta}_d - \cos\left(\theta\right)\dot{\theta}_d^2 & -\cos\left(\theta\right)\ddot{\theta}_d + \sin\left(\theta\right)\dot{\theta}_d^2 & 0 \\ \cos\left(\theta\right)\ddot{\theta}_d - \sin\left(\theta\right)\dot{\theta}_d^2 & -\sin\left(\theta\right)\ddot{\theta}_d - \cos\left(\theta\right)\dot{\theta}_d^2 & 0 \\ 0 & 0 & \ddot{\theta}_d \end{bmatrix}$$
$$\tag{2.38}$$

$$\phi_2 = \begin{bmatrix} mL\cos\beta & mL\sin\beta & I_{cm} + 2mL^2 \end{bmatrix}^T \tag{2.39}$$

and $Y_{df}\left(q, t\right) \in \mathbb{R}^{3\times 3}$ denotes a desired filtered regression matrix. A measurable[3] prediction error $\varepsilon(t) \in \mathbb{R}^3$ can now be defined as follows

$$\varepsilon = u_f - Y_{df}\hat{\phi}_2 = Y_f\phi - Y_{df}\hat{\phi}_2 \tag{2.40}$$

---

[3] Given that $f(t)$ and $\dot{f}(t)$ are known and that $q(t)$ and $\dot{q}(t)$ are measurable, it is clear from (2.35) and (2.40) that $\varepsilon(t)$ is measurable.

where (2.34) has been utilized. The parameter estimate vector $\hat{\phi}_2(t) \in \mathbb{R}^3$ given in (2.40) is computed according to the following composite adaptive update law

$$\dot{\hat{\phi}}_2 = PY_{df}^T \varepsilon + PY_d^T r \tag{2.41}$$

where the inverse of the time-varying gain matrix $P(t) \in \mathbb{R}^{3 \times 3}$ is generated according to the following differential expression

$$\dot{P}^{-1} = -\lambda P^{-1} + Y_{df}^T Y_{df} \tag{2.42}$$

which can also be expressed by the following integral expression

$$P^{-1}(t) = P^{-1}(0) \exp\left(-\int_0^t \lambda(\sigma)\, d\sigma\right)$$
$$+ \int_0^t \exp\left(-\int_\sigma^t \lambda(\eta)\, d\eta\right) Y_{df}^T(\sigma) Y_{df}(\sigma)\, d\sigma \tag{2.43}$$

where $Y_{df}(\cdot)$ is given in (2.37), and $\lambda(t) \in \mathbb{R}$ is a positive gain-adjusted forgetting factor [38]. To facilitate the subsequent control design and stability analysis, the expression given in (2.40) can be rewritten as follows

$$\varepsilon = \Omega + Y_{df}\tilde{\phi}_2 \tag{2.44}$$

where $\Omega(q, \dot{q}, t) \in \mathbb{R}^3$ quantifies the mismatch between the filtered torque and the desired filtered torque as follows

$$\Omega = Y_f \phi - Y_{df}\phi_2 \tag{2.45}$$

and the parameter estimation error $\tilde{\phi}_2(t) \in \mathbb{R}^3$ is defined as

$$\tilde{\phi}_2 = \phi_2 - \hat{\phi}_2. \tag{2.46}$$

Remark 2.2 *It is important to note that the* $\sin(\theta)$ *and* $\cos(\theta)$ *terms in (2.38) can be correctly calculated even if the measured value for* $\theta(t)$ *is reset every revolution. That is, since we are concerned with the angular velocity tracking problem, the control can be implemented by resetting* $\theta(t)$ *each revolution, and hence, we can ensure that* $\theta(t)$ *remains bounded. We also note that the regression matrix formulation given by (2.38) is crucial for developing the persistency of excitation arguments used in the subsequent stability analysis.*

Control Formulation

To develop the open-loop error system for $r(t)$, the same operations as given in the previous adaptive control design can be performed to obtain the expression given in (2.21). To facilitate the DCAL-based control development, the linear parameterization $Y_d(\cdot)\phi_2$ is added and subtracted to (2.47) to obtain the following expression

$$M\dot{r} = -V_m r + Y_d\phi_2 + \chi - u \tag{2.47}$$

where the desired regression matrix $Y_d(\cdot)$ is defined in (2.38), the parameter vector $\phi_2 \in \mathbb{R}^3$ is defined in (2.39), and the auxiliary term $\chi(q,\dot{q},t) \in \mathbb{R}^3$ is defined as follows

$$\chi = Y_1\phi_1 - Y_d\phi_2. \tag{2.48}$$

To facilitate the subsequent stability analysis, $\chi(q,\dot{q},t)$ is upper bounded as follows (see Lemma B.2 in Appendix B)

$$\|\chi\| \le \zeta_1 \|z\| \tag{2.49}$$

where $\zeta_1 \in \mathbb{R}$ is a known positive bounding constant, and $z(t) \in \mathbb{R}^6$ is defined as follows

$$z(t) = \begin{bmatrix} x(t) & y(t) & e(t) & r^T(t) \end{bmatrix}^T \tag{2.50}$$

where $x(t)$, $y(t)$, $e(t)$, and $r(t)$ are defined in (2.3), (2.18), and (2.19), respectively. Based on the open-loop error system given in (2.47) and (2.48), we design the control input $u(t)$ as follows

$$u = Y_d\hat{\phi}_2 + K_s r - \begin{bmatrix} k_{p1}x & k_{p2}y & 0 \end{bmatrix}^T + k_n\zeta_1^2 r \tag{2.51}$$

where $Y_d(\cdot)$ is defined in (2.38), $\hat{\phi}_2(t)$ is defined in (2.41), K_s and k_{p1}, k_{p2} are given in (2.23), $k_n \in \mathbb{R}$ is a positive control gain, and ζ_1 is defined in (2.49). After substituting (2.51) into (2.47), the following closed-loop error system for $r(t)$ can be obtained

$$M\dot{r} = -V_m r + Y_d\tilde{\phi}_2 + \chi - K_s r + \begin{bmatrix} k_{p1}x & k_{p2}y & 0 \end{bmatrix}^T - k_n\zeta_1^2 r. \tag{2.52}$$

After substituting (2.44) into (2.41) for $\varepsilon(t)$, the following closed-loop error system for $\tilde{\phi}_2(t)$ can be obtained

$$\dot{\tilde{\phi}}_2 = -PY_{df}^T\Omega - PY_{df}^T Y_{df}\tilde{\phi}_2 - PY_d^T r \tag{2.53}$$

where the following fact was used

$$\dot{\tilde{\phi}}_2 = -\dot{\hat{\phi}}_2. \tag{2.54}$$

The subsequent stability analysis exploits the fact that the mismatch term $\Omega(q,\dot{q},t)$ defined in (2.45) can be upper bounded as follows (see Lemma B.3 in Appendix B)

$$\|\Omega\| \le \gamma\zeta \|\psi\| \tag{2.55}$$

where γ was defined in (2.33), $\zeta \in \mathbb{R}$ is a known positive bounding constant, and $\psi(t) \in \mathbb{R}^9$ is defined as follows

$$\psi(t) = \left[\begin{array}{cc} z^T(t) & \tilde{\phi}_2^T(t) \end{array} \right]^T. \tag{2.56}$$

Moreover, the subsequent exponential stability result requires that $P^{-1}(t)$ given in (2.43) be upper and lower bounded by constants. Given the need to ensure that $P^{-1}(t)$ is lower bounded by a constant, the desired filtered regression matrix $Y_{df}(\cdot)$ given in (2.37) is required to satisfy the following PE condition

$$\int_{t_n}^{t_n + \delta_n} Y_{df}^T(\sigma) Y_{df}(\sigma) \, d\sigma \geq \mu I_3 \tag{2.57}$$

where I_3 denotes the standard 3×3 identity matrix, and $\mu \in \mathbb{R}$ is a positive bounding constant. Given the need to ensure that $P^{-1}(t)$ is upper bounded by a constant, the forgetting factor $\lambda(t)$ given in (2.42) is defined as follows [38]

$$\lambda = \frac{\lambda_1}{k_1} (k_1 - \|P\|_{i2}) \tag{2.58}$$

where λ_1, $k_1 \in \mathbb{R}$ are positive constants that represent the maximum forgetting rate and the prespecified bound for the magnitude of the gain matrix $P(t)$, respectively, and $\|\cdot\|_{i2}$ denotes the induced 2-norm.

Remark 2.3 *The time-varying design of $\lambda(t)$ given in (2.58) versus a simple positive constant is motivated by the desire to achieve the benefits of data forgetting while maintaining boundedness of the gain matrix $P(t)$ (see the discussion given in [38] regarding data forgetting and an example of gain unboundedness when $\lambda(t)$ is selected as a simple positive constant).*

Stability Analysis

Based on the closed-loop error systems given in (2.52) and (2.53), the following theorem defines the exponential envelope that confines the transient response of the disk tracking error and the parameter estimation error defined in (2.18) and (2.46), respectively.

Theorem 2.2 *The controller and the composite adaptive update law given in (2.41–2.58) ensure that the angular velocity tracking error, the disk center of geometry, and the parameter estimation error are exponentially regulated to zero in the following sense*

$$\|\psi(t)\| \leq \sqrt{\frac{\xi_2}{\xi_1} \|\psi(0)\|^2 \exp\left(-\frac{(\lambda_c - \gamma\zeta)}{\xi_2}t\right)} \tag{2.59}$$

where $\psi(t)$ was defined in (2.56). The positive constant parameters ξ_1, $\xi_2 \in \mathbb{R}$ given in (2.59) are defined as follows

$$\xi_1 = \frac{1}{2}\min\left\{m_1, k_{p1}, k_{p2}, \frac{1 + k_1\kappa}{k_1}\right\},$$

(2.60)

$$\xi_2 = \frac{1}{2}\max\left\{m_2, k_{p1}, k_{p2}, k_2\right\},$$

the filter control parameter γ is given in (2.33), ζ is the positive bounding constant given in (2.55), and $\lambda_c \in \mathbb{R}$ is a positive bounding constant defined as follows

$$\lambda_c = \min\left\{\lambda_b, \frac{\lambda_1\kappa}{2}\right\}$$

(2.61)

where λ_b, $\kappa \in \mathbb{R}$ are positive constants defined as follows

$$\lambda_b = \min\left\{\lambda_{\min}\left\{K_s\right\}, k_{p1}\alpha_1, k_{p2}\alpha_2\right\} - \frac{1}{k_n},$$

(2.62)

$$\kappa = \min\lambda_{\min}\left\{\left\{\left(P^{-1}(0) - \frac{1}{k_1}I_3\right)\exp\left(-\lambda_1\delta_0\right)\right\},$$

$$\cdot\exp\left(-\lambda_1\left(\delta_{i-1} + \delta_i\right)\right)\right\} \qquad \forall i = 2, 3, ..., n$$

(2.63)

where $\delta_i \in \mathbb{R}$ represents the length of n different time intervals between $[0, t]$ (i.e., length of the interval from t_0 to t_1 is δ_0, t_1 to t_2 is δ_1,...., t_n to t_{n+1} is δ_n) and $\lambda_{\min}\left\{\cdot\right\}$ denotes the minimum eigenvalue of a matrix. The result given in (2.59) can be proven provided that: (i) the desired filtered regression matrix $Y_{df}\left(\cdot\right)$ defined in (2.37) satisfies the PE condition given in (2.57), (ii) $P(0)$ is selected to be positive-definite symmetric and satisfies the following inequality

$$P(0) < k_1 I_3$$

(2.64)

where k_1 was defined in (2.58), and (iii) the control gain k_n of (2.51) and γ are selected to satisfy the following sufficient conditions

$$k_n > \frac{1}{\min\left\{\lambda_{\min}\left\{K_s\right\}, k_{p1}\alpha_1, k_{p2}\alpha_2\right\}}$$

(2.65)

$$\gamma < \min\left\{\frac{\lambda_c}{\zeta}, \frac{1}{2\zeta}\right\}.$$

(2.66)

Proof: To prove Theorem 2.2, we define a nonnegative function $V(t) \in \mathbb{R}$ as follows

$$V = \frac{1}{2}r^T M r + \frac{1}{2}k_{p1}x^2 + \frac{1}{2}k_{p2}y^2 + \frac{1}{2}\tilde{\phi}_2^T P^{-1}\tilde{\phi}_2.$$

(2.67)

Based on the structure of (2.67), we can use (2.13) and invoke Lemma B.5 of Appendix B to bound $V(t)$ by the following inequalities

$$\xi_1 \|\psi\|^2 \le V \le \xi_2 \|\psi\|^2 \tag{2.68}$$

where the positive constants ξ_1 and ξ_2 are defined in (2.60). After taking the time derivative of (2.67), substituting (2.52) into the resulting expression for $M(q)\dot{r}(t)$, and then cancelling common terms, the following expression is obtained

$$\dot{V} = r^T \left(\chi - K_s r + \begin{bmatrix} k_{p1}x & k_{p2}y & 0 \end{bmatrix}^T - k_n \zeta_1^2 r \right)$$

$$+ k_{p1}x(-r_1 - \alpha_1 x) + k_{p2}y(-r_2 - \alpha_2 y) \tag{2.69}$$

$$+ \tilde{\phi}_2^T \left(Y_d^T r + P^{-1} \dot{\tilde{\phi}}_2 + \frac{1}{2} \dot{P}^{-1} \tilde{\phi}_2 \right)$$

where (2.14) and (2.19) have been used. After substituting (2.53) and (2.42) into (2.69) for $\dot{P}^{-1}(t)$ and $\dot{\tilde{\phi}}_2(t)$, respectively, the following expression is obtained

$$\dot{V} \le -r^T K_s r - k_{p1}\alpha_1 x^2 - k_{p2}\alpha_2 y^2 - \frac{1}{2}\left\| Y_{df}\tilde{\phi}_2 \right\|^2 + \|r\| \|\chi\|$$

$$- k_n \zeta_1^2 r^T r + \left\| Y_{df}\tilde{\phi}_2 \right\| \|\Omega\| - \frac{1}{2}\tilde{\phi}_2^T \lambda P^{-1}\tilde{\phi}_2. \tag{2.70}$$

After substituting (2.55) and (2.49) into (2.70) for the bounds for $\|\Omega(\cdot)\|$ and $\|\chi(\cdot)\|$, respectively, the following expression is obtained

$$\dot{V} \le -r^T K_s r - k_{p1}\alpha_1 x^2 - k_{p2}\alpha_2 y^2$$

$$- \frac{1}{2}\left\| Y_{df}\tilde{\phi}_2 \right\|^2 + \left[\zeta_1 \|r\| \|z\| - k_n \zeta_1^2 \|r\|^2 \right] \tag{2.71}$$

$$+ \gamma\zeta \left(\left\| Y_{df}\tilde{\phi}_2 \right\|^2 + \|\psi\|^2 \right) - \frac{1}{2}\tilde{\phi}_2^T \lambda P^{-1}\tilde{\phi}_2$$

after utilizing the fact that

$$\|a\| \|b\| \le \|a\|^2 + \|b\|^2 \quad \forall a, b \in \mathbb{R}. \tag{2.72}$$

After invoking the nonlinear damping argument given in Lemma A.17 of Appendix A to the bracketed terms, an upper bound for (2.71) can be formulated as follows

$$
\dot{V} \leq \quad -\left(\min\left\{ \lambda_{\min}\left\{ K_s \right\}, k_{p1}\alpha_1, k_{p2}\alpha_2 \right\} - \frac{1}{k_n} \right) \|z\|^2
$$

$$
-\left(\frac{1}{2} - \gamma\zeta \right) \left\| Y_{df}\tilde{\phi}_2 \right\|^2 + \gamma\zeta \left\| \psi \right\|^2 - \left[\frac{1}{2}\tilde{\phi}_2^T \lambda P^{-1} \tilde{\phi}_2 \right].
$$

(2.73)

If the conditions given in (2.57) and (2.64) are satisfied, the following inequalities (see Lemma B.5 of Appendix B for the details regarding the development of the following inequalities)

$$
0 < \frac{\lambda_1 k_1 \kappa}{1 + k_1 \kappa} \leq \lambda(t)
$$

(2.74)

$$
\frac{1 + k_1 \kappa}{k_1} I_3 \leq P^{-1}(t) \leq k_2 I_3
$$

(2.75)

can be used to obtain the following expression

$$
\frac{1}{2}\tilde{\phi}_2^T \lambda P^{-1} \tilde{\phi}_2 \geq \frac{\lambda_1 \kappa}{2} \left\| \tilde{\phi}_2 \right\|^2
$$

(2.76)

where κ is defined in (2.63). Provided that the control parameters γ and k_n given in (2.33) and (2.51) are selected according to the conditions given in (2.66) and (2.65), we can now use (2.76) to rewrite (2.73) as follows

$$
\dot{V} \leq -\lambda_b \|z\|^2 + \gamma\zeta \|\psi\|^2 - \frac{\lambda_1 \kappa}{2} \left\| \tilde{\phi}_2 \right\|^2
$$

(2.77)

where λ_b was defined in (2.62). We can now express (2.77) in the following compact form

$$
\dot{V} \leq -(\lambda_c - \gamma\zeta) \|\psi\|^2
$$

(2.78)

after utilizing (2.56) and (2.61). If the control parameter γ of (2.33) is selected to satisfy the sufficient condition given in (2.66), then we can use (2.56), (2.67), and (2.78) to prove that $r(t)$, $x(t)$, $y(t)$, and $\tilde{\phi}_2(t) \in \mathcal{L}_\infty$ and that $\psi(t) \in \mathcal{L}_2$ (see Lemma A.11 of Appendix A). Hence, from (2.56), $r(t)$, $e(t)$, and $\tilde{\phi}_2(t) \in \mathcal{L}_2$. Since $r(t)$ and $\tilde{\phi}_2(t) \in \mathcal{L}_\infty$, (2.18), (2.19), and (2.46) can be used to determine that $e(t)$, $q(t)$, $\dot{q}(t)$, and $\hat{\phi}_2(t) \in \mathcal{L}_\infty$. Since the desired regression matrix $Y_d(\cdot)$ and its time derivative $\dot{Y}_d(\cdot)$ are made up of bounded arguments (i.e., $q(t)$, $\dot{q}(t)$, $\dot{q}_d(t)$, $\ddot{q}_d(t)$, $\dddot{q}_d(t)$), we can state that $Y_d(\cdot)$, $\dot{Y}_d(\cdot) \in \mathcal{L}_\infty$. Based on the previous boundedness arguments, (2.51) can now be used to prove that $u(t) \in \mathcal{L}_\infty$. From previous arguments and the definition of (2.48), we can also prove that $\chi(\cdot) \in \mathcal{L}_\infty$. It follows from (2.52) that $\dot{r}(t) \in \mathcal{L}_\infty$; hence, $\ddot{e}(t)$ and $\ddot{q}(t) \in \mathcal{L}_\infty$.

To prove the exponential tracking result given in (2.59), the inequalities given in (2.68) can be used to rewrite (2.78) as follows

$$\dot{V} \leq -\frac{(\lambda_c - \gamma\zeta)}{\xi_2}V. \tag{2.79}$$

After solving the differential equation given in (2.79) according to Lemma A.10 of Appendix A and then utilizing (2.68), the result given in (2.59) is obtained. □

Remark 2.4 *To obtain the inequality given by (2.77), $Y_{df}(\cdot)$ defined in (2.37) must satisfy the PE condition given in (2.57) (see Lemma B.5 of Appendix B for details). Based on the fact that $Y_d(\cdot)$, $\dot{Y}_d(\cdot) \in \mathcal{L}_\infty$, we can invoke Lemma A.8 of Appendix A to prove that if $Y_d(\cdot)$ defined in (2.38) satisfies the PE condition, then $Y_{df}(\cdot)$ will also be PE. Based on the knowledge that we would require $Y_d(\cdot)$ to satisfy the PE condition, particular attention was devoted to the manner in which (2.38) was constructed. Specifically, $Y_d(\cdot)$ was constructed such that the integrand of the PE condition takes the following form*

$$Y_d^T Y_d = \begin{bmatrix} \ddot{\theta}_d^2 + \dot{\theta}_d^4 & 0 & 0 \\ 0 & \ddot{\theta}_d^2 + \dot{\theta}_d^4 & 0 \\ 0 & 0 & \ddot{\theta}_d^2 \end{bmatrix}. \tag{2.80}$$

It is clear from (2.80) that the desired angular velocity trajectory $\dot{\theta}_d(t)$ can be designed to ensure that $Y_d(\cdot)$ satisfies the PE condition (e.g., $\dot{\theta}_d(t)$ can be constructed to be nonconstant for some time interval).

Remark 2.5 *If $\lambda(t)$ defined in (2.42) is selected simply to be a positive scalar constant, it is not difficult to show that $z(t)$ defined in (2.50) is asymptotically regulated to zero.*

Remark 2.6 *As indicated by the proof of Theorem 2.2, the exponential stability result given by (2.59) requires that $P^{-1}(t)$ be lower bounded by a constant (see (2.75)). In [38], a method is presented to develop a lower bound for $P^{-1}(t)$; however, it seems that a technical mistake was made in the derivation. Given that there does not seem to be a straightforward remedy to correct the mistake of [38], an alternative approach to lower bound $P^{-1}(t)$ is provided in Lemma B.5 of Appendix B.*

Remark 2.7 *As indicated by examining (2.60–2.64), the exponential envelope for the transient performance for $\psi(t)$ given in (2.59) can be adjusted through the selection of the various control parameters.*

2.2.5   Experimental Setup and Results

Description of the Experimental Testbed

The experimental testbed for the auto-balancing control design consists of an unbalanced disk with a center of geometry that does not coincide with the center of mass (see Figure 2.2). The disk is rotated by a DC motor via a belt-pulley transmission coupled with a universal joint and is contained within a circular space created by four horseshoe-shaped electromagnets forming a large air-gap magnetic bearing assembly (see Figure 2.3 for a close-up view of the universal joint and horseshoe-shaped electromagnetic assembly). The universal joint is used to ensure uninhibited planar disk vibrations. The horseshoe-shaped electromagnetic assembly is used to generate the vibration damping forces. An ultra-high intensity (5000 [mcd]) light-emitting diode (LED) with a viewing angle of 30 [steradians] is attached to the center of the free-end position of the rotating disk and is measured by two linear charge-coupled device (CCD) cameras which are placed at a distance of 0.86 [m] below the LED and are offset from each other by 90°. The data provided by the camera is sampled at 2 [kHz] and undergoes several stages of hardware and software decoding before the data is available as deflection in meters. A high-sensitivity slip-ring is placed at the clamped end of the rotor to provide the necessary excitation voltage to the LED.

A Pentium 266 MHz PC running QNX (a real-time micro-kernel-based operating system) hosts the control algorithm. The graphical user interface Qmotor [11] provides an environment to write the control algorithm in the C programming language. It also provides features such as on-line plotting and allows the user to vary control gains without recompiling the control program. The MultiQ I/O board provides for data transfer between the computer subsystem and the electrical interface. Five A/D channels are used to sense the currents flowing through the coils of the electromagnets and the DC motor. Five D/A channels output voltages that power the four electromagnets and drive the DC motor. These voltages go through two stages of amplification; the first stage consists of OP07C operational amplifiers, while in the second stage, Techron linear power amplifiers source a maximum current of 10 [amps] at 100 [volts].

A custom-designed software commutation strategy [16] ensures that the desired force commanded by the control law is applied to the rotating unbalanced disk. Roughly speaking, the commutation strategy involves translation of the desired force trajectory into desired current trajectories. To ensure that the actual magnetizing currents track the desired current trajectories, a high-gain current feedback loop is used to apply voltages to the four electromagnets. The four magnetizing currents and the motor current are measured using hall-effect current sensors. The planar and angular velocities of the disk were obtained by applying a backwards difference algorithm, in conjunction with a second-order digital low-pass filter, to the LED position measurements obtained from the cameras and to the motor angular position measurements obtained from the motor encoder, respectively. Each of the implemented controllers were executed with a sampling period of 0.5 [msec].

FIGURE 2.2. Autobalance experimental testbed.

Description of Experimental Results

The objective of the experiment presented in this section is to force the angular velocity of an unbalanced disk to track a desired velocity profile, while driving the planar disk displacements to zero. The desired velocity trajectory for each controller (see Figure 2.4) was designed as follows

$$\dot{\theta}_d(t) = \begin{cases} 400 \ [\text{rpm}] & 0 \le t < 27 \ [\text{sec}] \\ 400 - 10\cos(t) - 5\sin(20t) \ [\text{rpm}] & 27 \le t < 77 \ [\text{sec}] \\ 400 \ [\text{rpm}] & 77 \le t < 90 \ [\text{sec}]. \end{cases} \quad (2.81)$$

FIGURE 2.3. Close-up view of electromagnet assembly and rotating rod with universal joint.

During the interval $27 \leq t < 77$ [sec], the desired trajectory given in (2.81) ensures that the regression matrix $Y_d(\cdot)$ of (2.38) satisfies the PE condition of (2.57), thereby allowing a comparison of the parameter estimate convergence in the presence and absence of the PE condition.

Two experiments were performed to compare the performance of the DCAL-based controller with an open-loop controller. In the open-loop mode, the electromagnets were deenergized. That is, $F_x(t)$ and $F_y(t)$ defined in (2.5) were set to zero while the control voltage to the motor, denoted by $V_{cm}(t) \in \mathbb{R}$, was applied using the following high-gain current feedback approach

$$V_{cm} = \tau_z - k_\tau I_m \qquad (2.82)$$

where $\tau_z(t)$, defined in (2.5), is computed according to the following expression

$$\tau_z = \dot{\theta} - \dot{\theta}_d , \qquad (2.83)$$

$k_\tau \in \mathbb{R}$ is the torque constant of the motor, and $I_m(t) \in \mathbb{R}$ is the measured current in the motor. The open-loop response of the system is given in Figure 2.5. Rapid growth in the magnitude of the disk displacement can be seen until about 4 [sec] when the disk makes contact and is contained by the inside of the magnetic bearing assembly at about ± 14 [mm].

To contrast the open-loop response with that of the DCAL-based control law, the control design given in (2.51–2.58) was then implemented. Specifically, the control voltage $V_{cm}(t)$ was applied using the high-gain current feedback strategy given in (2.82), with the exception that $\tau_z(t)$ was now computed according to the DCAL-based control law given in (2.51–2.58). To develop the control voltage for the electromagnets, the desired force applied by each of the electromagnets was calculated as follows[4] [16]

$$f_{di} = \frac{1}{2}(-1)^{i+1} F_x + \sqrt{F_x^2 + \gamma_0^2} \quad \text{for } i = 1, 2 \tag{2.84}$$

where $F_x(t)$ is the desired control input along the x axis that is computed from (2.51) and $\gamma_0 \in \mathbb{R}$ is a small positive constant used to set the desired threshold winding current. Based on the desired electromagnetic force, the desired current through each of the electromagnets, denoted by $I_{d1}(t), I_{d2}(t) \in \mathbb{R}$, was developed as follows

$$I_{di} = \sqrt{\frac{2 f_{di}}{\beta L_1 \exp(\beta x)}} \quad \text{for } i = 1, 2 \tag{2.85}$$

where $\beta \in \mathbb{R}$ is an experimentally determined constant associated with the specific electromagnets utilized in the testbed and $L_1 \in \mathbb{R}$ represents the inductance of each of the electromagnets of the magnetic bearing assembly. The voltage applied to each of the electromagnets, denoted by $V_{o1}(t), V_{o2}(t) \in \mathbb{R}$, can then be applied using high-gain current feedback as follows

$$V_{oi} = k_e (I_{di} - I_i) \quad \text{for } i = 1, 2 \tag{2.86}$$

where $I_1(t), I_2(t) \in \mathbb{R}$ are the measured currents through each of the coils and $k_e \in \mathbb{R}$ is a positive control gain. The gains that resulted in the best

---

[4] Here, we have only presented the control implementation for the x-axis. The y-axis is similar.

closed-loop performance are given below

$$P(0) = \begin{bmatrix} 10.0 & 0 & 0 \\ 0 & 10.0 & 0 \\ 0 & 0 & 0.1 \end{bmatrix}$$

$$\gamma_0 = 0.001 \qquad \beta = 60 \qquad L_1 = 0.0002$$

$$\alpha = \text{diag}\,\{1.2, 1.2, 0\} \qquad K_p = \text{diag}\,\{0.1, 0.1, 0\}$$

$$K_s = \text{diag}\,\{2.8, 2.8, 1.0\} \qquad k_n = 2.0 \qquad \gamma = 2.5$$

$$k_e = 1.6 \qquad k_1 = 10.0 \qquad \lambda_1 = 0.1.$$

(2.87)

Figure 2.6 shows the disk displacements along the x and y axes. Notice that the peak closed-loop error with the adaptive controller is about ± 2.3 [mm], which is approximately 17% of the peak open-loop error. Figure 2.4 shows the angular velocity tracking error for the adaptive controller (the open-loop angular velocity tracking error response is similar to that of the adaptive controller and, hence, was not shown). As illustrated in Figure 2.7, during the time interval when the desired angular velocity satisfies the PE condition, the parameter estimates converge to different values.

The adaptive controller was also implemented with the following more aggressive desired velocity trajectory

$$\dot{\theta}_d(t) = \begin{cases} 900 \ [\text{rpm}] & 0 \le t < 27 \ [\text{sec}] \\ 900 - 50\cos(t) - 10\sin(20t) \ [\text{rpm}] & 27 \le t < 77 \ [\text{sec}] \\ 900 \ [\text{rpm}] & 77 \le t < 100 \ [\text{sec}], \end{cases}$$

(2.88)

with control gains similar to those used for the 400 [rpm] trajectory. The results for the trajectory given in (2.88) are given in Figures 2.8–2.10. As illustrated by these figures, the closed-loop response when the desired velocity trajectory is given by (2.88) is similar to that obtained with the trajectory given in (2.81).

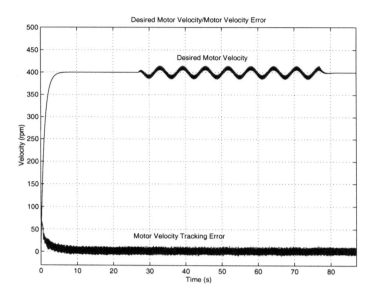

FIGURE 2.4. Desired motor velocity trajectory and motor velocity tracking error at 400 [rpm].

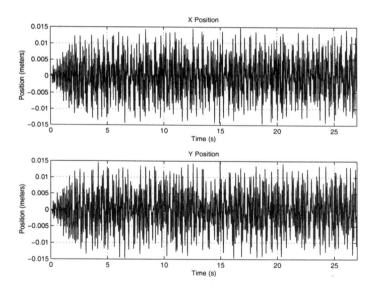

FIGURE 2.5. Open-loop performance of the disk displacement along the $x-$ and $y-$ axis at 400 [rpm].

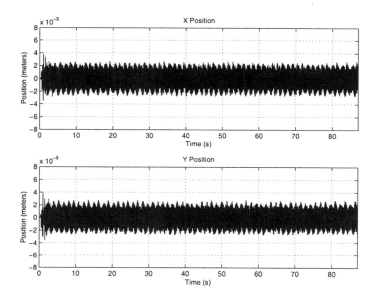

FIGURE 2.6. Adaptive controller performance of the disk displacement along the $x-$ and $y-$ axis at 400 [rpm].

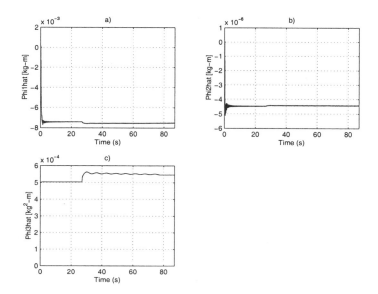

FIGURE 2.7. Parameter estimates for the adaptive controller at 400 [rpm] : (a) $mL\sin(\beta)$, (b) $mL\cos(\beta)$, and (c) $I_{cm} + 2mL^2$.

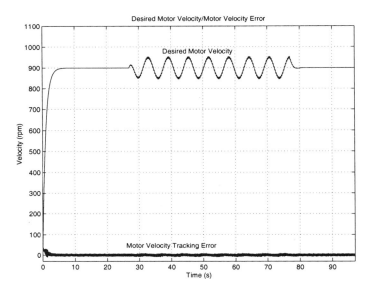

FIGURE 2.8. Desired motor velocity trajectory and motor velocity tracking error at 900 [rpm].

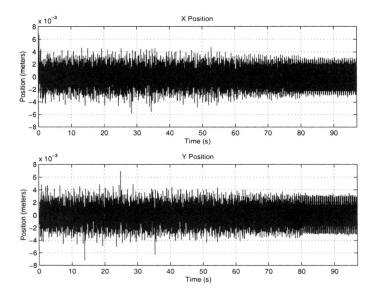

FIGURE 2.9. Adaptive controller performance of the disk displacement along the $x-$ and $y-$ axis at 900 [rpm].

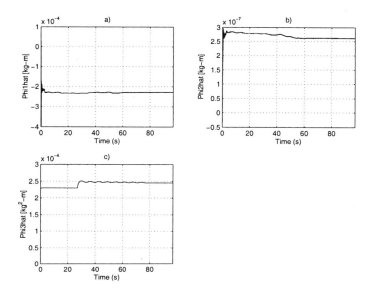

FIGURE 2.10. Parameter estimates for the adaptive controller at 900 [rpm] : (a) $mL\sin(\beta)$, (b) $mL\cos(\beta)$, and (c) $I_{cm} + 2mL^2$.

2.3 Dynamically Positioned Ships

In this section, two adaptive controllers are developed for dynamically positioned ships. The first controller uses a standard full-state feedback gradient-based adaptive update law. Based on the desire to eliminate velocity measurements, the second controller leverages off of recent results in the area of global adaptive output feedback control of robot manipulators [44, 46]. Specifically, the second controller is composed of an adaptive feedforward term that depends on the desired ship trajectory, a nonlinear feedback term, and a nonlinear filter that generates a surrogate velocity signal. Both controllers provide global asymptotic position tracking while compensating for parametric uncertainty in the ship dynamics. To prove the global adaptive output feedback tracking result for the second controller, an innovative nonquadratic Lyapunov function is used.

2.3.1 System Model

The dynamic and kinematic models for thruster-driven dynamically positioned ships can be written as follows [19, 29]

$$M\dot{\nu} + D\nu + K\eta = \tau \tag{2.89}$$

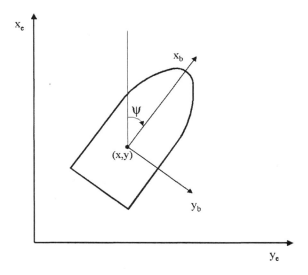

FIGURE 2.11. Ship coordinate frames.

$$\dot{\eta} = R(\psi)\nu \tag{2.90}$$

where $\nu(t)$, $\dot{\nu}(t) \in \mathbb{R}^3$ represent the ship velocity and acceleration, respectively, relative to the body-fixed coordinate frame $\{X_b, Y_b\}$, and $\eta(t) \in \mathbb{R}^3$ is defined in terms of the ship translational position, denoted by $x(t)$, $y(t) \in \mathbb{R}$, and the yaw angle, denoted by $\psi(t) \in \mathbb{R}$, relative to an Earth-fixed coordinate frame $\{X_e, Y_e\}$ as follows (see Figure 2.11)

$$\eta = \begin{bmatrix} x & y & \psi \end{bmatrix}^T. \tag{2.91}$$

The constant positive-definite and symmetric mass-inertia matrix $M \in \mathbb{R}^{3 \times 3}$ and the constant damping matrix $D \in \mathbb{R}^{3 \times 3}$ given in (2.89) are defined as follows

$$M = \begin{bmatrix} m_{11} & 0 & 0 \\ 0 & m_{22} & m_{23} \\ 0 & m_{23} & m_{33} \end{bmatrix} \quad D = \begin{bmatrix} d_{11} & 0 & 0 \\ 0 & d_{22} & d_{23} \\ 0 & d_{32} & d_{33} \end{bmatrix}, \tag{2.92}$$

$K = \text{diag}\{k_1, k_2, k_3\} \in \mathbb{R}^{3 \times 3}$ is a diagonal matrix representing the mooring forces, and $\tau(t) \in \mathbb{R}^3$ represents the control force/torque input vector provided by the thruster system. The matrix[5] $R(\psi) \in SO(3)$ given in (2.90)

---

[5] The set of all 3×3 rotation matrices is usually referred to as $SO(3)$ (Special Orthogonal group of order (3)) [25].

represents the rotation between the Earth and body-fixed coordinate frames that is defined as follows

$$R(\psi) = \begin{bmatrix} \cos(\psi) & -\sin(\psi) & 0 \\ \sin(\psi) & \cos(\psi) & 0 \\ 0 & 0 & 1 \end{bmatrix}. \tag{2.93}$$

To facilitate the subsequent control design and stability analysis, (2.89), (2.90), and the fact that

$$R^{-1}(\psi) = R^T(\psi) \tag{2.94}$$

are used to rewrite the dynamic model for the thruster-driven dynamically positioned ship in the following form

$$M^*\ddot{\eta} + V_m\dot{\eta} + F_1\dot{\eta} + F_2\eta = \tau^* \tag{2.95}$$

where the transformed dynamics given in (2.95) are related to (2.89) and (2.90) through the following expressions

$$M^*(\eta) = RMR^T \qquad V_m(\eta,\dot{\eta}) = RM\dot{R}^T$$

$$\tag{2.96}$$

$$F_1(\eta) = RDR^T \qquad F_2(\eta) = RK \qquad \tau^*(t) = R\tau.$$

Similar to the dynamic model of the rotating unbalanced disk, the structure of the dynamic model in (2.95) exhibits several properties that will be utilized in the subsequent control development and stability analysis.

Property 2.4: Symmetric and Positive-Definite Inertia Matrix

The transformed symmetric and positive-definite mass-inertia matrix $M^*(\eta)$ given in (2.96) satisfies the following inequalities

$$m_1 \|\xi\|^2 \le \xi^T M^*\xi \le m_2 \|\xi\|^2 \qquad \forall \xi \in \mathbb{R}^3 \tag{2.97}$$

where m_1, $m_2 \in \mathbb{R}$ are known positive constants, and $\|\cdot\|$ denotes the standard Euclidean norm.

Property 2.5: Skew-Symmetry

The transformed dynamic terms $M^*(\eta)$ and $V_m(\eta,\dot{\eta})$ given in (2.96) satisfy the following skew-symmetric relationship

$$\xi^T \left(\frac{1}{2}\dot{M}^* - V_m \right) \xi = 0 \qquad \forall \xi \in \mathbb{R}^3 \tag{2.98}$$

where $\dot{M}^*(q)$ denotes the time derivative of the transformed inertia matrix.

Property 2.6: Switching Property

The matrix $V_m(\eta, \dot{\eta})$ of (2.96) satisfies the following relationship

$$V_m(\eta, u)w = V_m(\eta, w)u \quad \forall u, w \in \mathbb{R}^3. \qquad (2.99)$$

Property 2.7: Bounding Inequalities

The norm of $V_m(\eta, \dot{\eta})$ and $F_1(\eta)$ can be upper bounded as follows

$$\|V_m(\eta, \dot{\eta})\|_{i\infty} \leq \zeta_{v1} \|\dot{\eta}\| \qquad (2.100)$$

$$\|F_1(\eta)\|_{i\infty} \leq \zeta_{f1} \qquad (2.101)$$

where $\zeta_{v1}, \zeta_{f1} \in \mathbb{R}$ are known positive bounding constants, and $\|\cdot\|_{i\infty}$ denotes the induced infinity norm. The following bounds are also valid for $\forall u, w \in \mathbb{R}^3$ (see Section B.1.5 of Appendix B for the proofs)

$$\|M^*(u) - M^*(w)\|_{i\infty} \leq \zeta_m \|Tanh\,(u - w)\|$$

$$\|V_m(u, \dot{\eta}) - V_m(w, \dot{\eta})\|_{i\infty} \leq \zeta_{v2} \|\dot{\eta}\| \, \|Tanh\,(u - w)\| \qquad (2.102)$$

$$\|F_1(u) - F_1(w)\| \leq \zeta_{f2} \|Tanh\,(u - w)\|$$

where ζ_m, ζ_{v2}, and $\zeta_{f2} \in \mathbb{R}$ are some positive bounding constants, and the vector function $Tanh(\cdot) \in \mathbb{R}^3$ is defined as follows

$$Tanh\,(a) \triangleq [\tanh(a_1), \tanh(a_2), \tanh(a_3)]^T \quad \forall a = [a_1, a_2, a_3]^T. \qquad (2.103)$$

Property 2.8 Linearity in the Parameters

The dynamics given in (2.96) can be linearly parameterized as follows

$$Y_s \phi = M^* \ddot{\eta} + V_m \dot{\eta} + F_1 \dot{\eta} \qquad (2.104)$$

where $Y_s(\eta, \dot{\eta}, \ddot{\eta}) \in \mathbb{R}^{3 \times 9}$ denotes a measurable regression matrix, and $\phi \in \mathbb{R}^9$ denotes the constant system parameter vector defined as

$$\phi = \begin{bmatrix} m_{11} & m_{22} & m_{23} & m_{33} & d_{11} & d_{22} & d_{23} & d_{32} & d_{33} \end{bmatrix}^T. \qquad (2.105)$$

2.3.2   Adaptive Full-State Feedback Control

The objective of this section is to design a global position tracking controller for the dynamically positioned ship model given by (2.95) despite parametric uncertainty in the nonlinear ship dynamics where $\eta(t)$, $\dot{\eta}(t)$ are assumed to be measurable. To quantify the ship position tracking performance, a position tracking error $e(t) \in \mathbb{R}^3$ is defined as follows

$$e = \eta_d - \eta \tag{2.106}$$

where $\eta_d(t) \in \mathbb{R}^3$ represents the ship's desired position trajectory that must be constructed such that $\eta_d(t), \dot{\eta}_d(t), \ddot{\eta}_d(t), \dddot{\eta}_d(t) \in \mathcal{L}_\infty$. In a similar manner as in the previous section, a filtered tracking error is defined as follows

$$r = \dot{e} + \alpha e \tag{2.107}$$

where $\alpha \in \mathbb{R}^{3 \times 3}$ is a control gain matrix given by the following expression

$$\alpha = \text{diag}\{\alpha_1, \alpha_2, \alpha_3\} \tag{2.108}$$

where $\alpha_1, \alpha_2, \alpha_3 \in \mathbb{R}$ are positive control gains.

Control Formulation

To develop the open-loop error system for $r(t)$, we take the time derivative of (2.107), premultiply the resulting expression by $M^*(\eta)$, and then substitute (2.95) for $M^*(\eta)\ddot{\eta}(t)$ to obtain the following expression

$$M^* \dot{r} = -V_m r + Y\phi - \tau^* \tag{2.109}$$

where $Y(\eta, \dot{\eta}, t) \in \mathbb{R}^{3 \times p}$ denotes a measurable regression matrix such that

$$Y\phi = M^* (\ddot{\eta}_d + \alpha\dot{e}) + V_m (\dot{\eta}_d + \alpha e) + F_1\dot{\eta} + F_2\eta \tag{2.110}$$

where ϕ is given in (2.105), and (2.106) and (2.107) have been used. Based on the structure of the open-loop error dynamics given in (2.109) and the subsequent stability analysis, we design the control input $\tau^*(t)$ as follows

$$\tau^* = Y\hat{\phi} + K_s r + K_p e \tag{2.111}$$

where $\hat{\phi}(t) \in \mathbb{R}^p$ denotes a parameter estimate vector for ϕ that is generated according to the following gradient update law

$$\dot{\hat{\phi}} = -\Gamma Y^T r \tag{2.112}$$

where Γ, K_s, and $K_p \in \mathbb{R}^{3 \times 3}$ denote positive-definite control gain matrices defined as follows

$$\begin{aligned} \Gamma &= \text{diag}\{\Gamma_1, \Gamma_2, \Gamma_3\} \\ K_s &= \text{diag}\{k_{s1}, k_{s2}, k_{s3}\} \\ K_p &= \text{diag}\{k_{p1}, k_{p2}, k_{p3}\}. \end{aligned} \tag{2.113}$$

After substituting (2.111) into (2.109), the following closed-loop error system is obtained

$$M^*\dot{r} = -V_m r + Y\tilde{\phi} - K_s r - K_p e \qquad (2.114)$$

where the parameter estimation error $\tilde{\phi}(t) \in \mathbb{R}^p$ is defined as follows

$$\tilde{\phi} = \phi - \hat{\phi}. \qquad (2.115)$$

Stability Analysis

The stability of the adaptive ship position tracking controller given in (2.111) and (2.112) can be examined through the following theorem.

Theorem 2.3 *The controller given by (2.111) along with the adaptive update law given in (2.112) ensures global asymptotic ship position tracking in the sense that*

$$\lim_{t \to \infty} e(t) = 0 \qquad (2.116)$$

where $e(t)$ was defined in (2.106).

Proof: To prove Theorem 2.3, we define a nonnegative function $V(t) \in \mathbb{R}$ as follows

$$V = \frac{1}{2}r^T M^* r + \frac{1}{2}e^T K_p e + \frac{1}{2}\tilde{\phi}^T \Gamma^{-1}\tilde{\phi}. \qquad (2.117)$$

After taking the time derivative of (2.117), substituting (2.114) into the resulting expression for $M^*(\eta)\dot{r}(t)$, substituting the time derivative of (2.115) for $\dot{\tilde{\phi}}(t)$, and then cancelling common terms, the following expression can be obtained

$$\dot{V} = -r^T K_s r - e^T K_p \alpha e \qquad (2.118)$$

where (2.98) and (2.107) have been used. From (2.117) and (2.118), it is clear that $r(t)$, $e(t)$, $\tilde{\phi}(t) \in \mathcal{L}_\infty$ and that $e(t)$, $r(t) \in \mathcal{L}_2$ (see Lemma A.11 of Appendix A). Since $r(t) \in \mathcal{L}_\infty$, Lemma A.13 of Appendix A can be used along with (2.106) and (2.107) to prove that $\dot{e}(t)$, $\eta(t)$, $\dot{\eta}(t) \in \mathcal{L}_\infty$; furthermore, (2.115) can be used to prove that $\dot{\hat{\phi}}(t) \in \mathcal{L}_\infty$. Since the regression matrix $Y(\cdot)$ is made up of bounded arguments, we can state that $Y(\cdot) \in \mathcal{L}_\infty$; therefore, (2.111) can be used to prove that the control input $u(t) \in \mathcal{L}_\infty$. It follows from (2.114) that $\dot{r}(t) \in \mathcal{L}_\infty$; hence, $\ddot{e}(t)$ and $\ddot{\eta}(t) \in \mathcal{L}_\infty$. Based on the facts that $r(t)$, $\dot{r}(t) \in \mathcal{L}_\infty$ and $r(t) \in \mathcal{L}_2$, Barbalat's Lemma (see Lemma A.16 of Appendix A) can now be invoked to prove that

$$\lim_{t \to \infty} r(t) = 0. \qquad (2.119)$$

The condition (2.119) and Lemma A.15 of Appendix A can now be utilized to prove the result given in (2.28). Alternatively, the expression in (2.119),

$e(t)$, $\dot{e}(t) \in \mathcal{L}_\infty$ and $e(t) \in \mathcal{L}_2$, and Barbalat's Lemma can now be utilized to prove the result given in (2.116). □

2.3.3   Adaptive Output Feedback Control

In the previous section, the fact that full-state feedback was available for the control design (i.e., $\tau^*(t)$ of (2.111) depends on $\eta(t), \dot{\eta}(t)$) was exploited. However, for many mechanical systems the requirement for velocity measurements is undesirable due to increased cost/complexity of an additional sensor and the added noise that is inherent to velocity measurements. Hence, motivation exists to develop output feedback controllers that only require the measurement of the output signal (e.g., the ship position $\eta(t)$ in this application). Given this motivation, the objective of this section is to design a global adaptive position tracking controller for the dynamically positioned ship given in (2.95) despite the additional constraint that only the ship position $\eta(t)$ is available for measurement. To achieve a global stability result for the subsequent adaptive output feedback controller, the dynamic model given in (2.95) must be simplified by neglecting the mooring effects. That is, if the mooring effects are included in the dynamic model then the result degrades to a semi-global stability result (see the discussion provided in the subsequent Remarks 2.10 and 2.13).

To facilitate the output feedback control design and the stability analysis, a new filtered tracking error signal, denoted by $r(t) \in \mathbb{R}^3$, is defined as follows

$$r = \dot{e} + Tanh\,(e) + z \qquad (2.120)$$

where $e(t)$ was defined in (2.106). The velocity-related tracking error signal $z(t) = [z_1, z_2, z_3]^T$ given in (2.120) is defined by the following dynamic nonlinear filter

$$z_i = p_i - k_c e_i \qquad (2.121)$$

$$
\begin{aligned}
\dot{p}_i = \;& -\left(1 - (p_i - k_c e_i)^2\right)^2 (p_i - k_c e_i - \tanh(e_i)) \\
& -k_c\,(\tanh(e_i) + p_i - k_c e_i)
\end{aligned}
\qquad (2.122)
$$

where $k_c \in \mathbb{R}$ denotes a positive constant filter gain, and $p_i(t) \in \mathbb{R}$ is a filter variable that is initialized to satisfy the following inequalities

$$-\frac{1}{\sqrt{3}} + k_c e_i(0) < p_i(0) < \frac{1}{\sqrt{3}} + k_c e_i(0) \quad \forall i = 1, 2, 3. \qquad (2.123)$$

Based on the development given in (2.120–2.123), the dynamics for the velocity-related tracking error signal can be obtained by taking the time derivative of (2.121) as follows

$$\dot{z}_i = -(1 - z_i^2)^2(z_i - \tanh(e_i)) - k_c r_i$$

$$|z_i(0)| < \frac{1}{\sqrt{3}} \quad \forall i = 1, 2, 3 \tag{2.124}$$

where (2.122) was utilized.

Remark 2.8 *Based on the definition for the filtered tracking error signal given in (2.120), it is clear that from (2.106) that $r(t)$ depends on $\dot{\eta}(t)$. The control objective in this section is targeted at the output feedback control problem, and hence, $\dot{\eta}(t)$, $r(t)$, and $\dot{z}_i(t)$ are not measurable signals. Although it cannot be directly used in the subsequent controller, the filtered tracking error signal facilitates the subsequent control development and stability analysis.*

Control Formulation

To develop the open-loop error system for $r(t)$, we take the time derivative of (2.120), premultiply the resulting expression by $M^*(\eta)$, and then substitute (2.95) for $M^*(\eta)\ddot{\eta}(t)$ (with the mooring effects neglected) to obtain the following expression

$$M^*\dot{r} = M^*\ddot{\eta}_d + V_m\dot{\eta} + F_1\dot{\eta} - \tau^* + M^* Cosh^{-2}(e)\dot{e} + M^*\dot{z} \tag{2.125}$$

where the matrix function $Cosh(\cdot) \in \mathbb{R}^{3\times3}$ is defined as follows

$$Cosh(e) = \text{diag}\{\cosh(e_1), \cosh(e_2), \cosh(e_3)\} \quad \forall e = [e_1,\ e_2,\ e_3]^T \in \mathbb{R}^3. \tag{2.126}$$

To facilitate the subsequent analysis, a desired linear parameterization is defined as follows

$$Y_d\phi = M^*(\eta_d)\ddot{\eta}_d + V_m(\eta_d, \dot{\eta}_d)\dot{\eta}_d + F_1(\eta_d)\dot{\eta}_d \tag{2.127}$$

where $Y_d(t) \in \mathbb{R}^{3\times p}$ denotes a desired regression matrix, and the constant unknown parameter vector ϕ is given in (2.105). After adding and subtracting (2.127) and the product $V_m(\eta, \dot{\eta})r(t)$ to (2.125), the open-loop dynamics for $r(t)$ can be formulated as follows

$$M^*\dot{r} = -V_m r + Y_d\phi - \tau^* - k_c M^* r + \tilde{Y} + \chi \tag{2.128}$$

where (2.106), (2.120), and (2.124) were used. In (2.128), the auxiliary terms $\tilde{Y}(e, z, \eta, t)$, $\chi(e, z, \eta, t) \in \mathbb{R}^3$ are defined as follows

$$\tilde{Y} = M^*(\eta)\ddot{\eta}_d + V_m(\eta, \dot{\eta}_d)\dot{\eta}_d + F_1(\eta)\dot{\eta} - Y_d\phi \tag{2.129}$$

and

$$\chi = M^*(\eta) \, Cosh^{-2}(e)(r - Tanh(e) - z) - M^*(\eta)T(z - Tanh(e))$$

$$+V_m(\eta, \dot{\eta}_d + Tanh(e) + z)(Tanh(e) + z)$$

$$+V_m(\eta, \dot{\eta}_d)(Tanh(e) + z) - V_m(\eta, r)(\dot{\eta}_d + Tanh(e) + z). \tag{2.130}$$

Based on the open-loop dynamics for $r(t)$ given in (2.128), the following adaptive output feedback tracking control law is developed

$$\tau^* = Y_d\hat{\phi} - k_cT^{-1}z + Tanh(e) \tag{2.131}$$

where $k_c \in \mathbb{R}$ is a positive control gain, and the matrix $T(z) \in \mathbb{R}^{3 \times 3}$ is defined as

$$T(z) = \text{diag}\left\{ \left(1 - z_1^2\right)^2, \left(1 - z_2^2\right)^2, \left(1 - z_3^2\right)^2 \right\} \tag{2.132}$$

where $z(t)$ is defined in (2.121). The parameter estimate $\hat{\phi}(t) \in \mathbb{R}^p$ given in (2.131) is generated by the following gradient adaptation law

$$\hat{\phi} = \Gamma \int_0^t Y_d^T(\eta_d(\sigma), \dot{\eta}_d(\sigma), \ddot{\eta}_d(\sigma))(Tanh(e(\sigma)) + z(\sigma)) \, d\sigma$$

$$-\Gamma \int_0^t \dot{Y}_d^T(\eta_d(\sigma), \dot{\eta}_d(\sigma), \ddot{\eta}_d(\sigma))e(\sigma)d\sigma + \Gamma Y_d^T e \tag{2.133}$$

where $\Gamma \in \mathbb{R}^{p \times p}$ denotes a constant diagonal positive-definite adaptation gain matrix.

After substituting the control law given in (2.131) into (2.128), the following closed-loop dynamics for $r(t)$ are obtained

$$M^*\dot{r} = -V_m r + Y_d\tilde{\phi} + k_cT^{-1}z - Tanh(e) - k_cM^*r + \tilde{Y} + \chi \tag{2.134}$$

where (2.115) was used. After differentiating (2.115) and (2.133), the following dynamic relationship for the parameter estimation error can also be developed

$$\dot{\tilde{\phi}} = -\Gamma Y_d^T r \tag{2.135}$$

where (2.120) has been utilized.

Remark 2.9 *As proven by Lemma B.9 of Appendix B, (2.97), (2.100), several trigonometric identities, and the boundedness properties of the desired trajectory can be used to prove that $\chi(\cdot)$ of (2.130) can be upper bounded as follows*

$$\|\chi\| \leq \zeta_1 \|\varrho\| + \zeta_2 \|z\|^2 + \zeta_3 \|z\|^3 + \zeta_4 \|z\|^4 + \zeta_5 \|z\|^5 + \zeta_6 \|r\| \|z\| \tag{2.136}$$

where the composite state vector $\varrho(t) \in \mathbb{R}^9$ is defined as

$$\varrho \triangleq \left[\ r^T \quad Tanh^T(e) \quad z^T \ \right]^T \tag{2.137}$$

and $\zeta_i \in \mathbb{R}$, $i = 1, ..., 6$ are known positive bounding constants that depend on the system parameters and the desired trajectory. Furthermore, by substituting (2.127) into (2.129) for $Y_d(\cdot)\phi$ and then using (2.101), (2.102), (2.106), and (2.120), it can be shown that $\tilde{Y}(\cdot)$ of (2.129) can be upper bounded as follows

$$\left\| \tilde{Y} \right\| \le \zeta_7 \left\| \varrho \right\| \tag{2.138}$$

where $\zeta_7 \in \mathbb{R}$ is a known positive constant depending on the system parameters and the desired trajectory.

Remark 2.10 *To develop the bound for $\tilde{Y}(e, z, \eta, t)$ given in (2.138), the mooring effects of (2.95) must be neglected because the mooring effects given by $F_2(\eta)\eta$ cannot be upper bounded by a function of $\varrho(t)$. Specifically, the norm of the mismatch between the desired mooring term and the actual mooring effect given below*

$$F_2(\eta)\eta - F_2(\eta_d)\eta_d \tag{2.139}$$

can only be upper bounded in terms of $e(t)$ as follows

$$\left\| F_2(\eta)\eta - F_2(\eta_d)\eta_d \right\| \le \zeta_8 \left\| e \right\| + \zeta_9 \left\| Tanh(e) \right\| \tag{2.140}$$

where ζ_8, $\zeta_9 \in \mathbb{R}$ are positive bounding constants that depend on the system parameters and the desired trajectory. That is, since

$$\left\| e \right\| \ge \left\| Tanh(e) \right\| , \tag{2.141}$$

the norm of the mismatch given in (2.140) can not be upper bounded in terms of $\varrho(t)$ as in (2.138).

Stability Analysis

The stability of the adaptive output feedback ship position tracking controller given in (2.121), (2.123), (2.131), and (2.133) can be examined through the following theorem.

Theorem 2.4 *Given the ship dynamics of (2.95), the adaptive output feedback tracking controller of (2.121), (2.123), (2.131), and (2.133) ensures global asymptotic tracking in the sense that*

$$\lim_{t \to \infty} r(t), z(t), e(t) = 0 \tag{2.142}$$

in the region

$$\left\{ \left(r, \tilde{\phi}, z, e \right) \in \mathbb{R}^3 \times \mathbb{R}^p \times \left(-\frac{1}{\sqrt{3}}, \frac{1}{\sqrt{3}} \right)^3 \times \mathbb{R}^3 \right\} \qquad (2.143)$$

provided the control gain k_c given in (2.131) is selected as follows

$$k_c = \frac{1}{m_1} \left(1 + k_n \left(\zeta_1 + \zeta_7 \right)^2 + 16\zeta_2^2 + 8\zeta_3^2 + 4\zeta_4^2 + 16\zeta_5^2 + \zeta_6 \right) \qquad (2.144)$$

where m_1 is given in (2.97), ζ_i, $i = 1, ..., 7$ were defined in (2.136) and (2.138), and k_n is an additional control gain selected to satisfy the following sufficient condition

$$k_n > 2. \qquad (2.145)$$

Proof: To prove Theorem 2.4, we define a nonnegative function $V(t) \in \mathbb{R}$ as follows

$$V = \frac{1}{2} r^T M^* r + \frac{1}{2} \sum_{i=1}^{3} \frac{z_i^2}{1 - z_i^2} + \sum_{i=1}^{3} \ln \left(\cosh(e_i) \right) + \frac{1}{2} \tilde{\phi}^T \Gamma^{-1} \tilde{\phi} \qquad (2.146)$$

where $\ln \left(\cosh(e_i) \right)$ is positive-definite and radially unbounded, and the function $\frac{z_i^2}{1 - z_i^2}$ is positive-definite and radially unbounded on the interval $[-1, 1]$; hence, $V(t)$ is a positive-definite radially unbounded function in the set

$$S = \left\{ \left(r, \tilde{\phi}, z, e \right) \in \mathbb{R}^3 \times \mathbb{R}^p \times [-1, 1]^3 \times \mathbb{R}^3 \right\}. \qquad (2.147)$$

After taking the time derivative of (2.146), utilizing (2.98), (2.120), (2.124), (2.134), and (2.135) and then cancelling the common terms, the following expression can be obtained

$$\dot{V} = r^T \left(-k_c M^* r + \tilde{Y} + \chi \right) - \sum_{i=1}^{3} z_i^2 - \sum_{i=1}^{3} \tanh^2(e_i). \qquad (2.148)$$

After using (2.136) and (2.138) and substituting (2.144) into (2.148) for k_c, the following upper bound can be obtained

$$\dot{V} \leq \; - \|r\|^2 - \|z\|^2 - \| Tanh(e) \|^2$$

$$+ \left[(\zeta_1 + \zeta_7) \|\varrho\| \, \|r\| - k_n \left(\zeta_1 + \zeta_7 \right)^2 \|r\|^2 \right]$$

$$+ \left[\zeta_2 \|z\|^2 \|r\| - 16\zeta_2^2 \|r\|^2 \right] + \left[\zeta_3 \|z\|^3 \|r\| - 8\zeta_3^2 \|r\|^2 \right] \qquad (2.149)$$

$$+ \left[\zeta_4 \|z\|^4 \|r\| - 4\zeta_4^2 \|r\|^2 \right] + \left[\zeta_5 \|z\|^5 \|r\| - 16\zeta_5^2 \|r\|^2 \right]$$

$$+ \zeta_6 \left(\|z\| - 1 \right) \|r\|^2.$$

After completing the squares on the bracketed terms of (2.149), we can further upper bound $\dot{V}(t)$ as follows

$$
\dot{V} \leq -\frac{1}{2}\|\varrho\|^2 + \frac{1}{k_n}\|\varrho\|^2 + \frac{1}{2}\|z\|^2 \left[-1 + \frac{1}{8}\|z\|^2 + \frac{1}{4}\|z\|^4\right]
$$

$$
+ \frac{1}{2}\|z\|^2 \left[\frac{1}{2}\|z\|^6 + \frac{1}{8}\|z\|^8\right] + \zeta_6\left[\|z\| - 1\right]\|r\|^2
\tag{2.150}
$$

where (2.137) was utilized. If k_n is selected according to (2.145), then (2.150) can be utilized to prove the following inequality

$$
\dot{V} \leq -\beta\|\varrho\|^2 \quad \text{if} \quad \|z(t)\| < 1 \; \forall t \geq 0
\tag{2.151}
$$

where β is a positive constant that satisfies the following inequality

$$
0 < \beta < \frac{1}{2}.
\tag{2.152}
$$

Based on (2.151), the following inequality can be utilized

$$
\|z\|^2 \leq 3 \max_i |z_i|^2
\tag{2.153}
$$

to define the following set

$$
S_1 = \left\{\left(r, \tilde{\phi}, z, e\right) \in \mathbb{R}^3 \times \mathbb{R}^p \times \left(-\frac{1}{\sqrt{3}}, \frac{1}{\sqrt{3}}\right)^3 \times \mathbb{R}^3\right\};
\tag{2.154}
$$

hence, (2.151) can be written as follows

$$
\dot{V} \leq -\beta\|\varrho\|^2 \quad \text{if} \quad \left(r, \tilde{\phi}, z, e\right) \in S_1.
\tag{2.155}
$$

Since $S_1 \subset S$, where S was defined in (2.147), the region of attraction will contain the largest level set of $V(t)$ inside the set S_1. Since all level sets of $V(t)$ are contained inside S, then S_1 is invariant and an estimate of the stability region. Hence, for initial conditions inside S_1, (2.146) and (2.155) can be used to prove that $\varrho(t)$, $e(t)$, $\tilde{\phi}(t) \in \mathcal{L}_\infty$ and that $\varrho(t) \in \mathcal{L}_2$ (see Lemma A.11 of Appendix A). Based on the fact that $\tilde{\phi}(t) \in \mathcal{L}_\infty$, (2.115) can be used to prove that $\hat{\phi}(t) \in \mathcal{L}_\infty$. Given that $\varrho(t) \in \mathcal{L}_\infty$, (2.137) can be used to prove that $r(t)$, $z(t) \in \mathcal{L}_\infty$. Since $e(t)$, $r(t)$, $z(t) \in \mathcal{L}_\infty$, (2.106), (2.120), (2.121), (2.124), and the assumption that $\eta_d(t) \in \mathcal{L}_\infty$ can be used to prove that $\eta(t)$, $p_i(t)$, $\dot{p}_i(t)$, $\dot{e}(t)$, $\dot{z}(t) \in \mathcal{L}_\infty$. The previous development indicates that

$$
|z_i(t)| \leq \frac{1}{\sqrt{3}}.
\tag{2.156}
$$

From (2.132) and (2.156), we can prove that $T(z)$, $T^{-1}(z) \in \mathcal{L}_\infty$. Based on the fact that $\eta_d(t), \dot{\eta}_d(t), \ddot{\eta}_d(t)$ are assumed to be bounded, (2.127–2.131) can be used to prove that $Y_d(t)$, $\tilde{Y}(\cdot)$, $\chi(\cdot)$, $\dot{r}(t)$, $\tau^*(t) \in \mathcal{L}_\infty$. Since $\dot{r}(t)$, $\dot{e}(t)$, $\dot{z}(t) \in \mathcal{L}_\infty$, the definition given in (2.137) can be used to prove that $\dot{\varrho}(t) \in \mathcal{L}_\infty$; hence, Barbalat's Lemma (see Lemma A.16 of Appendix A) can be invoked to prove that

$$\lim_{t \to \infty} \varrho(t) = 0 \quad \text{if} \quad \left(r, \tilde{\phi}, z, e \right) \in S_1. \tag{2.157}$$

Based on the definition of $\varrho(t)$ given in (2.137), (2.157) can be used to prove that

$$\lim_{t \to \infty} r(t),\, z(t),\, Tanh(e(t)) = 0 \tag{2.158}$$

in the region

$$\left\{ \left(r, \tilde{\phi}, z, e \right) \in \mathbb{R}^3 \times \mathbb{R}^p \times \left(-\frac{1}{\sqrt{3}}, \frac{1}{\sqrt{3}} \right)^3 \times \mathbb{R}^3 \right\}. \tag{2.159}$$

Based on the fact that

$$\lim_{t \to \infty} e(t) = 0 \tag{2.160}$$

is a necessary condition for

$$\lim_{t \to \infty} Tanh(e(t)) = 0, \tag{2.161}$$

the result given in (2.142) can now be obtained. $\square$

Remark 2.11 *Despite the initial condition restriction for the filter signal $z(t)$ given in (2.124), the stability result is still global for the position tracking error $e(t)$ since no restrictions are placed on the size of $\|e(0)\|$ (see (2.143)). Moreover, no restrictions are placed on the size of $\|r(0)\|$, $\left\|\tilde{\phi}(0)\right\|$, and $\|\dot{e}(0)\|$. Based on the control structure given by (2.121), (2.123), (2.131), and (2.133), we note that the initial conditions of the filter only need to be adjusted relative to the measurable quantity $e(0)$.*

Remark 2.12 *As in [29], the dynamic model presented in (2.89) (with the mooring effects neglected) can be modified to include a bias term representing drift, currents, and wave load as follows*

$$M\dot{\nu} + D\nu - R(\psi)b = \tau \tag{2.162}$$

where $b \in \mathbb{R}^3$ is an unknown constant vector. Given the model modification of (2.162), the adaptive output feedback tracking controller of (2.121),

*(2.123), (2.131), and (2.133) can be easily extended by adding an additional
term to $\tau^*(t)$ of (2.131) as follows*

$$\tau^* = Y_d\hat{\phi} - k_cT^{-1}z + Tanh(e) - \hat{b} \tag{2.163}$$

where the estimate $\hat{b}(t)$ is updated as follows

$$\dot{\hat{b}} = -\Gamma_2 \int_0^t (Tanh(e(\sigma)) + z(\sigma))\, d\sigma - \Gamma_2 e \tag{2.164}$$

*where $\Gamma_2 \in \mathbb{R}^{3\times3}$ is a constant diagonal positive-definite adaptation gain
matrix.*

Remark 2.13 *If the thruster-assisted mooring effects are included in the
dynamic model for the global adaptive output feedback control section, some
extra terms will be present that will change the bounds given in (2.138) for
$\tilde{Y}(t)$ (see the comments in Remark 2.10). If the thruster-assisted mooring
effects are incorporated in the model as in the previous full-state feedback design, a control design methodology similar to the one of [5] could be followed
to construct a semi-global adaptive output feedback tracking controller.*

2.3.4   Simulation Results

To simulate the adaptive output feedback controller, a ship model with the
following mass-inertia and damping matrices was used [20]

$$M = \begin{bmatrix} 1.0852 & 0 & 0 \\ 0 & 2.0575 & -0.4087 \\ 0 & -0.4087 & 0.2153 \end{bmatrix}$$

$$D = \begin{bmatrix} 0.08656 & 0 & 0 \\ 0 & 0.0762 & 0.1510 \\ 0 & 0.0151 & 0.0031 \end{bmatrix}. \tag{2.165}$$

The desired position trajectory for the ship was selected as follows

$$\eta_d(t) = [10\sin(0.2t)\ [\text{m}],\ 10\cos(0.2t)\ [\text{m}],\ 5\sin(0.2t)\ [\text{rad}]]^T. \tag{2.166}$$

The initial conditions for the ship position $\eta(0)$ were selected as

$$\eta(0) = [1\ [\text{m}],\ \text{-}1\ [\text{m}],\ 1\ [\text{rad}]]^T \tag{2.167}$$

and the initial condition for the ship velocity $\dot{\eta}(0)$, the velocity-related
tracking error signal $z(0)$, and parameter estimate $\hat{\phi}(0)$ were all set to zero.

The control and adaptation gains were selected as follows[6]

$$k_c = \text{diag}\{20, 40, 50\} \quad \Gamma = \text{diag}\{35, 50, 250, 280, 50, 100, 200, 400, 600\}.$$
$$(2.168)$$

Figure 2.12 illustrates the position tracking error performance while the control inputs are shown in Figure 2.13. The parameter estimates for the inertia and damping parameters are presented in Figures 2.14 and 2.15, respectively.

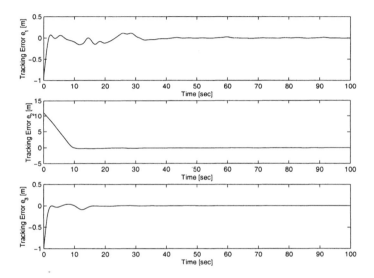

FIGURE 2.12. Position tracking errors.

---

[6]The stability analysis required that the gain k_c be defined as a scalar; however, k_c was defined as a matrix during the simulation. Although we cannot theoretically justify this modification, we have verified from experience that it usually improves the tracking performance in numerical simulations and real-time implementations.

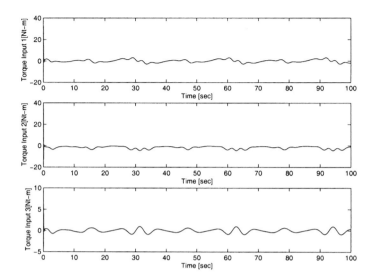

FIGURE 2.13. Control input torques.

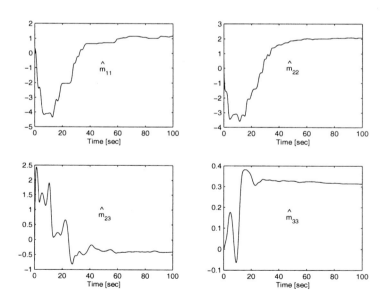

FIGURE 2.14. Parameter estimates for the entries of the inertia matrix.

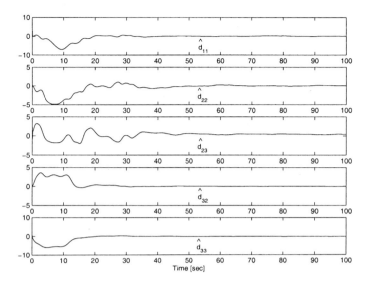

FIGURE 2.15. Parameter estimates for the entries of the damping matrix.

2.4 Euler-Lagrange Systems

A task space-based tracking controller for general Euler-Lagrange systems is formulated in this section. To eliminate singularity problems associated with some of the three-parameter task space formulations, as outlined in the introduction of this chapter, the orientation of a coordinate frame fixed to a mass is expressed in terms of the unit quaternion, and thereby, the orientation tracking error problem is formulated as commonly done in orientation (i.e., attitude) control problems[7] [3, 12, 27]. To demonstrate how the control design methodology for general Euler-Lagrange systems can be formulated to target a specific example, a unit quaternion-based controller for fully actuated UUVs is designed.

2.4.1 System Model

Kinematic Model

Let $\mathcal{E}$ and $\mathcal{B}$ be orthogonal coordinate frames attached to the body mass of a general Euler-Lagrange system and the inertial frame, respectively.

---

[7]Note that most task space robotic controllers formulate the orientation error according to the classical operational space approach [6] of taking the algebraic difference between the desired and actual Euler angles (see, for example, [45]).

The position and orientation of $\mathcal{E}$ relative to $\mathcal{B}$ are commonly represented by a homogeneous transformation matrix that is expressed in terms of a generalized coordinate system. From this homogeneous transformation matrix, several different representations can be used to develop the kinematic model, including three-parameter representations (e.g., the Euler angles, Gibbs vector, Cayley-Rodrigues parameters, and Modified Rodrigues parameters) and the constrained four-parameter unit quaternion representation. In this section, the unit quaternion parameterization is used to provide a global nonsingular parameterization of the orientation of $\mathcal{E}$ with respect to $\mathcal{B}$. To facilitate the unit quaternion parameterization, Euler's theorem is used to note that any rotation matrix can be uniquely represented by a rotation of angle $\varphi(t) \in \mathbb{R}$ about a suitable unit vector $k(t) \in \mathbb{R}^3$ (i.e., the axis-angle representation [26, 40]). Thus, given a rotation matrix in terms of a generalized coordinate system, the angle-axis parameters (φ, k) can be easily calculated (e.g., the algorithm given in [40] could be utilized). Given (φ, k), an alternative parameterization of the orientation of $\mathcal{E}$ with respect to $\mathcal{B}$ is provided by the unit quaternion vector $q(t) = [q_0(t), q_v^T(t)]^T \in \mathbb{R}^4$ with $q_0(t) \in \mathbb{R}$ and $q_v(t) \in \mathbb{R}^3$ as follows

$$q(t) \triangleq \begin{bmatrix} q_0(t) \\ q_v(t) \end{bmatrix} = \begin{bmatrix} \cos\left(\dfrac{\varphi(t)}{2}\right) \\ k(t)\sin\left(\dfrac{\varphi(t)}{2}\right) \end{bmatrix} \tag{2.169}$$

where it is clear from (2.169) that the unit quaternion is subject to the following constraint

$$q^T q = 1. \tag{2.170}$$

To develop the unit quaternion parameterization given in (2.169), a rotation matrix expressed in terms of a generalized coordinate system was utilized. Given the unit quaternion parameterization, the rotation matrix can be expressed in terms of the unit quaternion as follows [27]

$$R(q) = \left(q_o^2 - q_v^T q_v\right) I_3 + 2q_v q_v^T + 2q_o q_v^\times \tag{2.171}$$

where I_3 is the 3×3 identity matrix, and the notation $a^\times$ denotes the following skew-symmetric matrix

$$q_v^\times = \begin{bmatrix} 0 & -q_{v3} & q_{v2} \\ q_{v3} & 0 & -q_{v1} \\ -q_{v2} & q_{v1} & 0 \end{bmatrix} \qquad \forall q_v = [q_{v1}, \ q_{v2}, \ q_{v3}]^T. \tag{2.172}$$

Given the previous development, a relationship between the position and orientation of $\mathcal{E}$ relative to $\mathcal{B}$ can be developed as follows [27]

$$\begin{bmatrix} p \\ q \end{bmatrix} = \begin{bmatrix} f_p(\theta) \\ f_q(\theta) \end{bmatrix} \tag{2.173}$$

where $f_p(\theta) \in \mathbb{R}^3$ and $f_q(\theta) \in \mathbb{R}^4$ are kinematic functions, $\theta(t) \in \mathbb{R}^n$ denotes the position of $\mathcal{E}$ in a generalized coordinate system, $p(\theta) \in \mathbb{R}^3$ represents the position of the origin of $\mathcal{E}$ with respect to the origin of $\mathcal{B}$, and the unit quaternion was defined in (2.169). After differentiating (2.173), the following velocity relationships can be formulated [27]

$$\begin{bmatrix} \dot{p} \\ \dot{q} \end{bmatrix} = \begin{bmatrix} J_p(\theta) \\ J_q(\theta) \end{bmatrix} \dot{\theta} \tag{2.174}$$

where $\dot{\theta}(t) \in \mathbb{R}^n$ denotes the velocity of $\mathcal{E}$ in a generalized coordinate system, and $J_p(\theta) \in \mathbb{R}^{3 \times n}$, $J_q(\theta) \in \mathbb{R}^{4 \times n}$ denote the position and orientation Jacobian matrices, respectively, that are defined as follows

$$J_p = \frac{\partial f_p}{\partial \theta} \quad J_q = \frac{\partial f_q}{\partial \theta}. \tag{2.175}$$

To facilitate the subsequent control development and stability analysis, the fact that $q(t)$ is related to the angular velocity of $\mathcal{E}$ relative to $\mathcal{B}$, denoted by $w(t) \in \mathbb{R}^3$ with coordinates expressed in $\mathcal{B}$, via the following differential equation [3, 27] is exploited

$$\dot{q} = B(q)\omega \tag{2.176}$$

where the Jacobian-type matrix $B(q) \in \mathbb{R}^{4 \times 3}$ is defined as follows

$$B(q) = \frac{1}{2} \begin{bmatrix} -q_v^T \\ q_o I_3 - q_v^\times \end{bmatrix} \tag{2.177}$$

where $B(q)$ satisfies the following useful property (see Lemma B.11 of Appendix B)

$$B^T(q)B(q) = I_3. \tag{2.178}$$

Based on (2.178), we can rewrite (2.176) as follows

$$\omega = B^T(q)\dot{q}. \tag{2.179}$$

After combining (2.174) and (2.179), the final kinematic expression that relates the generalized Cartesian velocity to the generalized coordinate system is developed as follows

$$\begin{bmatrix} \dot{p} \\ \omega \end{bmatrix} = J(\theta)\dot{\theta} \tag{2.180}$$

where the Jacobian matrix $J(\theta) \in \mathbb{R}^{6 \times n}$ is defined by the following expression

$$J(\theta) = \begin{bmatrix} J_p(\theta) \\ B^T(q)J_q(\theta) \end{bmatrix}. \tag{2.181}$$

For some systems (e.g., redundant robot manipulators), the Jacobian given in (2.181) may have more columns than rows. Since the Jacobian may be nonsquare, we will exploit the use of the pseudo-inverse of the Jacobian in the subsequent development. Specifically, the pseudo-inverse of $J(\theta)$, denoted by $J^+(\theta) \in \mathbb{R}^{n \times 6}$, is defined as follows

$$J^+ = J^T \left(JJ^T \right)^{-1} \quad \text{such that} \quad JJ^+ = I_6. \tag{2.182}$$

As shown in [32], the pseudo-inverse given in (2.182) satisfies the Moore-Penrose Conditions given below

$$JJ^+J = J \qquad J^+JJ^+ = J^+$$

$$\left(J^+J \right)^T = J^+J \qquad \left(JJ^+ \right)^T = JJ^+. \tag{2.183}$$

In addition, the matrix $I_n - J^+J$, which projects vectors onto the null space of $J(\theta)$, satisfies the following properties

$$\begin{aligned}
\left(I_n - J^+J \right)\left(I_n - J^+J \right) &= I_n - J^+J & J\left(I_n - J^+J \right) &= 0 \\
\left(I_n - J^+J \right)^T &= \left(I_n - J^+J \right) & \left(I_n - J^+J \right) J^+ &= 0.
\end{aligned} \tag{2.184}$$

Remark 2.14 *During the subsequent control development, we assume that the minimum singular value of $J(\theta)$ is greater than a known small positive constant $\delta > 0$, such that $\sup_{\theta} \{\|J^+(\theta)\|\}$ is known a priori, and hence, all kinematic singularities are always avoided, where the notation $\sup_{\theta} \{\cdot\}$ is used to denote the supreme value over all θ.*

Remark 2.15 *As an example of how the previous development can be applied to a particular Euler-Lagrange system, consider an n-link robot manipulator. In the robotics literature, the position and orientation of the end-effector of a robot manipulator are commonly related to the manipulator base through a homogeneous transformation matrix. Specifically, by using the Denavit-Hartenberg representation, the transformation matrix $T(\theta) \in \mathbb{R}^{4 \times 4}$ can be calculated as follows [26, 40]*

$$T(\theta) = \begin{bmatrix} R(\theta) & p(\theta) \\ 0_{1 \times 3} & 1 \end{bmatrix} \tag{2.185}$$

where $0_{1 \times 3} = [0 \ 0 \ 0]$, $\theta(t) \in \mathbb{R}^n$ denotes a vector of manipulator link positions, $p(\theta) \in \mathbb{R}^3$ represents the position of the origin of the end-effector *with respect to the manipulator base, and $R(\theta) \in SO(3)$ represents the rotation of the coordinate frame attached to the end-effector with respect to the base coordinate frame. Given (2.185), three-parameter representations are typically utilized to develop the forward kinematic model of the manipulator. As demonstrated in the previous development, the homogeneous transformation in (2.185) can be utilized along with Euler's theorem to obtain (2.169), and hence, the forward kinematic model can be developed using the four-parameter unit quaternion representation as demonstrated in the previous development.*

Dynamic Model

The dynamic model for a generalized Euler-Lagrange system is assumed to have the following form

$$M(\theta)\ddot{\theta} + V_m(\theta, \dot{\theta})\dot{\theta} + G(\theta) + F_d\dot{\theta} = \tau \qquad (2.186)$$

where $\theta(t)$ was given in (2.173), and $\dot{\theta}(t), \ddot{\theta}(t) \in \mathbb{R}^n$ denote the velocity and acceleration of $\mathcal{E}$ in terms of the generalized coordinate system, respectively. For the dynamic model given in (2.186), $M(\theta) \in \mathbb{R}^{n \times n}$ represents the inertia matrix, $V_m(\theta, \dot{\theta}) \in \mathbb{R}^{n \times n}$ is the centripetal-Coriolis matrix, $G(\theta) \in \mathbb{R}^n$ represents the gravity effects, $F_d \in \mathbb{R}^{n \times n}$ is a diagonal matrix that contains the constant viscous friction coefficients, and $\tau(t) \in \mathbb{R}^n$ represents the control input vector. As in the previous sections, the subsequent control design and stability analysis will exploit the facts that the dynamic model given in (2.186) satisfies the following properties.

Property 2.9: Symmetric and Positive-Definite Inertia Matrix

The symmetric and positive-definite mass-inertia matrix $M(\theta)$ given in (2.186) satisfies the following inequalities

$$m_1 \|\xi\|^2 \leq \xi^T M \xi \leq m_2 \|\xi\|^2 \quad \forall \xi \in \mathbb{R}^n \qquad (2.187)$$

where $m_1, m_2 \in \mathbb{R}$ are known positive constants, and $\|\cdot\|$ denotes the standard Euclidean norm.

Property 2.10: Skew-Symmetry

The time derivative of the inertia matrix and the centripetal-Coriolis terms satisfy the following skew-symmetric relationship

$$\xi^T \left(\frac{1}{2}\dot{M} - V_m \right) \xi = 0 \quad \forall \xi \in \mathbb{R}^n \qquad (2.188)$$

where $\dot{M}(\theta)$ denotes the time derivative of the inertia matrix.

Property 2.11: Linearity in the Parameters

The dynamics given in (2.186) can be linearly parameterized as follows

$$Y_g \phi = M(\theta)\ddot{\theta} + V_m(\theta, \dot{\theta})\dot{\theta} + G(\theta) + F_d\dot{\theta} \qquad (2.189)$$

where $Y_g(\theta, \dot{\theta}, \ddot{\theta}) \in \mathbb{R}^{n \times p}$ denotes a regression matrix, and $\phi \in \mathbb{R}^p$ denotes the constant system parameter vector.

Remark 2.16 *The dynamic and kinematic terms for the general Euler Lagrange system, denoted above by $M(\theta)$, $V_m(\theta, \dot{\theta})$, $G(\theta)$, $J(\theta)$, and $J^+(\theta)$, are assumed to only depend on $\theta(t)$ as arguments of trigonometric functions, and hence, remain bounded for all possible $\theta(t)$.*

2.4.2   Control Objective

The objective in this section is to design a control input to ensure that the position and orientation of $\mathcal{E}$ track the position and orientation of a desired orthogonal coordinate frame $\mathcal{E}_d$ where $p_d(t) \in \mathbb{R}^3$ denotes the position of the origin of $\mathcal{E}_d$ relative to the origin of $\mathcal{B}$ and the rotation matrix from $\mathcal{E}_d$ to $\mathcal{B}$ is denoted by $R_d(t) \in SO(3)$. If the orientation of $\mathcal{E}_d$ relative to $\mathcal{B}$ is specified in terms of a desired unit quaternion $q_d(t) = \begin{bmatrix} q_{od}(t) & q_{vd}^T(t) \end{bmatrix}^T \in \mathbb{R}^4$, then similarly to (2.171), $R_d(q_d)$ can be calculated from $q_d(t)$ as follows

$$R_d(q_d) = \left(q_{od}^2 - q_{vd}^T q_{vd}\right) I_3 + 2q_{vd}q_{vd}^T + 2q_{od}q_{vd}^{\times}. \qquad (2.190)$$

As in (2.176), the time derivative of $q_d(t)$ is related to the angular velocity of $\mathcal{E}_d$ relative to $\mathcal{B}$, denoted by $\omega_d(t) \in \mathbb{R}^3$ with coordinates in $\mathcal{B}$, through the following kinematic equation

$$\dot{q}_d = B(q_d)\omega_d. \qquad (2.191)$$

The position tracking error $e_p(t) \in \mathbb{R}^3$ is defined as follows

$$e_p = p_d - p \qquad (2.192)$$

where the standard assumption is made that $p_d(t)$, $\dot{p}_d(t)$, and $\ddot{p}_d(t)$ are all bounded functions of time. To quantify the difference between $\mathcal{E}$ and $\mathcal{E}_d$, we define the rotation matrix $\tilde{R}(e_q) \in SO(3)$ from $\mathcal{E}$ to $\mathcal{E}_d$ as follows

$$\tilde{R} = R_d^T R = \left(e_o^2 - e_v^T e_v\right) I_3 + 2e_v e_v^T + 2e_o e_v^{\times}. \qquad (2.193)$$

The quaternion tracking error $e_q(t) = \begin{bmatrix} e_o(t) & e_v^T(t) \end{bmatrix}^T \in \mathbb{R}^4$ given in (2.193) can be explicitly calculated from $q(t)$ and $q_d(t)$ as follows [42]

$$e_q = \begin{bmatrix} e_o \\ e_v \end{bmatrix} = \begin{bmatrix} q_o q_{od} + q_v^T q_{vd} \\ q_{od}q_v - q_o q_{vd} + q_v^{\times} q_{vd} \end{bmatrix} \qquad (2.194)$$

by using the fact that the quaternion equivalent of (2.193) is given by the following quaternion product [42]

$$e_q = q q_d^*$$ (2.195)

where $q_d^*(t) \in \mathbb{R}^4$ is explicitly defined as follows

$$q_d^*(t) = \left[q_{od}(t), -q_{vd}^T(t) \right]^T.$$ (2.196)

Based on (2.169) and (2.196), it is clear that the unit quaternion tracking error satisfies the following constraint

$$e_q^T e_q = e_0^2 + e_v^T e_v = 1$$ (2.197)

where

$$0 \leq \|e_v(t)\| \leq 1 \qquad 0 \leq |e_0(t)| \leq 1.$$ (2.198)

Based on the previous definitions, the position and orientation tracking objectives can be stated as follows

$$\lim_{t \to \infty} e_p(t) = 0 \quad \text{and} \quad \lim_{t \to \infty} \tilde{R}(e_q) = I_3,$$ (2.199)

respectively. The orientation tracking objective given in (2.199) can also be stated in the terms of the unit quaternion error of (2.194). Specifically, from (2.197) it can be proven that

$$\text{if} \quad \lim_{t \to \infty} e_v(t) = 0, \quad \text{then} \quad \lim_{t \to \infty} |e_0(t)| = 1;$$ (2.200)

hence, (2.193) and (2.200) can be used to prove that

$$\text{if} \quad \lim_{t \to \infty} e_v(t) = 0 \quad \text{then} \quad \lim_{t \to \infty} \tilde{R}(e_q) = I_3$$ (2.201)

(i.e., the orientation tracking objective given in (2.199) is achieved).

2.4.3   Quaternion-Based Control

Tracking Error System Development

In this section, we develop the open-loop tracking error system that is utilized as the basis for the subsequent quaternion-based control design. To this end, the position tracking error is developed by taking the time derivative of (2.192) as follows

$$\dot{e}_p = \dot{p}_d - \dot{p}.$$ (2.202)

After utilizing (2.176), (2.191), and (2.194) the orientation error system can be formulated as follows

$$\dot{e}_o = -\frac{1}{2}e_v^T \tilde{\omega} \tag{2.203}$$

$$\dot{e}_v = \frac{1}{2}\left(e_o I_3 - e_v^\times\right)\tilde{\omega} \tag{2.204}$$

where $\tilde{\omega}(t) \in \mathbb{R}^3$ quantifies the difference between the angular velocity of $\mathcal{E}$ with respect to $\mathcal{E}_d$ (expressed in $\mathcal{E}_d$) and is defined as follows

$$\tilde{\omega} = R_d^T\left(\omega - \omega_d\right). \tag{2.205}$$

To facilitate the subsequent analysis, we define a nonnegative function $V_1(t) \in \mathbb{R}$ as follows

$$V_1 = \frac{1}{2}e_p^T e_p + (1 - e_o)^2 + e_v^T e_v. \tag{2.206}$$

After taking the time derivative of (2.206) and substituting (2.202–2.204) into the resulting expression for $\dot{e}_p(t)$, $\dot{e}_o(t)$, and $\dot{e}_v(t)$, respectively, the following expression is obtained

$$\dot{V}_1 = e_p^T\left(\dot{p}_d - \dot{p}\right) + e_v^T R_d^T\left(\omega - \omega_d\right) \tag{2.207}$$

where (2.205) was used along with Lemma A.6 of Appendix A. Adding and subtracting $e_p^T K_1 e_p$ and $e_v^T K_2 e_v$ to the right-hand side of (2.207) yields

$$\dot{V}_1 = -e_p^T K_1 e_p - e_v^T K_2 e_v \tag{2.208}$$

$$+ \begin{bmatrix} e_p^T & e_v^T \end{bmatrix} \left(\begin{bmatrix} \dot{p}_d + K_1 e_p \\ -R_d^T \omega_d + K_2 e_v \end{bmatrix} + \begin{bmatrix} -I_3 & 0_{3\times 3} \\ 0_{3\times 3} & R_d^T \end{bmatrix} \begin{bmatrix} \dot{p} \\ \omega \end{bmatrix} \right)$$

where $K_1, K_2 \in \mathbb{R}^{3\times 3}$ are positive-definite diagonal control gain matrices, and the notation $0_{\xi_1 \times \xi_2}$ is used to denote a $\xi_1 \times \xi_2$ matrix of zeros. By utilizing (2.180), the following expression can be obtained

$$\dot{V}_1 = -e_p^T K_1 e_p - e_v^T K_2 e_v + \begin{bmatrix} e_p^T & e_v^T \end{bmatrix} \Lambda J r \tag{2.209}$$

where the matrix $\Lambda(t) \in \mathbb{R}^{6\times 6}$ is defined as

$$\Lambda = \begin{bmatrix} -I_3 & 0_{3\times 3} \\ 0_{3\times 3} & R_d^T \end{bmatrix} \tag{2.210}$$

and the auxiliary signal $r(t) \in \mathbb{R}^n$ is defined as

$$r \triangleq J^+ \Lambda^{-1} \begin{bmatrix} \dot{p}_d + K_1 e_p \\ -R_d^T \omega_d + K_2 e_v \end{bmatrix} + \left(I_n - J^+ J\right) h + \dot{\theta} \tag{2.211}$$

where $h\left(\cdot\right) \in \mathbb{R}^n$ may be used to control the self-motion of the system.[8]

To obtain the open-loop dynamics for $r(t)$, we take the time derivative of (2.211), premultiply the resulting equation by $M(\theta)$, and then substitute (2.186) for $M(\theta)\ddot{\theta}(t)$ to obtain the following expression

$$M\dot{r} = -V_m r + Y\phi + \tau. \tag{2.212}$$

The linear parameterization $Y(\cdot)\phi$ given in (2.212) is defined as follows

$$
\begin{aligned}
Y\phi = \ & M\frac{d}{dt}\left\{ J^+\Lambda^{-1}\left[\begin{array}{c} \dot{p}_d + K_1 e_p \\ -R_d^T \omega_d + K_2 e_v \end{array}\right] + (I_n - J^+J)h\right\} \\[2mm]
& + V_m\left(J^+\Lambda^{-1}\left[\begin{array}{c} \dot{p}_d + K_1 e_p \\ -R_d^T \omega_d + K_2 e_v \end{array}\right] + (I_n - J^+J)h\right) \\[2mm]
& - G(\theta) - F_d\dot{\theta}
\end{aligned}
\tag{2.213}
$$

where $Y(p, q, \theta, \dot{\theta}, h, \dot{h}, t) \in \mathbb{R}^{n\times p}$ denotes a measurable regression matrix, and ϕ was defined in (2.189).

Control Formulation

Based on the open-loop kinematic tracking error systems given in (2.202–2.212) and the subsequent stability analysis, the control input is designed as follows

$$\tau = -Y\phi - K_r r - (\Lambda J)^T\left[\begin{array}{c} e_p \\ e_v \end{array}\right] \tag{2.214}$$

where $K_r \in \mathbb{R}^{n\times n}$ is a positive-definite diagonal control gain matrix, and exact knowledge of the dynamics (including ϕ given in (2.189)) and position/velocity measurements (i.e., $p(t)$, $q(t)$, $\theta(t)$, $h(t)$, $\dot{p}(t)$, $\dot{q}(t)$, $\dot{\theta}(t)$, $\dot{h}(t)$) are assumed to be available. After substituting (2.214) into (2.212), the closed-loop error system for $r(t)$ can be written in the following form

$$M\dot{r} = -V_m r - K_r r - (\Lambda J)^T\left[\begin{array}{c} e_p \\ e_v \end{array}\right]. \tag{2.215}$$

Stability Analysis

The stability of the quaternion-based tracking controller given in (2.214) can be examined through the following theorem.

---

[8] The subsequent stability analysis mandates that $h(t)$ be formulated in such a manner that both $h(t)$ and $\dot{h}(t)$ are bounded signals.

Theorem 2.5 *The control law given in (2.214) guarantees global asymptotic position and orientation tracking in the sense that*

$$\lim_{t \to \infty} e_p(t) = 0 \tag{2.216}$$

and

$$\lim_{t \to \infty} \tilde{R}(e_q(t)) = I_3. \tag{2.217}$$

Proof: To prove Theorem 2.5, we define a nonnegative function $V(t) \in \mathbb{R}$ as follows

$$V = \frac{1}{2} r^T M r + V_1 \tag{2.218}$$

where $V_1(t)$ was defined in (2.206). After taking the time derivative of (2.218) and then substituting for $\dot{V}_1(t)$ and $M(\theta)\dot{r}(t)$ from (2.209–2.211) and (2.215), respectively, the following expression is obtained

$$\dot{V} = r^T \left(-K_r r - (\Lambda J)^T \begin{bmatrix} e_p \\ e_v \end{bmatrix} \right)$$
$$-e_p^T K_1 e_p - e_v^T K_2 e_v + \begin{bmatrix} e_p^T & e_v^T \end{bmatrix} \Lambda J r \tag{2.219}$$

where (2.188) was used. After cancelling common terms in (2.219), the following expression is obtained

$$\dot{V} = -r^T K_r r - e_p^T K_1 e_p - e_v^T K_2 e_v. \tag{2.220}$$

From (2.218) and (2.220), it is clear that $e_p(t)$, $e_v(t)$, and $r(t) \in \mathcal{L}_\infty \cap \mathcal{L}_2$ (see Lemma A.11 of Appendix A). Based on these facts, along with the assumption that $p_d(t)$, $\dot{p}_d(t)$, $\ddot{p}_d(t)$, $h(t)$, and $\dot{h}(t) \in \mathcal{L}_\infty$ and the facts outlined in Remark 2.14, standard signal chasing arguments can be used to prove that all of the signals (with the exception of $\theta(t)$) remain bounded: the boundedness of $\theta(t)$ can not be proven due to the self-motion of the system. From the above boundedness statements, we can use (2.202), (2.204), and (2.215) to conclude that $\dot{e}_p(t)$, $\dot{e}_v(t)$, and $\dot{r}(t) \in \mathcal{L}_\infty$, which implies from Lemma A.9 of Appendix A that $e_p(t)$, $e_v(t)$, and $r(t)$ are uniformly continuous. Since $e_p(t)$, $e_v(t)$, and $r(t) \in \mathcal{L}_\infty \cap \mathcal{L}_2$ and uniformly continuous, Barbalat's Lemma (see Lemma A.16 of Appendix A) can be invoked to prove that

$$\lim_{t \to \infty} e_p(t), e_v(t), r(t) = 0 \tag{2.221}$$

and hence, (5.37) can be used to prove that

$$\lim_{t \to \infty} \tilde{R}(e_q(t)) = I_3. \tag{2.222}$$

Remark 2.17 *Due to the redundancy in the system, we cannot prove that $\theta(t) \in \mathcal{L}_\infty$; therefore, $h(t)$ defined in (2.211) should only depend on $\theta(t)$ as a function of bounded trigonometric signals. In addition, $h(t)$ cannot be a function of $\dot{\theta}(t)$ since $\dot{h}(t)$ would be a function of $\ddot{\theta}(t)$, which would lead to an algebraic loop in the control law of (2.214). An example of $h(t)$ that maximizes the robot manipulability while satisfying the above conditions is given by [22]*

$$h = \frac{\partial}{\partial \theta} \left(\det \left(J J^T \right) \right).$$
(2.223)

Remark 2.18 *Based on the structure of the open-loop dynamics given in (2.212), the previous assumption of exact model knowledge can be easily relaxed by designing an adaptive controller to compensate for parametric uncertainties. Specifically, the control input given in (2.214) can be redesigned as follows*

$$\tau = -Y\hat{\phi} - K_r r - (\Lambda J)^T \begin{bmatrix} e_p \\ e_v \end{bmatrix}.$$
(2.224)

The parameter estimate vector $\hat{\phi}(t) \in \mathbb{R}^r$ given in (2.224) is defined by the following gradient-based update law

$$\dot{\hat{\phi}} = \Gamma Y^T r$$
(2.225)

where $\Gamma \in \mathbb{R}^{r \times r}$ denotes a positive-definite diagonal adaptation gain matrix. Robust and sliding mode controllers could also be easily designed to compensate for modeling uncertainties not restricted to parametric uncertainties (e.g. see [14]). In addition to relaxing the assumption of exact model knowledge, the assumption that velocity measurements are available can also be relaxed. For example, in [41] an observer-based control scheme is used to achieve semi-global asymptotic position/orientation tracking.

2.4.4   UUV Extension

In this section, we discuss the application of the proposed control design methodology to fully actuated nonredundant UUVs. Specifically, we present the kinematic and dynamic models for UUVs and then transform the models into a similar form as the models given in the previous section. However, the following clarification is necessary regarding the kinematic notation used in this section. In the UUV (and aerospace) literature, rotation matrices are conventionally defined from the inertial frame to the moving frames [19, 23]. This convention is the opposite of the convention used in the previous section (and in the robotics literature) of defining rotation matrices from the moving coordinate frame to the inertial (base) frame.

This difference in notation will cause the UUV kinematic equations developed in this section to be slightly different than the kinematic equations presented in the previous section.

Kinematic and Dynamic Models

The attitude kinematic equations for the UUV can be written as follows [19]

$$\dot{q} = \begin{bmatrix} \dot{q}_o \\ \dot{q}_v \end{bmatrix} = \frac{1}{2} \begin{bmatrix} -q_v^T \\ q_o I_3 + q_v^\times \end{bmatrix} \omega \qquad (2.226)$$

where the unit quaternion $q(t) = \begin{bmatrix} q_o(t) & q_v^T(t) \end{bmatrix}^T \in \mathbb{R}^4$ describes the UUV attitude with coordinates in a fixed inertial orthogonal coordinate frame, denoted by $\mathcal{I}$, and $\omega(t) \in \mathbb{R}^3$ denotes the UUV angular velocity with coordinates in an orthogonal coordinate frame fixed to the UUV, denoted by $\mathcal{U}$. The translational kinematic equation is given as follows [19]

$$\dot{p} = R^T(q) v_L \qquad (2.227)$$

where $p(t) \in \mathbb{R}^3$ is the UUV position with coordinates in $\mathcal{I}$, $v_L(t) \in \mathbb{R}^3$ is the UUV linear velocity with coordinates in $\mathcal{U}$, and $R(q) \in SO(3)$ is a rotation matrix that brings $\mathcal{I}$ onto $\mathcal{U}$ that is defined in terms of the unit quaternion as follows

$$R(q) = \left(q_o^2 - q_v^T q_v \right) I_3 + 2 q_v q_v^T - 2 q_o q_v^\times. \qquad (2.228)$$

The dynamic model for the UUV can be written as follows [19]

$$M\dot{\nu} + V_m \nu + D\nu + G = u \qquad (2.229)$$

where $\nu(t) \in \mathbb{R}^6$ is composed of the linear and angular velocity vectors as follows

$$\nu = \begin{bmatrix} v_L^T & \omega^T \end{bmatrix}^T. \qquad (2.230)$$

In the dynamic model given in (2.229), $M \in \mathbb{R}^{6\times6}$ denotes the constant positive-definite inertia matrix, $V_m(\nu) \in \mathbb{R}^{6\times6}$ denotes the skew-symmetric, centripetal-Coriolis matrix, $D(\nu) \in \mathbb{R}^{6\times6}$ is the diagonal positive-definite hydrodynamic damping matrix, $G(q) \in \mathbb{R}^6$ is the vector of gravitational and buoyant forces/torques, and $u(t) \in \mathbb{R}^6$ represents the force/torque control input. The inertia and centripetal-Coriolis matrices include the added mass effects [19].

Control Objective

Let $\mathcal{U}_d$ be a desired orthogonal coordinate frame such that $p_d(t) \in \mathbb{R}^3$ and $q_d(t) \in \mathbb{R}^4$ denote the desired UUV position and attitude, respectively, where

$$q_d(t) = \begin{bmatrix} q_{od}(t) & q_{vd}^T(t) \end{bmatrix}^T. \qquad (2.231)$$

Given $q_d(t)$, the desired rotation matrix $R_d(q_d) \in SO(3)$ from $\mathcal{I}$ to $\mathcal{U}_d$ can be calculated from an equation similar in structure to (2.228). Furthermore, the time derivative of $q_d(t)$ is related to the desired UUV angular velocity $\omega_d(t) \in \mathbb{R}^3$ with coordinates in $\mathcal{U}_d$ by an equation similar in structure to (2.226). The UUV position tracking error $e_p(t) \in \mathbb{R}^3$ is defined as in (2.192), while the attitude tracking error is represented by

$$\tilde{R} = R R_d^T \qquad (2.232)$$

where $\tilde{R}(\cdot) \in SO(3)$ is the rotation matrix from $\mathcal{U}_d$ to $\mathcal{U}$. The quaternion tracking error $e_q(t) = \begin{bmatrix} e_o(t) & e_v^T(t) \end{bmatrix}^T \in \mathbb{R}^4$ can then be calculated from (2.228), (2.232), and the definition of $R_d(q_d)$. The angular velocity tracking error $\tilde{\omega}(t) \in \mathbb{R}^3$ is defined as

$$\tilde{\omega} = \omega - \tilde{R}\omega_d. \qquad (2.233)$$

Based on the previous definitions, the control objective is to design the control input $u(t)$ to ensure UUV position and attitude tracking in the sense that

$$\lim_{t \to \infty} e_p(t) = 0 \quad \text{and} \quad \lim_{t \to \infty} \tilde{R}(e_q(t)) = I_3 \qquad (2.234)$$

or as shown in Section 2.4.2

$$\lim_{t \to \infty} e_p(t) = 0 \quad \text{and} \quad \lim_{t \to \infty} e_v(t) = 0. \qquad (2.235)$$

Control Development and Analysis

The UUV kinematic error system can be developed in a similar manner as in Section 2.4.3 as follows

$$\begin{aligned} \dot{e}_p &= \dot{p}_d - R^T v_L \\ \dot{e}_o &= -\frac{1}{2} e_v^T \tilde{\omega} \\ \dot{e}_v &= \frac{1}{2} \left(e_o I_3 + e_v^\times \right) \tilde{\omega}. \end{aligned} \qquad (2.236)$$

Based on the subsequent stability analysis, an auxiliary signal $r(t) \in \mathbb{R}^6$ is defined as follows

$$r = \begin{bmatrix} R\dot{p}_d + RK_1 e_p \\ \tilde{R}\omega_d + K_2 e_v \end{bmatrix} - \nu \qquad (2.237)$$

where $\nu(t)$ was defined in (2.230). To develop the open-loop error system for $r(t)$, we take the time derivative of (2.237) and premultiply the resulting expression by the inertia matrix to obtain the following expression

$$M\dot{r} = Y\phi - u \qquad (2.238)$$

where (2.229) was utilized. The linear parametrization $Y(q, e_q, \nu, t)\phi$ given in (2.238) is defined as follows

$$Y\phi = M\frac{d}{dt}\left(\left[\begin{array}{c} R(\dot{p}_d + K_1 e_p) \\ \tilde{R}\omega_d + K_2 e_v \end{array}\right]\right) + V_m\nu + D\nu + G \qquad (2.239)$$

where $Y(\cdot) \in \mathbb{R}^{6 \times r}$ denotes a measurable regression matrix, and $\phi \in \mathbb{R}^r$ represents a vector containing the constant coefficients from the UUV dynamic model given in (2.229). Based on the structure of (2.238) and the subsequent stability analysis, the control input $u(t)$ is designed as follows

$$u = Y\phi + K_r r + \left[\begin{array}{cc} R & 0 \\ 0 & I_3 \end{array}\right]\left[\begin{array}{c} e_p \\ e_v \end{array}\right] \qquad (2.240)$$

where $K_r \in \mathbb{R}^{6 \times 6}$ denotes a positive-definite, diagonal gain matrix. After substituting the controller given in (2.240) into the open-loop error system dynamics given in (2.238), the following closed-loop dynamics can be developed for $r(t)$

$$M\dot{r} = -K_r r - \left[\begin{array}{cc} R & 0 \\ 0 & I_3 \end{array}\right]\left[\begin{array}{c} e_p \\ e_v \end{array}\right]. \qquad (2.241)$$

To analyze the stability of the controller given in (2.240), the same nonnegative function given in (2.218) can be utilized where $r(t)$ has been redefined as in (2.237). Specifically, after differentiating the nonnegative function defined in (2.218) and making the appropriate substitutions from (2.236) and (2.241), the following expression can be obtained

$$\begin{aligned}
\dot{V} &= -e_p^T\left(R^T v_L - \dot{p}_d\right) - e_v^T\left(\omega - \tilde{R}\omega_d\right) \\
&\quad -r^T\left(K_r r + \left[\begin{array}{cc} R & 0 \\ 0 & I_3 \end{array}\right]\left[\begin{array}{c} e_p \\ e_v \end{array}\right]\right) \\
&= \left[\begin{array}{cc} e_p^T & e_v^T \end{array}\right]\left[\begin{array}{cc} R^T & 0 \\ 0 & I_3 \end{array}\right]\left[\begin{array}{c} R\dot{p}_d - v_L \\ \tilde{R}\omega_d - \omega \end{array}\right]. \\
&\quad -r^T\left(K_r r + \left[\begin{array}{cc} R & 0 \\ 0 & I_3 \end{array}\right]\left[\begin{array}{c} e_p \\ e_v \end{array}\right]\right)
\end{aligned} \qquad (2.242)$$

where (2.233), Lemma A.6 of Appendix A, and the fact that $R^T(q)R(q) = I_3$ were used. After adding and subtracting the error terms $e_p^T(t)K_1e_p(t)$ and $e_v^T(t)K_2e_v(t)$ to (2.242), utilizing (2.237), and then cancelling common terms, the following expression can be obtained

$$\dot{V} = -e_p^T K_1 e_p - e_v^T K_2 e_v - r^T K_r r. \tag{2.243}$$

From (2.218) and (2.243), it is clear that $e_p(t)$, $e_v(t)$, and $r(t) \in \mathcal{L}_\infty \cap \mathcal{L}_2$ (see Lemma A.11 of Appendix A). Standard signal chasing arguments can be utilized to prove that all of the signals remain bounded and that $e_p(t)$, $e_v(t)$, and $r(t)$ are uniformly continuous. Since $e_p(t)$, $e_v(t)$, and $r(t) \in \mathcal{L}_\infty \cap \mathcal{L}_2$ and are uniformly continuous, Barbalat's Lemma (see Lemma A.16 of Appendix A) can be invoked to prove that

$$\lim_{t\to\infty} e_p(t), e_v(t), r(t) = 0 \tag{2.244}$$

and hence, (5.37) can be used to prove that

$$\lim_{t\to\infty} \tilde{R}(e_q(t)) = I_3. \tag{2.245}$$

Remark 2.19 *An adaptive controller could be designed as described in Remark 2.18 for the case where ϕ is uncertain. Moreover, an output feedback controller could also be designed according to the framework presented in [41] that produces semi-global asymptotic position/attitude tracking.*

2.4.5   Simulation Results

The model-based full-state feedback controller of (2.240) and the adaptive full-state feedback controller (see Remark 2.19) were simulated based on the dynamic model given in [33] as follows

$$M = \text{diag}\,(m_1, m_2, m_3, m_4, m_5, m_6)$$

$$D(\nu) = \text{diag}\,(d_{11} + d_{12}\,|\nu_1|, d_{21} + d_{22}\,|\nu_2|, d_{31} + d_{32}\,|\nu_3|,$$

$$d_{41} + d_{42}\,|\nu_4|, d_{51} + d_{52}\,|\nu_5|, d_{61} + d_{62}\,|\nu_6|)$$

$$V_m(\nu) = \begin{bmatrix} 0 & 0 & 0 & 0 & m_3\nu_3 & -m_2\nu_2 \\ 0 & 0 & 0 & -m_3\nu_3 & 0 & m_1\nu_1 \\ 0 & 0 & 0 & m_2\nu_2 & -m_1\nu_1 & 0 \\ 0 & m_3\nu_3 & -m_2\nu_2 & 0 & m_6\nu_6 & -m_5\nu_5 \\ -m_3\nu_3 & 0 & m_1\nu_1 & -m_6\nu_6 & 0 & m_4\nu_4 \\ m_2\nu_2 & -m_1\nu_1 & 0 & m_5\nu_5 & -m_4\nu_4 & 0 \end{bmatrix}$$

$$\tag{2.246}$$

where ν_i denotes the ith element of the velocity vector ν. The UUV is assumed to be neutrally buoyant with the origin of the $\mathcal{U}$ frame located at the center of gravity; thus, $G(q) = 0$. The following numerical values of the mass, inertia, and damping parameters in (2.246) were used in the simulation

$$
\begin{array}{lll}
m_1 = 215 \text{ [kg]} & d_{11} = 70 \text{ [Nm} \cdot \text{sec]} & d_{41} = 30 \text{ [Nm} \cdot \text{sec]} \\
m_2 = 265 \text{ [kg]} & d_{12} = 100 \text{ [N} \cdot \text{sec}^2\text{]} & d_{42} = 50 \text{ [N} \cdot \text{sec}^2\text{]} \\
m_3 = 265 \text{ [kg]} & d_{21} = 100 \text{ [Nm} \cdot \text{sec]} & d_{51} = 50 \text{ [Nm. sec]} \\
m_4 = 40 \text{ [kg} \cdot \text{m}^2\text{]} & d_{22} = 200 \text{ [N} \cdot \text{sec}^2\text{]} & d_{52} = 100 \text{ [N} \cdot \text{sec}^2\text{]} \quad (2.247) \\
m_5 = 80 \text{ [kg} \cdot \text{m}^2\text{]} & d_{31} = 200 \text{ [N} \cdot \text{sec}^2\text{]} & d_{61} = 50 \text{ [Nm} \cdot \text{sec]} \\
m_6 = 80 \text{ [kg} \cdot \text{m}^2\text{]} & d_{32} = 50 \text{ [N} \cdot \text{sec}^2\text{]} & d_{62} = 100 \text{ [N} \cdot \text{sec}^2\text{]}.
\end{array}
$$

From (2.246), the parameter vector ϕ in (2.239) can be constructed as follows

$$
\phi = \begin{bmatrix} m_1 & m_2 & m_3 & m_4 & m_5 & m_6 & d_{11} & d_{12} & d_{21} \\
d_{22} & d_{31} & d_{32} & d_{41} & d_{42} & d_{51} & d_{52} & d_{61} & d_{62} \end{bmatrix}^T. \quad (2.248)
$$

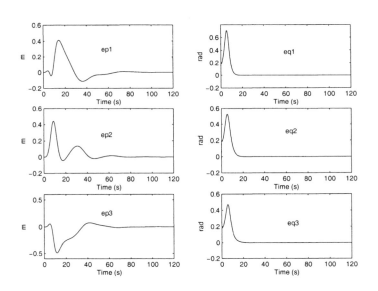

FIGURE 2.16. Position and attitude tracking errors for the model-based full-state feedback UUV controller.

For each controller, the initial attitude of the vehicle was selected as follows

$$
q(0) = \begin{bmatrix} 0.9486, & 0.1826, & 0.1826, & 0.1826 \end{bmatrix}^T \quad (2.249)
$$

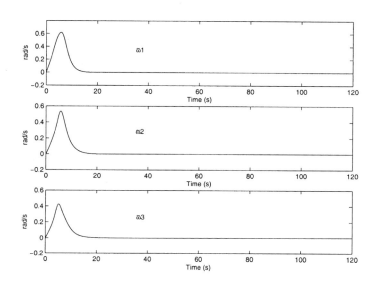

FIGURE 2.17. Angular velocity tracking error for the model-based full-state feedback UUV controller.

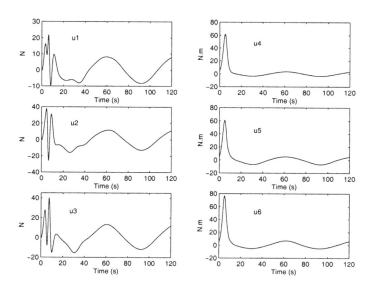

FIGURE 2.18. Force and torque control inputs for the model-based full-state feedback UUV controller.

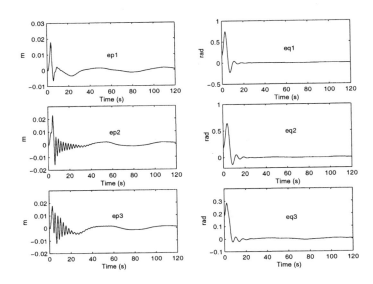

FIGURE 2.19. Position and attitude tracking errors for the adaptive full-state feedback UUV controller.

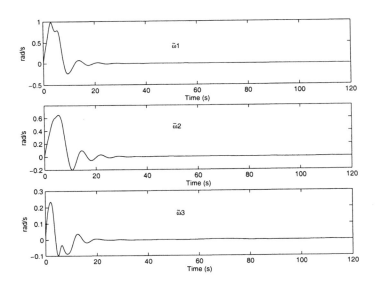

FIGURE 2.20. Angular velocity tracking errors for the adaptive full-state feedback UUV controller.

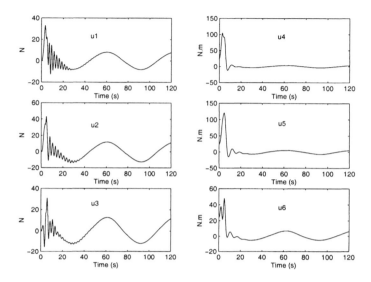

FIGURE 2.21. Force and torque control inputs for the adaptive full-state feedback UUV controller.

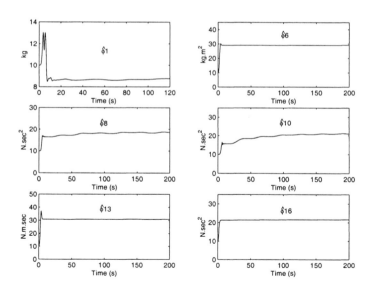

FIGURE 2.22. Selected parameter estimates for the adaptive full-state feedback UUV controller.

and the desired position trajectory was selected as

$$p_d(t) = \begin{bmatrix} \sin(0.1t)(1 - e^{-0.01t^2}) \\ \sin(0.1t)(1 - e^{-0.01t^2}) \\ \sin(0.1t)(1 - e^{-0.01t^2}) \end{bmatrix} \text{ [m]}. \tag{2.250}$$

Moreover, the desired attitude trajectory was generated using the following differential equation

$$\dot{q}_d(t) = \frac{1}{2} \begin{bmatrix} -q_{vd}^T(t) \\ q_{od}(t)I_3 + q_{vd}^\times(t) \end{bmatrix} \omega_d(t) \tag{2.251}$$

where the initial condition was selected as follows

$$q_d(0) = \begin{bmatrix} 1, & 0, & 0, & 0 \end{bmatrix}^T. \tag{2.252}$$

The desired angular velocity selected as

$$\omega_d(t) = \begin{bmatrix} 0.1\cos(0.1t)(1 - e^{-0.01t^2}) + 0.03t^2 \sin(0.1t)e^{-0.01t^2} \\ 0.1\cos(0.1t)(1 - e^{-0.01t^2}) + 0.03t^2 \sin(0.1t)e^{-0.01t^2} \\ 0.1\cos(0.1t)(1 - e^{-0.01t^2}) + 0.03t^2 \sin(0.1t)e^{-0.01t^2} \end{bmatrix} \text{ [rad/sec]}. \tag{2.253}$$

For the model-based full-state feedback controller, the following control gains were used

$$K_1 = \text{diag}(1,1,1) \quad K_2 = \text{diag}(1,1,1) \tag{2.254}$$

$$K_r = \text{diag}(25, 25, 25, 25, 25, 25).$$

Figures 2.16–2.18 illustrate the position and attitude tracking errors, angular velocity tracking errors, and the exact model knowledge full-state feedback control input, respectively. For the adaptive full-state feedback controller, the following control and adaptation gains were utilized

$$K_1 = \text{diag}(30, 30, 30) \quad K_2 = \text{diag}(4, 4, 4)$$

$$K_r = \text{diag}(30, 30, 30, 30, 30, 30) \tag{2.255}$$

$$\Gamma = \text{diag}(10, 100, 100, 10, 10, 10, 300, 300, 300, 300, 300, 300,$$
$$10, 10, 10, 10, 100, 100)$$

where the parameter estimates were initialized as follows

$$\hat{\phi}_i(0) = 10 \qquad \forall i = 1, 2, \dots 18. \tag{2.256}$$

Figures 2.19–2.21 show the position and attitude tracking errors, angular velocity tracking errors, and control inputs of the adaptive full-state feedback controller, respectively. Figure 2.22 illustrates selected parameter estimates.

2.5   Background and Further Reading

Motivated by the desire to suppress vibrational effects and due to the problems associated with passive damping, the design of active controllers for balancing high-speed rotational structures has been investigated. For example, in [30], Lum, Coppola, and Bernstein presented a new approach referred to as adaptive autocentering to stabilize the motion of the center of mass by attenuating the transmitted forces under varying rotational speed. This approach differs from autobalancing in that the autobalancing control objective is to stabilize the geometric center. The aforementioned work based on force attenuation was then expanded in [31] to facilitate imbalance cancellation. In [28], Long, Carroll, and Mukundan designed an adaptive controller for a single axis, constant speed rotor, active magnetic bearing system using integrator backstepping techniques to deal with a combination of an unknown static load change and an unbalanced sinusoidal load disturbance. In [47], a nonlinear controller was proposed to damp out vibration for a two-dimensional rotating flexible body-beam model with an unbalanced load attached to its free end. More recently, in [13] a DCAL-based adaptive autobalancing controller is designed to achieve a global exponential result under a persistency of excitation condition.

Motivated by various potential industrial applications, several researchers have investigated automatic control systems for dynamically positioned ships. These control systems are most commonly designed after linearizing the system dynamic equation about a set of prespecified yaw angles [20]. This procedure enables the application of linear control methods along with gain scheduling techniques. For example, early ship control systems used commercial PID controllers in cascade with a low-pass filter [4]. Later, linear optimal control laws in conjunction with Kalman filtering techniques were proposed in [21, 39]. With the goal of overcoming the linearization procedure, several control algorithms that take the nonlinear ship dynamics into account have also been proposed in [9, 18, 34]. In [18], a class of nonlinear PD-type control laws for position regulation was developed; however, robustness against parametric uncertainties could not be guaranteed. A robust nonlinear control law using singular perturbation theory was presented in [9] that accounts for parametric uncertainties and external disturbances, and a sliding-mode control approach was presented in [34].

Since most commercially available marine vessels only contain position sensors, control schemes that do not require direct velocity measurements are highly desirable (i.e., output feedback controllers as opposed to full-state feedback controllers). Motivated by this fact, [20] presented the design of a nonlinear output feedback controller using an observer backstepping method. Specifically, a nonlinear model-based observer-controller was used to eliminate the need for velocity measurements while achieving global exponential position tracking. More recently in [29], a globally asymptotically stabilizing controller for ship regulation was developed that only used position measurements. However, similar to [20], the observer-controller of [29] required exact knowledge of the nonlinear ship dynamics. Recently, the authors of [1] presented a global output tracking controller for a class of Euler-Lagrange systems that is directly applicable to the dynamically positioned surface vessel problem; however, the controller requires that the system parameters be known. The result in [1] was extended in [2] by imposing a monotone damping condition on the nonlinearities of the unmeasured states to remove the condition that the nonlinearities be globally Lipschitz. More recently, a global adaptive output feedback tracking controller for dynamically positioned ships was designed in [17].

Some past work that deals with task space control formulation can be found in [6, 24, 42]. Specifically, one of the first results in task space control of robot manipulators was presented in [24]. Resolved-rate and resolved-acceleration task space controllers using the quaternion parameterization were proposed in [42]. An experimental assessment of different end-effector orientation parameterizations for task space robot control was provided in [6]. Recently in [8], a stability proof using a strict Lyapunov function was presented for a quaternion-based resolved-acceleration controller. Examples of quaternion-based task space controllers that do not require velocity measurements are given in [7, 41]. Specifically, the controller proposed in [7] yields a local stability result, whereas the result given in [41], achieves a semi-global stability result. For additional work related to controllers for redundant robots, see [10, 22, 24, 35, 37, 43, 45] and the references therein. The mechanical system models and the control development in this chapter are based on ordinary differential equations (ODEs). Some mechanical systems such as flexible structures are modeled by partial differential equations (PDEs) and the corresponding control designs will vary from the approaches demonstrated in this chapter. For a treatment of mechanical systems that are modeled by ODEs and PDEs, see [15].

References

[1] O. M. Aamo, M. Arcak, T. I. Fossen, and P. V. Kokotović, "Global Output Tracking Control of a Class of Euler-Lagrange Systems," *Proceedings of the IEEE Conference on Decision and Control*, Dec. 2000, pp. 2478–2483.

[2] O. M. Aamo, M. Arcak, T. I. Fossen, and P. V. Kokotović, "Global Output Tracking Control of a Class of Euler-Lagrange Systems with Monotonic Nonlinearities in the Velocities," *International Journal of Control*, Vol. 74, No. 7, May 2001, pp. 649–658.

[3] J. Ahmed, V. T. Coppola, and D. S. Bernstein, "Adaptive Asymptotic Tracking of Spacecraft Attitude Motion with Inertia Matrix Identification," *Journal of Guidance, Control, and Dynamics*, Vol. 21, No. 5, Sept.–Oct. 1998, pp. 684–691.

[4] J. G. Balchen, N. A. Jenssen, and S. Saellid, "Dynamic Positioning of Floating Vessels Based on Kalman Filtering and Optimal Control," *Proceedings of the IEEE Conference on Decision and Control*, 1980, pp. 852–864.

[5] T. Burg, D. Dawson, and P. Vedagarbha, "A Redesigned DCAL Controller without Velocity Measurements: Theory and Demonstration," *Robotica*, Vol. 15, 1997, pp. 337–346.

[6] F. Caccavale, C. Natale, B. Siciliano, and L. Villani, "Resolved-Acceleration Control of Robot Manipulators: A Critical Review with Experiments," *Robotica*, Vol. 16, 1998, pp. 565–573.

[7] F. Caccavale, C. Natale, and L. Villani, "Task-space Control Without Velocity Measurements," *Proceedings of the IEEE International Conference on Robotics and Automation*, 1999, pp. 512–517.

[8] R. Campa, R. Kelly, and E. Garcia, "On Stability of the Resolved Acceleration Control," *Proceedings of the IEEE International Conference on Robotics and Automation*, 2001, pp. 3523–3528.

[9] C. Canudas De Wit, E. Olguin, D. Perrier, and M. Perrier, "Robust Nonlinear Control of an Underwater Vehicle/Manipulator System with Composite Dynamics," *Proceedings of the IEEE International Conference on Robotics and Automation*, 1998, pp. 452–457.

[10] R. Colbaugh and K. Glass, "Robust Adaptive Control of Redundant Manipulators," *Journal of Intelligent and Robotic Systems*, Vol. 14, 1995, pp. 68–88.

[11] N. Costescu, D. M. Dawson, and M. Loffler, "Qmotor 2.0-A PC Based Real-Time Multitasking Graphical Control Environment," *IEEE Control Systems Magazine*, Vol. 19, No. 3, June 1999, pp. 68–76.

[12] B. T. Costic, D. M. Dawson, M. S. de Queiroz, and V. Kapila, "A Quaternion-Based Adaptive Attitude Tracking Controller Without Velocity Measurements," *Proceedings of the IEEE Conference on Decision and Control*, Dec. 2000, pp. 2424–2429.

[13] B. T. Costic, S. P. Nagarkatti, D. M. Dawson, and M. S. de Queiroz, "Autobalancing DCAL Controller for Rotating Unbalanced Disk," *Mechatronics: An International Journal*, Vol. 12, No. 5, June 2002, pp. 685–712.

[14] D. M. Dawson, M. M. Bridges, and Z. Qu, *Nonlinear Control of Robotic Systems for Environmental Waste and Restoration*, Englewood Cliffs, NJ: Prentice-Hall, 1995.

[15] M. S. de Queiroz, D. M. Dawson, S. Nagarkatti, and F. Zhang, *Lyapunov-Based Control of Mechanical Systems*, Boston, MA: Birkhäuser, 2000.

[16] M. S. de Queiroz and D. M. Dawson, "Nonlinear Control of Active Magnetic Bearings: A Backstepping Approach," *IEEE Transactions on Control Systems Technology* (Special Issue on Magnetic Bearing Control), Vol. 4, No. 5, Sept. 1996, pp. 545–552.

[17] Y. Fang, E. Zergeroglu, M. S. de Queiroz, and D. M. Dawson, "Global Output Feedback Control of Dynamically Positioned Surface Vessels: An Adaptive Control Approach," *Proceedings of the American Control Conference*, June 2001, pp. 3109–3114.

[18] O. Fjellstad and T. I. Fossen, "Quaternion Feedback Regulation of Underwater Vehicles," *Proceedings of the IEEE Conference on Control Applications*, Dec. 1994, pp. 857–862.

[19] T. I. Fossen, *Guidance and Control of Ocean Vehicles*, Chichester, UK: John Wiley, 1994.

[20] T. I. Fossen and Å. Grøvlen, "Nonlinear Output Feedback Control of Dynamically Positioned Ships Using Vectorial Observer Backstepping," *IEEE Transactions on Control Systems Technology*, Vol. 6, No. 1, Jan. 1998, pp. 121–128.

[21] M. J. Grimble, R. J. Patton, and D. A. Wise, "The Design of Dynamic Positioning Control Systems Using Stochastic Optimal Control Theory," *Optimal Control Application Methods*, Vol. 1, 1980, pp. 167–202.

[22] P. Hsu, J. Hauser, and S. Sastry, "Dynamic Control of Redundant Manipulators," *Journal of Robotic Systems*, Vol. 6, 1989, pp. 133–148.

[23] P. C. Hughes, *Spacecraft Attitude Dynamics*, New York, NY: John Wiley, 1994.

[24] O. Khatib, "Dynamic Control of Manipulators in Operational Space," *IFTOMM Congress Theory of Machines and Mechanisms*, 1983, pp. 1–10.

[25] J. B. Kuipers, *Quaternions and Rotation Sequences*, Princeton, NJ: Princeton University Press, 1999.

[26] F. Lewis, C. Abdallah, and D. M. Dawson, *Control of Robot Manipulators*, New York: MacMillan Publishing Co., 1993.

[27] F. Lizarralde and J. T. Wen, "Attitude Control Without Angular Velocity Measurement: A Passivity Approach," *IEEE Transactions on Automatic Control*, Vol. 41, No. 3, Mar. 1996, pp. 468–472.

[28] M. L. Long, J. J. Carroll, and R. Mukundan, "Adaptive Control of Active Magnetic Bearings Under Unknown Static Load Change and Unbalance," *Proceedings of the IEEE Conference on Control Applications*, Dearborn, MI, Sept. 1996, pp. 876–881.

[29] A. Loria, T. I. Fossen, and E. Panteley, "A Separation Principle for Dynamic Positioning of Ships: Theoretical and Experimental Results," *IEEE Transactions on Control Systems Technology*, Vol., 8, No. 2, Mar. 2000, pp. 322–343.

[30] K. Lum, V. Coppola, and D. Bernstein, "Adaptive Autocentering Control for an Active Magnetic Bearing Supporting a Rotor with Unknown Mass Imbalance," *IEEE Transactions on Control Systems Technology*, Vol. 4, No. 5, Sept. 1996, pp. 587–597.

[31] K. Lum, V. Coppola, and D. Bernstein, "Adaptive Virtual Autobalancing for a Rigid Rotor with Unknown Mass Imbalance Supported by Magnetic Bearings," *ASME Journal of Vibration and Acoustics*, Vol. 120, Apr. 1998, pp. 557–570.

[32] Y. Nakamura, *Advanced Robotics Redundancy and Optimization*, Reading, MA: Addison-Wesley, 1991.

[33] Ola-Erik Fjellstad and Thor I. Fossen, "Quaternion Feedback Regulation of Underwater Vehicles," *Proceedings of the IEEE Conference on Control Applications*, Strathclyde University, Glasgow, Aug. 1994, pp. 857–862.

[34] F. A. Papoulias and A. J. Healey, "Path Control of Surface Ships Using Sliding Modes," *Journal of Ship Research*, Vol. 36, No. 2, 1992, pp. 141–153.

[35] Z. X. Peng and N. Adachi, "Compliant Motion Control of Kinematically Redundant Manipulators," *IEEE Transactions on Robotics and Automation*, Vol. 9, No. 6, Dec. 1993, pp. 831–837.

[36] N. Sadegh and R. Horowitz, "Stability and Robustness Analysis of a Class of Adaptive Controllers for Robotic Manipulators," *International Journal of Robotic Research*, Vol. 9, No. 9, June 1990, pp. 74–92.

[37] H. Seraji, "Configuration Control of Redundant Manipulators: Theory and Implementation," *IEEE Transactions on Robotics and Automation*, Vol. 5, No. 4, Aug. 1989, pp. 472–490.

[38] J. -J. E. Slotine and W. Li, *Applied Nonlinear Control*, Englewood Cliffs, NJ: Prentice-Hall, 1991.

[39] A. J. Sørensen, S. I. Sagatun, and T. I. Fossen, "Design of a Dynamic Positioning System Using Model-Based Control," *Control Engineering Practice*, Vol. 4, No. 3, 1996, pp. 359–368.

[40] M. W. Spong and M. Vidyasagar, *Robot Dynamics and Control*, New York, NY: John Wiley, 1989.

[41] B. Xian, M. S. de Queiroz, D. M. Dawson, and I. D. Walker, "Task-Space Tracking Control of Redundant Robot Manipulators via Quaternion Feedback," *Proceedings of the IEEE Conference on Control Applications*, Sept. 2001, pp. 363–368.

[42] J. S. C. Yuan, "Closed-Loop Manipulator Control Using Quaternion Feedback," *IEEE Transactions on Robotics and Automation*, Vol. 4, No. 4, Aug. 1988, pp. 434–440.

[43] T. Yoshikawa, "Analysis and Control of Robot Manipulators with Redundancy," *Robotics Research-The First International Symposium*, Cambridge, MA, 1984, pp. 735–747.

[44] E. Zergeroglu, D. M. Dawson, M.S. de Queiroz, and M. Krstić, "On Global Output Feedback Tracking Control of Robot Manipulators," *Proceedings of the IEEE Conference on Decision and Control*, 2000, pp. 5073–5078.

[45] E. Zergeroglu, D. M. Dawson, I. Walker, and A. Behal, "Nonlinear Tracking Control of Kinematically Redundant Robot Manipulators," *Proceedings of the American Control Conference*, June 2000, pp. 2513–2517.

[46] F. Zhang, D. M. Dawson, M. S. de Queiroz, and W. Dixon, "Global Adaptive Output Feedback Tracking Control of Robot Manipulators," *IEEE Transactions on Automatic Control*, Vol. 45, No. 6, June 2000, pp. 1203–1208.

[47] F. Zhang, S. P. Nagarkatti, B. Costic, and D. M. Dawson, "Adaptive Autobalancing Boundary Control for a Flexible Rotor," *Proceedings of the American Control Conference*, June 1999, pp. 2220–2224.

3
Electric Machines

3.1 Introduction

In this chapter, model-based control designs are developed for electric machines that are typically described by (i) electrical subsystem dynamics that include all of the relevant electrical effects, (ii) an algebraic torque coupling that represents the electrical to mechanical energy conversion, and (iii) mechanical subsystem dynamics that may include the rotor and position dependent load dynamics. A block diagram that illustrates the interconnection of these coupled nonlinear dynamics is given in Figure 3.1.
Since the electrical subsystem dynamics have a faster time constant than

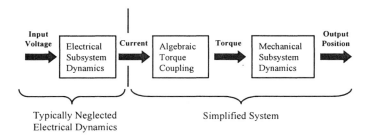

FIGURE 3.1. Block diagram for electric machines.

the associated mechanical dynamics, the electrical dynamics are often ne-

glected, especially for many industrial speed applications that are not required to be high-precision. For example, induction motors are widely used
in industrial speed control applications because of their simple design and
ruggedness. The popular field-oriented control (FOC) strategy controls the
induction motor by decoupling the current control inputs that appear bilin-
early in the mechanical subsystem dynamics. Specifically, the FOC strategy
is based on a rotational transformation that allows the system dynamics to
be expressed in a frame that aligns with the direction of the rotating magnetic field vector in the air gap of the motor. This transformation breaks
up the bilinear structure of the electromagnetic torque generation, and new
control inputs (in the transformed frame) can be designed independently to
provide speed tracking control as well as the establishment of a desired rotor flux magnitude. Based on the standard FOC strategy, researchers have
demonstrated several benefits of an indirect field-oriented control (IFOC)
strategy where the system is transformed to a frame that aligns with the
desired rotor flux vector as opposed to aligning with the actual rotor flux
vector in the standard FOC approach. In the first section of this chapter,
a brief review of the FOC and IFOC schemes is provided. A limitation of
the IFOC strategy is that it is constructed based on the standard FOC
legacy and does not integrate well with nonlinear analysis techniques. To
overcome this barrier an improved IFOC strategy is developed that is constructed via Lyapunov-based design and analysis techniques. Specifically,
the thrust will be to present simple alterations to the popular IFOC method
in a systematic manner. The modifications to the IFOC scheme involve the
injection of additional nonlinear terms into the current control input and
the so-called desired rotor flux angle dynamics. These additional nonlinear
terms facilitate an exponential stability result via the direct cancellation of
mechanical/electrical subsystem coupling terms during the Lyapunov stability argument. Furthermore, Lyapunov-based design and analysis tools
facilitate the design of adaptive controllers that compensate for the parametric uncertainty associated with the mechanical load.

For some positioning applications, high-precision electric machines are
required. An electric machine that is typically used in high-precision applications is the switched reluctance motor (SRM). The SRM can be considered to be an AC machine with an iron core rotor. Similar to the induction
motor, there are no mechanical brushes that require expensive maintenance
as required for permanent magnet brushed DC motors. In addition to the
reduction in maintenance, there are several advantages to using SRMs as
actuators in high-performance motion control applications. These advantages include (i) low cost and high reliability through simple design and

construction, (ii) increased capability to withstand high temperature environments, (iii) significant reduction in friction, and (iv) the ability to increase torque production through electromagnetic gearing. Since SRMs are typically used in high-precision applications, control designs for these machines incorporate the electrical dynamics. Moreover, SRMs are often used in applications that demand high torque (i.e., high current), requiring the motor to operate "high" on the saturation curve. Hence, many motor control experts also believe that saturation effects must be included in the SRM model to develop high-precision controllers. Typical approaches, which rely on feedback linearization techniques, exhibit control singularities when the stator current goes to zero. While it has been suggested that the singularity issue is a minor nuisance, the presence of singularities indicates that improvements in the control structure can still be accomplished. In pursuit of the aforementioned objectives, the second section of this chapter uses a general nonlinear model of the SRM to develop an adaptive controller for the full-order electromechanical model. Specifically, the developed SRM controller (i) is based on the full-order model (i.e., the electrical dynamics are not neglected), (ii) compensates for uncertainty in electromechanical model, (iii) utilizes a flux linkage model that includes magnetic saturation effects, and (iv) eliminates all control singularities (i.e., the voltage control input remains bounded for all operating conditions). The design assumes that a commutation strategy[1] can be developed to ensure the desired magnetic forces are delivered to the system. As an extension, a commutation design is presented to satisfy the assumed conditions required for the stability analysis.

Vibrational effects induced on a motor rotor (or long transmission links) can have a significant impact on many high-performance applications. Conventional bearings can be used to damp out these vibrations; however, frictional effects can result in negative system performance. Since magnetic bearings allow contact-free levitation, they exhibit several advantages over conventional bearings. For example, since magnetic bearings do not require lubrication, they can easily operate under environmental constraints that prevent the use of lubrication [29]. While allowing high circumferential speeds at high loads, magnetic bearings do not suffer frictional wear and tear; therefore, they offer a virtually unlimited lifetime with little or no maintenance. The term active magnetic bearing (AMB) is used to denote a set of electromagnets serving as a magnetic bearing by producing desired

---

[1] The term commutation strategy is often used to describe a step in the control design procedure that involves utilizing the desired torque trajectory and the structure of the torque transmission model to construct the desired current trajectory signals.

magnetic forces through an active controller. Through an AMB, magnetic forces can be developed to (i) eliminate vibration through active damping, (ii) adjust the stiffness of the suspended load, (iii) provide an automatic balancing capability, or (iv) deliberately excite vibrations [30]. Because of these advantages, AMBs are used in an increasing number of commercial high-performance applications in the domain of rotating machinery. These applications include ultra-high vacuum pumps, canned pipeline compressors and expanders, high-speed milling and grinding spindles, flywheels for energy storage, gyroscopes for space navigation, and spinning spindles. In a typical AMB application, the magnetic forces, which are applied by sets of stator electromagnets, must be adjusted online to ensure that the rotor is accurately positioned. Unfortunately, the ascendancy of AMBs over traditional bearings has been stymied because the control problem is complicated by the inherent nonlinearities associated with the electromechanical dynamics. In the third section of this chapter, the nonlinear model of a six degree of freedom (DOF) active magnetic bearing system is developed. The AMB model can be represented by the same block diagram as depicted in Figure 3.1. Based on this model, the AMB control problem is stated and a nonlinear tracking controller is developed. In a similar manner as for the SRM, a commutation design is presented as an extension to satisfy the assumed conditions required for the stability analysis.

3.2 Induction Motor

As stated previously, many industrial manufacturers have opted to use the induction motor as the desired actuator for many constant speed drive applications because of the its relative inexpensive cost, rugged construction, and inherent operating reliability. However, many of the previously designed velocity setpoint control methodologies may prove to be woefully inadequate for high-performance rotor velocity tracking applications using the induction motor. In an effort to use the induction motor for demanding tracking applications while still benefitting from its desirable features, nonlinear control strategies have been developed to overcome the numerous difficulties associated with the induction motor's complex coupled high-order electrical dynamics. Due to its extensive industrial use, the IFOC algorithm is one of the more popular strategies employed for current-fed induction motor control applications. In this section, a modified version of the standard IFOC is derived to achieve global exponential rotor velocity/rotor flux tracking. An extension is also provided that proves how a simple modification to the controller can yield global asymptotic tracking despite the parametric uncertainty associated with the mechanical dynamics and load.

3.2.1   System Model

Based on the common assumptions of equal mutual inductances and a linear magnetic circuit, the electromechanical model of a current-fed induction motor driving a mechanical subsystem in the rotating rotor reference frame can be written as follows [22]

$$J_m \dot{\omega} + B\omega + T_L(\omega) = \alpha_1 u^T J \psi \tag{3.1}$$

$$\dot{\psi} = -\beta_1 \psi + \beta_2 u \tag{3.2}$$

where $\omega(t)$, $\dot{\omega}(t) \in \mathbb{R}$ represent the rotor velocity and acceleration, respectively, and $\psi(t) = [\ \psi_a(t) \quad \psi_b(t)\ ]^T \in \mathbb{R}^2$ denotes the rotor flux. In the electromechanical model given in (3.1) and (3.2), the coefficients J_m, B, $T_L(\omega) \in \mathbb{R}$ denote the system inertia (including rotor inertia), the coefficient of viscous friction, and a known nonlinear load that is assumed to be bounded if $\omega(t)$ is bounded, respectively, $J \in \mathbb{R}^{2 \times 2}$ denotes the following skew-symmetric matrix

$$J = \begin{bmatrix} 0 & -1 \\ 1 & 0 \end{bmatrix}, \tag{3.3}$$

and $u(t) = [\ u_a(t) \quad u_b(t)\ ]^T \in \mathbb{R}^2$ denotes the stator current control input. The positive constants α_1, β_1, $\beta_2 \in \mathbb{R}$ given in (3.1) and (3.2) are related to electrical circuit parameters as follows

$$\alpha_1 = \frac{n_p M}{L_r} \qquad \beta_1 = \frac{R_r}{L_r} \qquad \beta_2 = \frac{R_r M}{L_r} \tag{3.4}$$

where L_r, $M \in \mathbb{R}$ denote the per-phase rotor self-inductance and mutual inductance, respectively, $R_r \in \mathbb{R}$ denotes the rotor resistance, and $n_p \in \mathbb{R}$ is the number of motor pole pairs.

Remark 3.1 *The electromagnetic torque transmission term $\alpha_1 u^T(t) J \psi(t)$ of (3.1) is bilinear (i.e., the system state vector $\psi(t)$ multiplies the control input vector $u(t)$). The presence of this bilinear term presents a design challenge because it couples the control inputs. Hence, to design the control inputs $u_a(t)$ and $u_b(t)$ independently, a control strategy must target the bilinear structure of (3.1). The desire to decouple the control inputs has motivated the development of previous FOC and IFOC strategies.*

3.2.2   Control Objective

The primary objective in this section is to develop a controller for the current-fed induction motor that ensures rotor velocity tracking under the constraint that rotor velocity is the only measurable signal. To this end, a standard IFOC rotor velocity *regulation* controller is first presented. Based on the structure of the standard IFOC design, an improved IFOC rotor velocity *tracking* controller is designed. The improved IFOC design is also redeveloped with a structure that facilitates Lyapunov-based analysis techniques. To quantify the objective of these controllers, a rotor velocity tracking error, denoted by $e(t) \in \mathbb{R}$, is defined as follows

$$e = \omega_d - \omega \tag{3.5}$$

where $\omega_d(t) \in \mathbb{R}$ represents the desired rotor velocity trajectory (for the standard IFOC regulation controller, ω_d is defined as a constant). The subsequent development is based on the typical assumption that the desired rotor trajectory is selected so that $\omega_d(t), \dot{\omega}_d(t) \in \mathcal{L}_\infty$. To provide flexibility in the design of feedback control laws for the mechanical subsystem dynamics given by (3.1), a filtered tracking error signal, denoted by $r(t) \in \mathbb{R}$, is defined as follows [23]

$$r = e + \mathcal{L}^{-1}\left\{ \frac{1}{s} K_F(s) e(s) \right\} \tag{3.6}$$

where $\mathcal{L}^{-1}\{\cdot\}$ denotes the inverse Laplace Transform operation, s denotes the Laplace Transform variable, and $K_F(s) \in \mathbb{R}$ denotes a linear filter selected to ensure that the following transfer function is proper and exponentially stable [19]

$$\frac{e(s)}{r(s)} = \frac{s}{s + K_F(s)}. \tag{3.7}$$

Motivation for the definition of the filtered tracking signal introduced in (3.6) is given by the fact that the subsequent control development and stability proof can be accommodated without restricting the feedback control structure. For example, by redefining the linear filter $K_F(s)$ given in (3.6), the control structure can be changed from proportional feedback to proportional-integral (PI) feedback without restructuring the controller and requiring additional stability analysis. Specifically, a proportional feedback controller can be obtained by defining the linear filter as follows

$$K_F(s) = 0 \Longrightarrow k_s r(t) = k_s e(t) \tag{3.8}$$

where $k_s \in \mathbb{R}$ is a constant control gain. To change from proportional feedback to PI feedback, the linear filter can be redefined as follows

$$K_F(s) = k_i \implies k_s r(t) = k_s e(t) + k_s k_i \int_0^t e(\tau)\, d\tau \qquad (3.9)$$

where $k_i \in \mathbb{R}$ is a constant control gain. Given that the transfer function defined in (3.7) is proper and exponentially stable, Lemmas A.14 and A.15 of Appendix A can be invoked to prove that if $r(t)$ is exponentially (or asymptotically) driven to zero, then $e(t)$ will be exponentially (or asymptotically) driven to zero.

In addition to the rotor velocity tracking control objective, a secondary control objective is to force the magnitude of the rotor flux to track a desired trajectory (or a constant setpoint for the standard IFOC controller). To quantify this secondary objective, the rotor flux magnitude tracking error, denoted by $\eta_\delta(t) \in \mathbb{R}$, is defined as follows

$$\eta_\delta = \delta_d^2 - \|\psi\|^2 \qquad (3.10)$$

where $\|\cdot\|$ denotes the standard Euclidean norm, and $\delta_d(t) \in \mathbb{R}$ denotes the desired magnitude of the rotor flux (for the standard IFOC regulation controller, δ_d is defined as a constant). The desired rotor flux magnitude is assumed to be selected so that $\delta_d(t)$, $\dot{\delta}_d(t) \in \mathcal{L}_\infty$, and that $\delta_d(t)$ is selected as a strictly positive function in order to prevent a potential singularity in the subsequent control development.

3.2.3 Standard IFOC Control

The electromechanical model for the induction motor given in (3.1) and (3.2) is expressed in terms of a rotating rotor reference frame. The motivation for an IFOC scheme is to transform the electromechanical model from the rotating rotor reference frame to a frame that is aligned with the direction of the desired rotor flux vector. That is, IFOC is aimed at transforming the dynamics from a Cartesian coordinate system representation to a polar coordinate system representation. By performing this transformation, the bilinearity in the torque transmission relationship of (3.1) can be eliminated, and hence, the design of the two stator current control inputs can be decoupled. Specifically, to transform the IFOC stator current input vector to the stator current input vector expressed in the rotating rotor reference frame, the following transformation can be applied

$$u = \begin{bmatrix} \cos\rho_d & -\sin\rho_d \\ \sin\rho_d & \cos\rho_d \end{bmatrix} u_{foc} \qquad (3.11)$$

where $u_{foc}(t) = \begin{bmatrix} u_{foc1}(t) & u_{foc2}(t) \end{bmatrix}^T \in \mathbb{R}^2$ is the standard stator current IFOC input expressed in the field-oriented reference frame as follows

$$u_{foc} = \begin{bmatrix} \dfrac{\beta_1}{\beta_2}\delta_d & \dfrac{\tau_d}{\alpha_1\delta_d} \end{bmatrix}^T \qquad (3.12)$$

where α_1, β_1, β_2 are defined in (3.4). For the control input given in (3.11), $\rho_d(t) \in \mathbb{R}$ denotes the desired rotor flux angle whose dynamics are governed by the following differential equation and initial condition

$$\dot{\rho}_d = \frac{\beta_2\tau_d}{\alpha_1\delta_d^2} \qquad \rho_d(0) = 0 \qquad (3.13)$$

where $\tau_d(t) \in \mathbb{R}$ denotes the following desired torque signal for the mechanical subsystem

$$\tau_d = y + k_s r. \qquad (3.14)$$

In (3.14), the auxiliary signal $y(\omega, e) \in \mathbb{R}$ denotes a feedforward signal that is defined as follows

$$y(\omega, e) = J_m\left(\mathcal{L}^{-1}\left\{K_F(s)\,e(s)\right\}\right) + B\omega + T_L(\omega) \qquad (3.15)$$

and $k_s \in \mathbb{R}$ denotes a positive constant control gain. Roughly speaking, $u_{foc1}(t)$ (the so-called direct-axis current) is designed to force the rotor flux magnitude to converge to δ_d, while $u_{foc2}(t)$ (the so-called quadrature-axis current) is designed to force the rotor angular velocity to converge to ω_d. As stated previously, the standard IFOC scheme targets the rotor velocity regulation problem, and hence, assumes that δ_d and ω_d are selected as constants. The IFOC design given in (3.11–3.15) can be proven to yield global asymptotic rotor velocity regulation (see [34]).

3.2.4   Improved IFOC Control

IFOC-Based Control

In this section, modifications to the previous IFOC design are provided as a means to achieve improved performance (e.g., faster transient response and improved robustness). That is, the improved IFOC strategy developed in this section fosters global exponential[2] rotor velocity tracking rather than

---

[2] With the achievement of exponential rotor velocity tracking as will be seen in (3.41), a certain degree of robustness is acquired for the improved IFOC strategy. That is, exponentially stable systems inherently have the ability to tolerate a greater degree of uncertainty in the form of unknown parameters, external disturbances, unmodeled dynamics, etc., as compared to asymptotically stable systems. For a more detailed discussion on the theorems and analysis concerning the robustness of exponentially stable systems, see [37] and the references therein.

the asymptotic regulation results associated with standard IFOC designs. Specifically, to develop the improved IFOC strategy, the controller given in (3.12) is modified as follows

$$u_{foc} = \left[\begin{array}{cc} \dfrac{\beta_1}{\beta_2}\delta_d + \dfrac{\dot{\delta}_d}{\beta_2} + \Omega_0 & \dfrac{\tau_d}{\alpha_1\delta_d} \end{array} \right]^T \tag{3.16}$$

where the auxiliary signal $\Omega_0(r,t) \in \mathbb{R}$ is injected to sever the electrical/mechanical subsystem interconnections and is explicitly given as

$$\Omega_0 = \frac{\tau_d r}{\beta_2 \delta_d}. \tag{3.17}$$

In addition to the modification of (3.12), the differential equation given in (3.13) is also modified as follows

$$\dot{\rho}_d = \frac{\beta_2 \tau_d}{\alpha_1 \delta_d^2} + \alpha_1 r \Omega_1 \qquad \rho_d(0) = 0 \tag{3.18}$$

where the auxiliary signal $\Omega_1(r,t) \in \mathbb{R}$ is defined as

$$\Omega_1 = \frac{\beta_1}{\beta_2} + \frac{\dot{\delta}_d}{\beta_2 \delta_d} + \frac{\Omega_0}{\delta_d} \tag{3.19}$$

and the auxiliary signal $y(\omega, e, t) \in \mathbb{R}$ is redefined as follows

$$y(\omega, e, t) = J_m \left(\dot{\omega}_d + \mathcal{L}^{-1} \left\{ K_F(s) e(s) \right\} \right) + B\omega + T_L(\omega). \tag{3.20}$$

Lyapunov-Based Control

The development of the control structure of the improved IFOC design given in (3.11), (3.14), and (3.16–3.20) is based on the legacy of the standard IFOC control structures. Unfortunately, the standard IFOC structure does not readily accommodate Lyapunov-based analysis techniques. In this section, the improved IFOC design is written in a structure that is more conducive to a Lyapunov-based analysis. To this end, a desired rotor flux trajectory signal, denoted by $\psi_d(t) = [\ \psi_{da}(t) \quad \psi_{db}(t)\ ]^T \in \mathbb{R}^2$, is defined as follows [32]

$$\psi_d = \delta_d \left[\begin{array}{c} \cos \rho_d \\ \sin \rho_d \end{array} \right]. \tag{3.21}$$

The dynamic response of $\psi_d(t)$ can be determined by taking the time derivative of (3.21) as follows

$$\dot{\psi}_d = \frac{\dot{\delta}_d}{\delta_d} \left(\delta_d \left[\begin{array}{c} \cos \rho_d \\ \sin \rho_d \end{array} \right] \right) + \dot{\rho}_d J \left(\delta_d \left[\begin{array}{c} \cos \rho_d \\ \sin \rho_d \end{array} \right] \right) \tag{3.22}$$

where (3.3) has been utilized. After using the definition for (3.21) and then substituting (3.18) into the resulting expression for $\dot{\rho}_d(t)$, the dynamics for $\dot{\psi}_d(t)$ can be rewritten in the following form

$$\dot{\psi}_d = \frac{\dot{\delta}_d}{\delta_d}\psi_d + \left(\frac{\beta_2\tau_d}{\alpha_1\delta_d^2} + \alpha_1 r\Omega_1\right)J\psi_d \tag{3.23}$$

where J was defined in (3.3). After substituting (3.16) into (3.11) for $u_{foc}(t)$ the following expression can be obtained

$$u = \left(\frac{\tau_d}{\alpha_1\delta_d^2}\right)J\left(\delta_d\begin{bmatrix}\cos\rho_d\\\sin\rho_d\end{bmatrix}\right)$$

$$+ \left(\frac{\beta_1}{\beta_2} + \frac{\dot{\delta}_d}{\delta_d\beta_2} + \frac{\Omega_0}{\delta_d}\right)\left(\delta_d\begin{bmatrix}\cos\rho_d\\\sin\rho_d\end{bmatrix}\right) \tag{3.24}$$

where (3.3) was used. The commutation strategy given in (3.24) can also be written in the following compact form

$$u = \left(\frac{\tau_d}{\alpha_1\delta_d^2}\right)J\psi_d + \Omega_1\psi_d \tag{3.25}$$

after utilizing (3.19) and (3.21).

Remark 3.2 *Based on the structure of (3.21), the following relationship can be determined*

$$\|\psi_d(t)\|^2 = \delta_d^2\begin{bmatrix}\cos\rho_d & \sin\rho_d\end{bmatrix}\begin{bmatrix}\cos\rho_d\\\sin\rho_d\end{bmatrix} = \delta_d^2(t). \tag{3.26}$$

Given (3.26), the rotor flux magnitude tracking error introduced in (3.10) can be written as follows

$$\eta_\delta = \|\psi_d(t)\|^2 - \|\psi(t)\|^2 = (\psi_d + \psi)^T\eta_\psi \tag{3.27}$$

where the rotor flux tracking error, denoted by $\eta_\psi(t) = [\ \eta_{\psi a}(t)\ \ \eta_{\psi b}(t)\ ]^T \in \mathbb{R}^2$, is defined as follows

$$\eta_\psi = \psi_d - \psi. \tag{3.28}$$

From (3.27), it can be shown that if $\eta_\psi(t)$ is exponentially stable and if both $\psi_d(t)$, $\psi(t) \in \mathcal{L}_\infty$, then $\eta_\delta(t)$ will be exponentially stable. Hence, the secondary control objective of (3.10) will be achieved. This fact will be exploited in the subsequent closed-loop development and stability analysis.

Closed-Loop Error System

To construct the closed-loop rotor flux tracking error system, we take the time derivative of (3.28) and substitute (3.2) into the resulting expression as follows

$$\dot{\eta}_\psi = \dot{\psi}_d + \beta_1 \psi - \beta_2 u. \qquad (3.29)$$

After substituting (3.23) and (3.25) into (3.29) for $\dot{\psi}_d(t)$ and $u(t)$, respectively, the following closed-loop dynamics for $\dot{\eta}_\psi(t)$ can be obtained

$$\dot{\eta}_\psi = \frac{\dot{\delta}_d}{\delta_d} \psi_d + \left(\frac{\beta_2 \tau_d}{\alpha_1 \delta_d^2}\right) J\psi_d + (\alpha_1 r \Omega_1) J\psi_d + \beta_1 (\psi_d - \eta_\psi)$$

$$- \beta_2 \left(\left(\frac{\tau_d}{\alpha_1 \delta_d^2}\right) J\psi_d + \left(\frac{\beta_1}{\beta_2} + \frac{\dot{\delta}_d}{\beta_2 \delta_d} + \frac{\tau_d r}{\beta_2 \delta_d^2}\right) \psi_d\right) \qquad (3.30)$$

where (3.19) and (3.28) have been utilized. After cancelling common terms in (3.30) and noting that $JJ = -I_2$ (where I_2 denotes the 2×2 identity matrix), the closed-loop rotor flux tracking error system can be rewritten in the following form

$$\dot{\eta}_\psi = -\beta_1 \eta_\psi + \alpha_1 r J \left[\left(\frac{\tau_d}{\alpha_1 \delta_d^2}\right) J\psi_d + \Omega_1 \psi_d\right]. \qquad (3.31)$$

After utilizing (3.25), (3.31) can be written as follows

$$\dot{\eta}_\psi = -\beta_1 \eta_\psi + \alpha_1 r J u. \qquad (3.32)$$

To formulate the closed-loop tracking error system for the mechanical subsystem, we take the time derivative of the rotor velocity tracking error given in (3.6), premultiply the resulting expression by J_m, and use (3.1) to obtain the following expression

$$J_m \dot{r} = y - \alpha_1 u^T J\psi_d + \alpha_1 u^T J\psi_d - \alpha_1 u^T J\psi \qquad (3.33)$$

where $y(\omega, e, t)$ was defined in (3.20), and the product $\alpha_1 u^T(t) J\psi_d(t)$ has been added and subtracted to (3.33). By substituting (3.25) into (3.33) for only the first occurrence of $u(t)$, the following expression can be obtained

$$J_m \dot{r} = y - \tau_d + \left(\tau_d - \left(\frac{\tau_d}{\delta_d^2}\right) \psi_d^T J^T J\psi_d + \Omega_1 \psi_d^T J\psi_d\right)$$

$$+ \alpha_1 u^T J\psi_d - \alpha_1 u^T J\psi \qquad (3.34)$$

where $\tau_d(t)$ has been added and subtracted to (3.34). By exploiting the facts that $J^T J = I_2$ and $\psi_d^T(t) J\psi_d(t) = 0$, the closed-loop dynamics for the rotor velocity tracking error can be obtained as follows

$$J_m \dot{r} = -k_s r + \alpha_1 u^T J \eta_\psi \qquad (3.35)$$

where (3.14), (3.26), and (3.28) have been utilized.

Stability Analysis

Given the improved IFOC controller of (3.11), (3.14), and (3.16–3.20), the following theorem can be used to prove that the rotor velocity/flux magnitude tracking errors for the current-fed induction motor are confined to an exponentially decaying envelope.

Theorem 3.1 *The improved IFOC structure of (3.11), (3.14), and (3.16–3.20) ensures global exponential rotor velocity/rotor flux tracking in the sense that*

$$\|z(t)\| \le \sqrt{\frac{\lambda_2}{\lambda_1}} \|z(0)\| \exp\left(-\frac{\lambda_3}{2\lambda_2} t\right). \qquad (3.36)$$

For the exponential envelope given in (3.36), $z(t) = \begin{bmatrix} r(t) & \eta_\psi^T(t) \end{bmatrix}^T \in \mathbb{R}^3$, *where* $r(t)$ *and* $\eta_\psi(t)$ *are defined in (3.6) and (3.28), respectively, and the coefficients* λ_1, λ_2, $\lambda_3 \in \mathbb{R}$ *are positive constants defined as follows*

$$\lambda_1 = \frac{1}{2} \min\{J_m, 1\}, \; \lambda_2 = \frac{1}{2} \max\{J_m, 1\}, \qquad (3.37)$$

and $\lambda_3 = \min\{k_s, \beta_1\}.$

Proof: To prove Theorem 3.1, we define the nonnegative function $V(t) \in \mathbb{R}$ as follows

$$V = \frac{1}{2} J_m r^2 + \frac{1}{2} \eta_\psi^T \eta_\psi. \qquad (3.38)$$

Based on the structure of (3.38), the following inequalities can be formulated

$$\lambda_1 \|z\|^2 \le V \le \lambda_2 \|z\|^2 \qquad (3.39)$$

where $z(t)$ was introduced in (3.36), and λ_1 and λ_2 were defined in (3.37). After taking the time derivative of (3.38) and utilizing (3.32) and (3.35), the following expression can be obtained

$$\dot{V} = -k_s r^2 - \beta_1 \left(\eta_\psi^T \eta_\psi\right) + \alpha_1 r \left(u^T J \eta_\psi + \eta_\psi^T J u\right). \qquad (3.40)$$

After using the fact that $J^T = -J$, (3.40) can be upper bounded as follows

$$\dot{V} \le -\lambda_3 \|z\|^2 \qquad (3.41)$$

where λ_3 was defined in (3.37). To prove the exponential tracking result given in (3.36), the inequalities given in (3.39) can be used to rewrite (3.41) as follows

$$\dot{V} \leq -\frac{\lambda_3}{\lambda_2} V. \tag{3.42}$$

After solving the differential equation given in (3.42) according to Lemma A.10 of Appendix A and then utilizing (3.39), the result given in (3.36) can be obtained; hence, $r(t)$ and $\eta_\psi(t)$ are also contained in the exponential envelope given in (3.36).

Since $\psi_d(t) \in \mathcal{L}_\infty$, (3.28) can be used to prove that $\psi(t) \in \mathcal{L}_\infty$. From the restrictions placed on $H(s)$ defined in (3.7) and the fact that $r(t)$ is bounded by the exponential envelope given in (3.36), Lemma A.14 of Appendix A can be used to prove that $e(t)$ is also confined to an exponentially decaying envelope. Specifically, for the PI compensator (i.e., $K_F(s) = k_i$), the following exponential envelope for $e(t)$ can be obtained

$$
\begin{aligned}
|e(t)| \leq \quad & k_i |e(0)| \exp(-k_i t) + \sqrt{\frac{\lambda_2}{\lambda_1}} \|z(0)\| \exp\left(-\frac{\lambda_3}{2\lambda_2} t\right) \\
& + \sqrt{\frac{\lambda_2}{\lambda_1}} \frac{2k_i \lambda_2 \|z(0)\|}{(2\lambda_2 k_i - \lambda_3)} \left(\exp\left(-\frac{\lambda_3}{2\lambda_2} t\right) - \exp\left(-k_i t\right)\right).
\end{aligned}
\tag{3.43}
$$

After using (3.26–3.28), and the fact that $\eta_\psi(t)$ is contained in the exponentially decaying envelope given in (3.36), the following exponential envelope can be developed for the rotor flux magnitude

$$|\eta_\delta| \leq 2\delta_d \sqrt{\frac{\lambda_2}{\lambda_1}} \|z(0)\| \exp\left(-\frac{\lambda_3}{2\lambda_2} t\right) + \frac{\lambda_2}{\lambda_1} \|z(0)\|^2 \exp\left(-\frac{\lambda_3}{\lambda_2} t\right). \tag{3.44}$$

$\square$

3.2.5   Adaptive Extension

To obtain the result given in (3.36), exact model knowledge of the mechanical subsystem was assumed. In practice, the constant parameters of the mechanical subsystem may not be exactly known (i.e., the parameters J_m, B, and the constant coefficients of the load function $T_L(\omega)$ may be unknown). In this extension, an adaptive update law is developed that can be incorporated with the previous improved IFOC design to compensate for parametric uncertainty in the mechanical subsystem. To this end, the function $y(\omega, e, t)$ defined in (3.20) is assumed to satisfy the following linear parameterization property

$$y = Y\theta \tag{3.45}$$

where $Y(\omega, e, t) \in \mathbb{R}^{1 \times p}$ is a known regression vector, and $\theta \in \mathbb{R}^p$ denotes the unknown constant parameter vector. To compensate for the parametric uncertainty contained in the unknown vector θ, the desired torque signal of (3.14) is redesigned as follows

$$\tau_d = Y\hat{\theta} + k_s r \tag{3.46}$$

where $\hat{\theta}(t) \in \mathbb{R}^p$ denotes a parameter estimate vector that is calculated on-line via the following adaptive gradient update law

$$\dot{\hat{\theta}} = \Gamma Y^T r \tag{3.47}$$

where $\Gamma \in \mathbb{R}^{p \times p}$ is a constant positive diagonal adaptation gain matrix.

Because (3.14) is redesigned as in (3.46) due to parametric uncertainty, the closed-loop error system for $r(t)$ must be modified. Specifically, after substituting the new expression for $\tau_d(t)$ given in (3.46) into (3.34) and exploiting the facts that $J^T J = I_2$ and $\psi_d^T J \psi_d = 0_2$, the closed-loop dynamics for the rotor velocity tracking error given in (3.35) are now given by the following expression

$$J_m \dot{r} = -k_s r + Y\tilde{\theta} + \alpha_1 u^T J \eta_\psi \tag{3.48}$$

where (3.26) and (3.28) have been used. The parameter estimation error $\tilde{\theta}(t) \in \mathbb{R}^p$ introduced in (3.48) is defined as follows

$$\tilde{\theta} = \theta - \hat{\theta}. \tag{3.49}$$

The following theorem can now be used to examine the stability of the adaptive extension of the improved IFOC design.

Theorem 3.2 *The adaptive improved IFOC structure of (3.11), (3.16–3.20), and (3.46) ensures global asymptotic rotor velocity/rotor flux tracking in the sense that*

$$\lim_{t \to \infty} e(t), \eta_\delta(t) = 0. \tag{3.50}$$

Proof: To prove Theorem 3.2, we define the nonnegative function $V(t) \in \mathbb{R}$ as follows

$$V = \frac{1}{2} J_m r^2 + \frac{1}{2} \eta_\psi^T \eta_\psi + \tilde{\theta}^T \Gamma^{-1} \tilde{\theta}. \tag{3.51}$$

After taking the time derivative of (3.51) and utilizing (3.32), (3.35), and (3.47), the following expression is obtained

$$\dot{V} = -k_s r^2 + rY\tilde{\theta} - \beta_1 \left(\eta_\psi^T \eta_\psi \right)$$
$$+ \alpha_1 r \left(u^T J \eta_\psi + \eta_\psi^T J u \right) - \tilde{\theta}^T Y^T r. \tag{3.52}$$

After cancelling common terms and using the fact that $J^T = -J$, (3.52) can be expressed as follows

$$\dot{V} = -k_s r^2 - \beta_1 \left\| \eta_\psi \right\|^2 . \tag{3.53}$$

Since $V(t)$ is a nonnegative function, and $\dot{V}(t)$ is negative semi-definite, $V(t)$ is upper bounded by $V(0)$; hence, $r(t)$, $\eta_\psi(t)$, $\tilde{\theta}(t)$, $\hat{\theta}(t) \in \mathcal{L}_\infty$. By invoking Lemmas A.11 and A.13 of Appendix A, we can prove that $r(t)$, $\eta_\psi(t) \in \mathcal{L}_2$ and that $e(t)$, $\dot{e}(t) \in \mathcal{L}_\infty$. Since $\psi_d(t) \in \mathcal{L}_\infty$, (3.28) can be used to prove that $\psi(t) \in \mathcal{L}_\infty$. Since $\eta_\psi(t)$, $\psi(t) \in \mathcal{L}_\infty$, (3.27) can be used to show that $\eta_\delta(t) \in \mathcal{L}_\infty$. By Definition A.1 in Appendix A, the product of an $\mathcal{L}_\infty$ function with an $\mathcal{L}_2$ function is also $\mathcal{L}_2$; hence, (3.27) can be utilized to show that $\eta_\delta(t) \in \mathcal{L}_2$. Since $\omega_d(t)$ and $\dot{\omega}_d(t)$ are selected as bounded functions of time, $\omega(t)$, $\dot{\omega}(t) \in \mathcal{L}_\infty$. Given the preceding facts, (3.46) and (3.47) can be used to prove that $\tau_d(t)$, $\dot{\hat{\theta}}(t) \in \mathcal{L}_\infty$. The expressions given in (3.17) and (3.19) can be used to prove that $\Omega_0(t)$, $\Omega_1(t) \in \mathcal{L}_\infty$, and from (3.25), $u(t) \in \mathcal{L}_\infty$. From (3.32) and (3.48), $\dot{r}(t)$, $\dot{\eta}_\psi(t) \in \mathcal{L}_\infty$, and hence, $\dot{\psi}(t) \in \mathcal{L}_\infty$. After taking the time derivative of (3.27), it can be seen that $\dot{\eta}_\delta(t) \in \mathcal{L}_\infty$. Since $\eta_\delta(t)$, $r(t) \in \mathcal{L}_\infty \cap \mathcal{L}_2$ and $\dot{r}(t)$, $\dot{\eta}_\delta(t) \in \mathcal{L}_\infty$, Barbalat's Lemma (see Lemma A.16 of Appendix A) can be invoked to prove that

$$\lim_{t \to \infty} r(t), \eta_\delta(t) = 0. \tag{3.54}$$

By invoking Lemma A.15 of Appendix A, the result given in (3.50) can now be obtained. □

3.2.6 Experimental Setup and Results

Experiments were conducted on a three-phase induction motor (Baldor Electric Co., Model M3541) powered by three linear amplifiers (Techron, Model 7570-60) to compare the performance of the standard IFOC scheme to that of the improved IFOC scheme for a rotor velocity tracking application.[3] The rotor position was measured using a 10,240 line shaft-mounted encoder (BEI Inc.). The rotor velocity was obtained using a backward difference algorithm applied to the rotor position signal with the resulting signal being passed through a second-order digital filter. The control algorithm was computed on a Pentium processor running under the QNX operating system at a sampling frequency of 2.0 [kHz]. Qmotor 3.0 [24] was used as the graphical user interface. The Quanser Consulting MultiQ I/O

---

[3] The standard IFOC design of [7] can be easily augmented for velocity tracking applications through simple modifications to the desired torque command.

board was used to deliver the three-phase voltages to the induction motor. To achieve the command current of (3.11), a high-gain current feedback control algorithm applied to the stator windings of the motor was implemented. That is, the stator voltage applied to the terminals of the motor is given by the following expression

$$V_s = k(u - I_s). \tag{3.55}$$

In (3.55), $V_s(t) \in \mathbb{R}^2$ denotes the stator voltage input in the rotating rotor reference frame, $u(t) \in \mathbb{R}^2$ represents the command input current of (3.11), $I_s(t) \in \mathbb{R}^2$ represents the measured stator current in the rotating rotor reference frame, and $k \in \mathbb{R}$ denotes a constant positive control gain. Three Hall-effect sensors (Microswitch, Model CSLB1AD) were used to measure the stator phase currents. Figure 3.2 depicts a block diagram of the experimental setup.

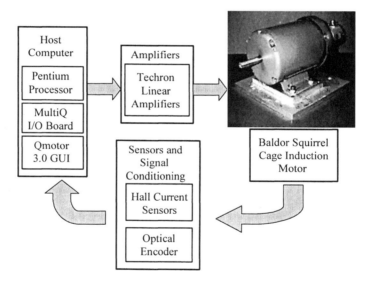

FIGURE 3.2. Block diagram configuration of the induction motor experimental setup.

An inertia wheel attached to the rotor formed the mechanical load for the system. The total system inertia J_m including the rotor inertia was calculated to be 0.044 [kg · m²]. The coefficient of viscous friction B was experimentally determined to have a nominal value of $B = 0.007$ [Nm · s]. The shaft of the induction motor was directly coupled to a separately excited direct current motor (Baldor Electric Co., Model CD3433) to provide a constant load torque of $T_L(\omega) = 0.75$ [Nm]. By using standard measurement procedures and the motor data sheet, the nominal values of the remaining system parameters were determined to have the following values:

$$L_r = 0.14 \ [\text{H}], \quad R_r = 1.99 \ [\Omega],$$

$$M = 0.12 \ [\text{H}], \quad \text{and} \quad n_p = 1.0. \tag{3.56}$$

For the experiment, the desired rotor velocity trajectory was selected as the following soft-start trajectory

$$\omega_d = 500(1 - \exp(-0.015t^3)) \ [\text{rpm}] \tag{3.57}$$

and the desired rotor flux norm $\delta_d(t)$ was selected as the following constant value $\delta_d = 0.62$ [Wb] (i.e., $\dot{\delta}_d = 0$). The standard IFOC scheme given in (3.11–3.15) with the following torque trajectory structure was implemented (i.e., a PI-based feedback control law)

$$\tau_d = y + k_s e + k_s k_i \int_0^t e(\sigma) \, d\sigma. \tag{3.58}$$

In (3.58), the control gains were selected as follows

$$k_s = 14.0, \ k_i = 5.18, \ \text{and} \ k = 3.0. \tag{3.59}$$

The improved IFOC scheme of (3.11), (3.16–3.20), and (3.58) was also implemented using the same control gains as in (3.59). To illustrate the improved transient response, the rotor velocity tracking errors of the standard IFOC scheme as well as the improved IFOC scheme are depicted in Figure 3.3. Plots of the actual and reference stator phase 1 currents are shown in Figure 3.4. To compare the performance of the two control strategies, the $\mathcal{L}_2$ norm of the rotor velocity tracking error $e(t)$ was calculated as in Table 3.1. From Table 3.1, a greater than 50% reduction in the $\mathcal{L}_2$ norm of the rotor velocity tracking error can be observed. It is important to note that the improved IFOC scheme achieves better performance via a different control structure as opposed to a different set of control gains (i.e., the two schemes use the same control gains).

Remark 3.3 *During the experiment, a mismatch between the reference and actual stator currents was observed. This mismatch may be observed with any strategy that neglects the effect of the stator current dynamics. For control strategies for the full-order model of the induction motor that consider stator current dynamics in the control synthesis, see [15] and [26]. In the next section, the stator current dynamics are incorporated in the control design for precision position tracking applications.*

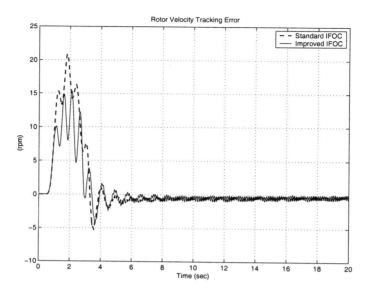

FIGURE 3.3. Rotor velocity tracking error: standard IFOC scheme vs. improved IFOC scheme.

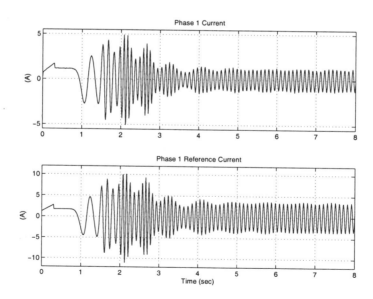

FIGURE 3.4. Phase 1 stator reference and actual currents for the improved IFOC scheme.

TABLE 3.1. Rotor velocity tracking error comparison

Implemented Controller	$\|e(t)\|_2$
Standard IFOC scheme	$5.025 \text{ s}^{-0.5}$
Improved IFOC scheme	$2.425 \text{ s}^{-0.5}$

3.3 Switched Reluctance Motor

As in the previous section, most existing control strategies for electric machines assume that the stator current dynamics can be neglected (i.e., the motor is treated as a current-fed motor), and the flux linkage characteristics can be modeled by a linear magnetic circuit. Since these assumptions may not always be justified, degraded tracking performance may result. Since the SRM is typically the electric machine of choice for high-precision position tracking applications, this section will focus on the development of a control design for the SRM that accounts for the nonlinearities in the flux linkage as well as the effects of the stator current dynamics. Specifically, a general nonlinear model of the SRM is used to develop an adaptive controller for the full-order electromechanical dynamic model. During the control formulation, a design condition related to the structure of the algebraic torque transmission relationship is developed which is used to construct a commutation strategy. Specifically, this design condition facilitates the modification of the commutation strategy originally proposed in [12] to ensure that the desired torque is delivered to the mechanical subsystem. Provided this design condition can be satisfied and the desired current trajectory signals satisfy some mild conditions,[4] the controller achieves global asymptotic position/velocity tracking. The proposed controller requires measurement of the rotor position, rotor velocity, and stator current, does not exhibit any control singularities, and compensates for uncertain electromechanical parameters that are independent of the flux linkage model. As an extension to the general nonlinear model, a commutation strategy is also designed to account for magnetic saturation associated with a proposed flux linkage model.

---

[4]Roughly speaking, the backstepping control design procedure requires that the desired current and its time derivative be bounded if the rotor position, velocity, and acceleration are bounded.

3.3.1   System Model

The mechanical subsystem for the SRM is assumed to be characterized by
the following differential expression

$$M_m\ddot{q} + W_m(q,\dot{q})\theta_m = \sum_{j=1}^{m} \tau_j(q, I_j) \qquad (3.60)$$

where the subsystem states $q(t)$, $\dot{q}(t)$, and $\ddot{q}(t) \in \mathbb{R}$ represent the rotor
position, velocity, and acceleration, respectively. In (3.60), $M_m \in \mathbb{R}$ denotes the unknown constant lumped inertia of the rotor-load, the linear
parameterization $W_m(q,\dot{q})\theta_m$ represents friction and loading effects, where
$W_m(q,\dot{q}) \in \mathbb{R}^{1 \times p}$ is assumed to be a known first-order differentiable regression vector and $\theta_m \in \mathbb{R}^p$ represents the unknown constant mechanical parameter vector, $m \in \mathbb{R}$ represents the number of phases $(m \geq 3)$,
$I_j(t) \in \mathbb{R}$ represents the phase current in the jth winding, and $\tau_j(q, I_j) \in \mathbb{R}$
denotes the torque produced by the jth electrical phase.[5] The per-phase
torque of the mechanical subsystem is directly related to the flux linkage
of the system. That is, from the flux linkage model the per-phase torque
can be calculated as follows

$$\tau_j(q, I_j) = \frac{\partial}{\partial q} \int_0^{I_j} \lambda_j(q, \bar{I}_j) d\bar{I}_j \qquad (3.61)$$

where $\lambda_j(q, \bar{I}_j) \in \mathbb{R}$ represents the known flux linkage model [43]. The flux
linkage model can also be used to compute the inductance and back-EMF
functions associated with the electrical subsystem dynamics. That is, for
the following electrical dynamics

$$L_j(q, I_j)\dot{I}_j + RI_j + B_j(q, I_j)\dot{q} = v_j \qquad (3.62)$$

the per-phase incremental inductance, denoted by $L_j(q, I_j) \in \mathbb{R}$, and per-
phase back-emf, denoted by $B_j(q, I_j) \in \mathbb{R}$, can be calculated from the flux
linkage as follows [43]

$$L_j(q, I_j) = \frac{\partial \lambda_j(q, I_j)}{\partial I_j} \quad \text{and} \quad B_j(q, I_j) = \frac{\partial \lambda_j(q, I_j)}{\partial q} \qquad (3.63)$$

where $R \in \mathbb{R}$ denotes the unknown electrical resistance in each matched
phase, and $v_j(t) \in \mathbb{R}$ denotes the input voltage of the jth phase.

---

[5] Since the stator phases are symmetrically wound on pole pairs, the common assumption is made for SRM systems that the mutual inductance between the phases can
be neglected [43].

Remark 3.4 *To facilitate the subsequent analysis, M_m is assumed to be lower bounded by a known positive constant (i.e., $\underline{M}_m < M_m$ where $\underline{M}_m$ is a known positive constant). In addition, it is assumed that $W_m(\cdot), \dot{W}_m(\cdot) \in L_\infty$ if $q(t), \dot{q}(t), \ddot{q}(t) \in L_\infty$, and $L_j(q, I_j), B_j(q, I_j), \tau_j(q, I_j), \dfrac{\partial \tau_j(q, I_j)}{\partial I_j}, \dfrac{\partial \tau_j(q, I_j)}{\partial q} \in L_\infty$ if $q(t), I_j(t) \in L_\infty$.*

3.3.2 Control Objective

The primary objective in this section is to develop a controller to ensure that the rotor tracks a desired position/velocity trajectory despite parametric uncertainty in the mechanical and electrical dynamics given in (3.60–3.63). To facilitate the control development for the primary objective, secondary objectives of phase currents tracking and torque tracking are also defined. To quantify the rotor tracking objective, a rotor position tracking error, denoted by $e(t) \in \mathbb{R}$, and a filtered tracking error, denoted by $r(t) \in \mathbb{R}$, are defined as follows

$$e = q_d - q \qquad r = \dot{e} + \alpha e \tag{3.64}$$

where $q_d(t)$ represents the desired rotor position trajectory, and $\alpha \in \mathbb{R}$ is a positive constant control gain. To quantify the secondary tracking objectives, a phase current tracking error, denoted by $\eta_j(t) \in \mathbb{R}$, and a torque tracking error, denoted by $\eta_\tau(t) \in \mathbb{R}$, are defined as follows

$$\eta_j = I_{dj} - I_j \qquad \eta_\tau = \sum_{j=1}^{m} \tau_j(q, I_{dj}) - \sum_{j=1}^{m} \tau_j(q, I_j) \tag{3.65}$$

where $I_{dj}(\cdot) \in \mathbb{R}$ represents a subsequently designed desired stator current trajectory. For the subsequent analysis, the standard assumption is made that $q_d(t), \dot{q}_d(t), \ddot{q}_d(t), \dddot{q}_d(t)$ are selected to be bounded functions of time. It is also assumed that $q(t), \dot{q}(t), I_j(t)$ are measurable.

Remark 3.5 *Based on the definition of the current and torque tracking errors given in (3.65) and some mild smoothness restrictions on the structure of $\tau_j(q, I_j)$, it is clear that if the current tracking objective is met, then the torque tracking objective will also be met (i.e., if $\eta_j(t) \to 0$, then $\eta_\tau(t) \to 0$).*

3.3.3 Adaptive Tracking Control

Based on the structure of the coupled mechanical and electrical dynamics, the following control design will exploit integrator backstepping techniques.

That is, these techniques will be used to develop a voltage control input for the electrical dynamics that produces a desired torque (current) input that achieves the rotor position/velocity tracking objective.

Filtered Tracking Error

To develop the open-loop error system for the mechanical dynamics, we take the time derivative of $r(t)$ of (3.64) and multiply the resulting expression by M_m as follows

$$M_m \dot{r} = M_m \left(\ddot{q}_d + \alpha \dot{e} \right) + W_m \theta_m - \sum_{j=1}^{m} \tau_j(q, I_{dj}) + \eta_\tau \qquad (3.66)$$

where (3.60), (3.65), and the definition of $e(t)$ given in (3.64) were used. To facilitate the subsequent analysis, it is assumed that the desired phase currents are designed according to the following commutation strategy

$$\sum_{j=1}^{m} \tau_j \left(q, I_{dj} \left(q, \tau_d \right) \right) = \tau_d \qquad (3.67)$$

where $\tau_d(t) \in \mathbb{R}$ denotes a desired torque trajectory defined as follows

$$\tau_d = \hat{M}_m \left(\ddot{q}_d + \alpha \dot{e} \right) + W_m \hat{\theta}_m + K_s r. \qquad (3.68)$$

For the desired torque trajectory introduced in (3.68), $K_s \in \mathbb{R}$ denotes a positive constant control gain, and $\hat{M}_m(t) \in \mathbb{R}$, $\hat{\theta}_m(t) \in \mathbb{R}^p$ denote subsequently designed dynamic estimates of the unknown inertia term M_m and unknown parameter vector θ_m, respectively. After substituting (3.67) and (3.68) into (3.66) and then simplifying the resulting expression, the closed-loop dynamics for the filtered tracking error can be obtained as follows

$$M_m \dot{r} = -K_s r + \tilde{M}_m \left(\ddot{q}_d + \alpha \dot{e} \right) + W_m \tilde{\theta}_m + \eta_\tau \qquad (3.69)$$

where the parameter estimate error vectors $\tilde{M}_m(t) \in \mathbb{R}$ and $\tilde{\theta}_m(t) \in \mathbb{R}^p$ are defined as follows

$$\tilde{M}_m = M_m - \hat{M}_m \quad \text{and} \quad \tilde{\theta}_m = \theta_m - \hat{\theta}_m. \qquad (3.70)$$

Remark 3.6 *The development of the closed-loop filtered tracking error dynamics assumes that a commutation strategy can be employed to satisfy (3.67) and (3.68). Given the structure of the coupled mechanical and electrical dynamics, we are motivated to employ integrator backstepping techniques to develop the voltage control input. However, the use of this technique places additional constraints on the commutation strategy. Specifically, in addition to satisfying the relationship given by (3.67), the use of* *an integrator backstepping design strategy will further require that $I_{dj}(\cdot)$, $\dot{I}_{dj}(\cdot) \in L_\infty$ provided $q(t)$, $\dot{q}(t)$, $\tau_d(t)$, $\dot{\tau}_d(t) \in L_\infty$. Given these requirements, not all commutation strategies (i.e., desired current trajectories) can be fused into the subsequent backstepping procedure. That is, a flux linkage model could be formulated that would render the subsequent design procedure invalid. However, there are many possible flux linkage models that are amenable to the design procedure, and after the subsequent stability analysis, a commutation strategy is presented that could be used to satisfy the assumptions on $I_{dj}(\cdot)$.*

Current Tracking Error

Based on the closed-loop error system given in (3.69) and the fact that a vanishing current tracking error implies a vanishing torque tracking error (see Remark 3.5), we are now motivated to develop the current tracking error dynamics. To this end, we take the time derivative of the current tracking error term defined in (3.65) and substitute the current dynamics of (3.62) for $\dot{I}_j(t)$ to obtain the following expression

$$\dot{\eta}_j = \frac{dI_{dj}}{d\tau_d}\dot{\tau}_d + \frac{dI_{dj}}{dq}\dot{q} + \frac{1}{L_j(q,I_j)}\left(RI_j + B_j(q,I_j)\dot{q} - v_j\right). \qquad (3.71)$$

Given the definition of the desired torque signal in (3.68), $\tau_d(t)$ can be expressed as a function of $\hat{M}_m(t)$, $\hat{\theta}_m(t)$, $q(t)$, $\dot{q}(t)$, $q_d(t)$, $\dot{q}_d(t)$, and $\ddot{q}_d(t)$; therefore, $\dot{\tau}_d(t)$ can be expressed as follows

$$\begin{aligned}
\dot{\tau}_d = {} & \hat{M}_m \dddot{q}_d + (\hat{M}_m\alpha + k_s)\ddot{q}_d + k_s\alpha\dot{q}_d \\
& + \frac{\partial \tau_d}{\partial \hat{M}_m}\dot{\hat{M}}_m + \frac{\partial \tau_d}{\partial \hat{\theta}_m}\dot{\hat{\theta}}_m + \frac{\partial \tau_d}{\partial q}\dot{q} + \frac{\partial \tau_d}{\partial \dot{q}}\ddot{q}
\end{aligned} \qquad (3.72)$$

where

$$\begin{aligned}
\frac{\partial \tau_d}{\partial \hat{M}_m} &= \ddot{q}_d + \alpha\dot{e} & \frac{\partial \tau_d}{\partial \hat{\theta}_m} &= W_m(q,\dot{q}) \\
\frac{\partial \tau_d}{\partial q} &= \frac{\partial W_m}{\partial q}\hat{\theta}_m - \alpha k_s & \frac{\partial \tau_d}{\partial \dot{q}} &= -\hat{M}_m\alpha + \frac{\partial W_m}{\partial \dot{q}}\hat{\theta}_m - k_s.
\end{aligned} \qquad (3.73)$$

To express (3.72) in terms of measurable signals, (3.60) is used to develop the following relationship for the rotor acceleration

$$\ddot{q} = -\frac{1}{M_m}W_m\theta_m + \frac{1}{M_m}\sum_{j=1}^{m}\tau_j(q,I_j). \qquad (3.74)$$

After substituting (3.74) into (3.72), substituting the resulting expression for $\dot{\tau}_d(t)$ into (3.71), and then multiplying both sides of the resulting expression by M_m, the following open-loop current tracking dynamics for $\eta_j(t)$ can be obtained

$$M_m \dot{\eta}_j = M_m \Omega_{1j} + \Omega_{2j} W_m \theta_m + M_m \Omega_{3j} R + \Omega_{4j} - \frac{M_m v_j}{L_j(q, I_j)} \qquad (3.75)$$

where $\Omega_{1j}(q, \dot{q}, I_j, t)$, $\Omega_{2j}(t)$, $\Omega_{3j}(q, I_j)$, $\Omega_{4j}(q, I_j, t) \in \mathbb{R}$ are measurable functions (see Definition B.1 of Appendix B). Based on the open-loop dynamics of (3.75) and the subsequent stability analysis, the voltage control input is designed as follows

$$\begin{aligned} v_j = \ & L_j(q, I_j) \left(K_{ej} \eta_j + \Omega_{1j} + \Omega_{3j} \hat{R} \right) \\ & + \hat{M}_m^{-1} L_j(q, I_j) \left(\Omega_{4j} + \Omega_{2j} W_m \hat{\theta}_m + u_j \right) \end{aligned} \qquad (3.76)$$

where $K_{ej} \in \mathbb{R}$ is a positive constant control gain, $\hat{R}(t) \in \mathbb{R}$ is a dynamic estimate of R, and $u_j(t) \in \mathbb{R}$ is an additional control term injected to sever the interconnection between the mechanical and electrical subsystem that is explicitly defined as follows[6] [39]

$$u_j = \begin{cases} \dfrac{\tau_j(q, I_{dj}) - \tau_j(q, I_j)}{\eta_j} r & \text{if } \eta_j \neq 0 \\[4mm] \dfrac{\partial \tau_j(q, I_{dj})}{\partial I_{dj}} r & \text{if } \eta_j = 0. \end{cases} \qquad (3.77)$$

After substituting the control input given by (3.76) into (3.75), the closed-loop dynamics for $\eta_j(t)$ can be written as follows

$$\begin{aligned} M_m \dot{\eta}_j = \ & -K_{ej} M_m \eta_j + \Omega_{2j} W_m \tilde{\theta}_m + M_m \Omega_{3j} \tilde{R} \\ & - \hat{M}_m^{-1} \left(\Omega_{4j} + \Omega_{2j} W_m \hat{\theta}_m + u_j \right) \tilde{M}_m - u_j \end{aligned} \qquad (3.78)$$

where $\tilde{R}(t) \in \mathbb{R}$ denotes the following parameter estimation error

$$\tilde{R} = R - \hat{R}. \qquad (3.79)$$

---

[6] It is important to note that the definition of $u_j(t)$ for $\eta_j(t) \neq 0$ reduces to the definition of $u_j(t)$ for $\eta_j(t) = 0$ via L'Hospital's rule. Hence, the auxiliary control input $u_j(t)$ does not exhibit a singularity or a discontinuity.

Based on the structure of the closed-loop dynamics given in (3.69) and (3.78), the parameter update laws for $\hat{R}_r$ (t), $\hat{\theta}_m$ (t), and $\hat{M}_m$ (t) can be designed as follows

$$\dot{\hat{R}} = \Gamma_3 \sum_{j=1}^{m} \Omega_{3j} \eta_j \quad \dot{\hat{\theta}}_m = \Gamma_2 \left(W_m^T r + \sum_{j=1}^{m} \left(\Omega_{2j} W_m^T \eta_j \right) \right) \qquad (3.80)$$

$$\dot{\hat{M}}_m = \begin{cases} \Omega_m & \text{if} \quad \hat{M}_m > \underline{M}_m \\ \Omega_m & \text{if} \quad \hat{M}_m = \underline{M}_m \quad \text{and} \quad \Omega_m \geq 0 \quad \hat{M}_m(0) = \underline{M}_m \\ 0 & \text{if} \quad \hat{M}_m = \underline{M}_m \quad \text{and} \quad \Omega_m < 0 \end{cases}$$

$$(3.81)$$

where the auxiliary term Ω_m $(t) \in \mathbb{R}$ is defined as follows

$$\Omega_m = \Gamma_1 \left(\sum_{j=1}^{m} \left(-\hat{M}_m^{-1} \left(\Omega_{4j} + \Omega_{2j} W_m \hat{\theta}_m + u_j \right) \eta_j \right) + (\ddot{q}_d + \alpha \dot{e}) r \right)$$

$$(3.82)$$

where Γ_1, $\Gamma_3 \in \mathbb{R}$ are positive constant gains, and $\Gamma_2 \in \mathbb{R}^{p \times p}$ is a positive-definite diagonal constant gain matrix.

Remark 3.7 *To avoid singularities in (3.76) and (3.82), the projection algorithm [37] given in (3.81) can be employed to ensure that $\hat{M}_m$ (t) is always positive. For further details regarding the use of a projection algorithm, see [9], [25], and [37].*

Stability Analysis

Given the adaptive tracking controller of (3.67), (3.68), (3.76), (3.77), and (3.80-3.82), global asymptotic rotor position/velocity tracking can be obtained as described by the following theorem.

Theorem 3.3 *Given the commutation design of (3.67) and (3.68) (assuming the conditions on $I_{dj}(q, \tau_d)$ given in Remark 3.6 are satisfied), the voltage control input given in (3.76), (3.77), and (3.80–3.82), ensures global asymptotic rotor position/velocity tracking in the sense that*

$$\lim_{t \to \infty} e(t), \dot{e}(t) = 0. \qquad (3.83)$$

Proof: To prove Theorem 3.3, we define the nonnegative function $V(t) \in \mathbb{R}$ as follows

$$V = \frac{1}{2} M_m r^2 + \frac{1}{2} \sum_{j=1}^{m} M_m \eta_j^2$$

$$+ \frac{1}{2} \Gamma_1^{-1} \tilde{M}_m^2 + \frac{1}{2} \tilde{\theta}_m^T \Gamma_2^{-1} \tilde{\theta}_m + \frac{1}{2} M_m \Gamma_3^{-1} \tilde{R}^2. \tag{3.84}$$

Based on the structure of $V(t)$ given in (3.84), the following inequalities can be developed

$$\lambda_1 \|z\|^2 \le V \le \lambda_2 \|z\|^2 \tag{3.85}$$

where $\lambda_1, \lambda_2 \in \mathbb{R}$ denote positive bounding constants defined as follows

$$\lambda_1 = \frac{1}{2} \min \left\{ M_m, \Gamma_1^{-1}, \lambda_{\min} \left\{ \Gamma_2^{-1} \right\}, M_m \Gamma_3^{-1} \right\} \tag{3.86}$$

$$\lambda_2 = \frac{1}{2} \max \left\{ M_m, \Gamma_1^{-1}, \lambda_{\max} \left\{ \Gamma_2^{-1} \right\}, M_m \Gamma_3^{-1} \right\} \tag{3.87}$$

and the composite state vector $z(t) \in \mathbb{R}^{3+p+m}$ is defined as

$$z = \begin{bmatrix} r & \eta_1 ... \eta_m & \tilde{R} & \tilde{M}_m & \tilde{\theta}_m^T \end{bmatrix}^T \tag{3.88}$$

where $\lambda_{\min} \{\cdot\}, \lambda_{\max} \{\cdot\}$ denote the minimum and maximum eigenvalues of a matrix, respectively. After taking the time derivative of (3.84) and substituting the mechanical and electrical closed-loop error systems from (3.69) and (3.78), the following expression can be obtained

$$\dot{V} = -K_s r^2 - \sum_{j=1}^{m} K_{ej} M_m \eta_j^2 + \left[\eta_\tau r - \sum_{j=1}^{m} \eta_j u_j \right]$$

$$+ \tilde{\theta}_m^T \left(W_m^T r + \sum_{j=1}^{m} \left(\Omega_{2j} W_m^T \eta_j \right) - \Gamma_2^{-1} \dot{\hat{\theta}}_m \right)$$

$$+ M_m \tilde{R} \left(\sum_{j=1}^{m} \left(\Omega_{3j} \eta_j \right) - \Gamma_3^{-1} \dot{\hat{R}} \right) \tag{3.89}$$

$$+ \tilde{M}_m \left(\sum_{j=1}^{m} \left(-\hat{M}_m^{-1} \left(\Omega_{4j} + \Omega_{2j} W_m \hat{\theta}_m + u_j \right) \eta_j \right) \right.$$

$$\left. + (\ddot{q}_d + \alpha \dot{e}) r - \Gamma_1^{-1} \dot{\hat{M}}_m \right).$$

After substituting (3.65) and (3.77) into (3.89) for $\eta_\tau(t)$ and $u_j(t)$, respectively, and substituting (3.80–3.82) into (3.89) for the parameter estimates $\dot{\hat{R}}(t)$, $\dot{\hat{\theta}}_m(t)$, and $\dot{\hat{M}}_m(t)$, respectively, the following upper bound for $\dot{V}(t)$ can be obtained[7]

$$\dot{V} \leq -\lambda_3 \|x\|^2 \tag{3.90}$$

where the positive bounding constant $\lambda_3 \in \mathbb{R}$ is defined as follows

$$\lambda_3 = \min\left\{K_s, \min_j \{K_{ej} M_m\}\right\} \tag{3.91}$$

and $x(t) \in \mathbb{R}^{1+m}$ is defined as

$$x = \begin{bmatrix} r & \eta_1 \cdots \eta_m \end{bmatrix}^T. \tag{3.92}$$

Based on (3.84) and (3.90), $z(t)$, $x(t) \in \mathcal{L}_\infty$ and $x(t) \in \mathcal{L}_2$ (see Lemma A.11 of Appendix A). Given that $z(t) \in \mathcal{L}_\infty$, (3.88) can be used to prove that $r(t)$, $\eta_j(t)$, $\tilde{R}(t)$, $\tilde{M}_m(t)$, $\tilde{\theta}_m(t) \in \mathcal{L}_\infty$. Since $r(t) \in \mathcal{L}_\infty$, Lemma A.13 of Appendix A can be used in conjunction with (3.64) to prove that $e(t)$, $\dot{e}(t)$, $q(t)$, $\dot{q}(t) \in \mathcal{L}_\infty$. Based on the fact that $\tilde{R}(t)$, $\tilde{M}_m(t)$, $\tilde{\theta}_m(t) \in \mathcal{L}_\infty$, (3.70) and (3.79) can be used to prove that $\hat{R}(t)$, $\hat{M}_m(t)$, $\hat{\theta}_m(t) \in \mathcal{L}_\infty$. Since $\eta_j(t) \in \mathcal{L}_\infty$, the structural information regarding the flux linkage dependent quantities discussed in Remark 3.4 as well as the restrictions on the desired current trajectories discussed in Remark 3.6 can be used to prove that $I_j(t)$, $v_j(t) \in \mathcal{L}_\infty$. Standard signal chasing arguments can also be used along with (3.69) and (3.78) to prove that $\dot{x}(t) \in \mathcal{L}_\infty$. Since $x(t) \in \mathcal{L}_\infty \cap \mathcal{L}_2$, and $\dot{x}(t) \in \mathcal{L}_\infty$, Barbalat's Lemma (see Lemma A.16 of Appendix A) can be used to prove that

$$\lim_{t\to\infty} x(t) = 0. \tag{3.93}$$

The definition of $x(t)$ given in (3.92) can now be used to prove the phase current tracking error and the filtered tracking error are asymptotically driven to zero as follows

$$\lim_{t\to\infty} r(t), \ \eta_j(t) = 0. \tag{3.94}$$

Given the definition of the filtered tracking error of (3.64) and the result given by (3.94), Lemma A.15 of Appendix A can be invoked to prove (3.83). $\square$

---

[7] See Lemma B.12 of Appendix B for an examination of the inequality given in (3.90) for the different cases of the projection algorithm given in (3.81).

3.3.4   Commutation Strategy

An inherent feature of the SRM is that it must be electronically commutated to produce motion. Based on the motor construction, the individual phases are energized as a function of the rotor position in order to turn the rotor. The commutation can be achieved through dedicated electronics or integrated into the controller design via a mathematical commutation strategy. Carroll, Dawson, and Leviner [12] developed a commutation strategy for the torque transmission equation given by $\lambda_j = L(x_j)I_j$, which shared the control responsibilities in a continuously differentiable fashion while also satisfying the desired current conditions described in Remark 3.6. Specifically, each phase is assigned the responsibility of producing some fraction of the desired torque based on the position of the rotor. At certain rotor positions this fraction is small, while at other rotor positions this fraction may be nearly 100% of the desired torque.

Based on the development in [12], a commutation strategy is now presented for constructing the desired current trajectories as described by (3.67). The first requirement is that the structure of the torque transmission function (i.e., $\tau_j (q, I_j)$ in (3.61)) allow the construction of differentiable desired current trajectories (i.e., $I_{dj} (q, \tau_d)$) which satisfy condition in Remark 3.6 while also allowing (3.67) to be simplified as follows

$$\sum_{j=1}^{m} g(x_j)\gamma_{dj} = \tau_d. \tag{3.95}$$

In (3.95), $x_j(q) \in \mathbb{R}$ denotes an auxiliary commutation signal defined as

$$x_j = \left(N_r q - \frac{2\pi(j-1)}{m} \right) \tag{3.96}$$

where $N_r \in \mathbb{R}$ denotes the number of rotor saliencies, $\gamma_{dj} (q, \tau_d) \in \mathbb{R}$ is an auxiliary strictly positive function that is embedded inside of the desired current trajectory (i.e., $I_{dj} \left(q, \gamma_{dj} (q, \tau_d)\right)$), and $g(x_j)$ is an odd differentiable periodic function that satisfies the following properties: (i) $g(x_j) \in \mathcal{L}_\infty$ if $x_j(t) \in \mathcal{L}_\infty$, (ii) $\sum_{j=1}^{m} g(x_j) = 0$ for $m \geq 3$, and (iii) $g(x_j)$ has a periodicity of 2π.

Similar to the linear flux linkage model used in [12], the auxiliary function $\gamma_{dj} (\cdot)$ introduced in (3.95) can be designed as follows

$$\gamma_{dj} = \frac{g(x_j)\tau_d S\left(g(x_j)\tau_d\right)}{S_T} + \gamma_{c0}. \tag{3.97}$$

In (3.97), the positive design parameter $\gamma_{c0} \in \mathbb{R}$ is used to set the desired per phase threshold winding current (note the parameter γ_{co} is used to prevent control singularities), $S(\cdot) \in \mathbb{R}$ is defined to be the following differentiable function

$$S(z) = \begin{cases} 0 & \text{for } z \leq 0 \\ 1 - \exp\left(-\epsilon_0 z^2\right) & \text{for } z > 0 \end{cases} \qquad \forall z(t) \in \mathbb{R} \qquad (3.98)$$

where $\epsilon_0 \in \mathbb{R}$ is a positive design parameter that determines how closely $S(\cdot)$ approximates the unit-step function, and $S_T(x_j, \tau_d) \in \mathbb{R}$ in (3.97) is also defined to be a differentiable function as follows

$$S_T = \sum_{j=1}^{m} g^2(x_j) S\left(g(x_j)\tau_d\right). \qquad (3.99)$$

To clarify the motivation for the structure of (3.97–3.99), the expression given in (3.97) can be substituted into (3.95) as follows

$$\sum_{j=1}^{m} g(x_j)\left(\frac{g(x_j)\tau_d S\left(g(x_j)\tau_d\right)}{S_T} + \gamma_{c0}\right) = \tau_d. \qquad (3.100)$$

After substituting (3.99) into (3.100) for $S_T(\cdot)$ and then simplifying the result, the following expression can be obtained

$$\tau_d + \gamma_{c0} \sum_{j=1}^{m} g(x_j) = \tau_d. \qquad (3.101)$$

Since $\sum\limits_{j=1}^{m} g(x_j) = 0$ for $m \geq 3$, (3.101) can be used to prove that the commutation strategy has been developed such that the desired torque trajectory is delivered to the mechanical subsystem. That is, the desired current trajectory signals have been designed to satisfy the relationship given by (3.67).

Remark 3.8 *At first glance, it seems as if (3.97) is undefined for $S_T(\cdot) = 0$. However, given the definition of $x_j(q)$ and the constraints previously imposed on the structure of $g(x_j)$, it is obvious that $g(x_j) \neq 0$ for all j simultaneously. Since $\sum\limits_{j=1}^{m} g(x_j) = 0$ for $m \geq 3$, this implies that $g(x_j) \not< 0$ for all j simultaneously. From the preceding observations and the definitions of $S(\cdot)$ and $S_T(\cdot)$, it is easy to see that $S_T(\cdot) = 0$ only when $\tau_d(t) = 0$. After evaluating the limit of the suspect term in (3.97), it can be shown that*

$$\lim_{\tau_d \to 0}\left[\frac{g(x_j)\tau_d S\left(g(x_j)\tau_d\right)}{S_T}\right] = 0 \qquad \forall x_j(t) \qquad (3.102)$$

since the numerator of the expression is a higher order of $\tau_d(t)$. Based on the result given in (3.102), it follows from (3.97) that

$$\lim_{\tau_d \to 0} \gamma_{dj} = \gamma_{c0}; \qquad (3.103)$$

hence, given the form of (3.97), $\gamma_{dj}(\cdot) \in \mathcal{L}_\infty$ given that $q(t)$, $\tau_d(t) \in \mathcal{L}_\infty$ (as required by condition in Remark 3.6).

Remark 3.9 *During the formulation of the current tracking error dynamics, the time derivative of the desired current trajectory is required. Given the fact that the above commutation strategy embeds the auxiliary variable $\gamma_{dj}(\cdot)$ inside of $I_{dj}(\cdot)$, the partial derivative terms required to compute the time derivative of derivative $I_{dj}(\cdot)$ can be expressed as follows*

$$\frac{dI_{dj}}{dq} = \frac{\partial I_{dj}}{\partial q} + \frac{\partial I_{dj}}{\partial \gamma_{dj}} \frac{\partial \gamma_{dj}}{\partial q} \qquad \frac{dI_{dj}}{d\tau_d} = \frac{\partial I_{dj}}{\partial \gamma_{dj}} \frac{\partial \gamma_{dj}}{\partial \tau_d} \qquad (3.104)$$

where $\dfrac{\partial \gamma_{dj}(\cdot)}{\partial \tau_d}$ and $\dfrac{\partial \gamma_{dj}(\cdot)}{\partial q}$ can be determined as follows

$$\frac{\partial \gamma_{dj}}{\partial \tau_d} = \frac{g(x_j)S(g(x_j)\tau_d)}{S_T} + \frac{\tau_d g(x_j)}{S_T}\frac{\partial S(\cdot)}{\partial \tau_d} - \frac{\tau_d g(x_j)S(g(x_j)\tau_d)}{S_T^2}\frac{\partial S_T}{\partial \tau_d}$$
$$(3.105)$$

$$\frac{\partial \gamma_{dj}}{\partial q} = \frac{\tau_d S(g(x_j)\tau_d)}{S_T}\frac{\partial g(x_j)}{\partial x_j}\frac{\partial x_j}{\partial q} + \frac{\tau_d g(x_j)}{S_T}\frac{\partial S(\cdot)}{\partial x_j}\frac{\partial x_j}{\partial q}$$
$$(3.106)$$

$$- \frac{\tau_d g(x_j)S(g(x_j)\tau_d)}{S_T^2}\frac{\partial S_T}{\partial x_j}\frac{\partial x_j}{\partial q}.$$

In (3.105) and (3.106) the partial derivative terms $\dfrac{\partial S(\cdot)}{\partial \tau_d}$, $\dfrac{\partial S_T(\cdot)}{\partial \tau_d}$, $\dfrac{\partial S(\cdot)}{\partial x_j}$, $\dfrac{\partial S_T(\cdot)}{\partial x_j}$ can be determined as follows

$$\frac{\partial S(\cdot)}{\partial \tau_d} = \begin{cases} 2\epsilon_0 g^2(x_j)\tau_d \exp(-\epsilon_0 g^2(x_j)\tau_d^2) & g(x_j)\tau_d > 0 \\[2mm] 0 & g(x_j)\tau_d \le 0 \end{cases} \qquad (3.107)$$

$$\frac{\partial S_T(\cdot)}{\partial \tau_d} = \sum_{j=1}^{m} \frac{\partial S(\cdot)}{\partial \tau_d} g^2(x_j) \qquad (3.108)$$

$$\frac{\partial S(\cdot)}{\partial x_j} = \begin{cases} 2\epsilon_0 g(x_j)\tau_d^2 \exp(-\epsilon_0 g^2(x_j)\tau_d^2)\dfrac{\partial g(x_j)}{\partial x_j} & g(x_j)\tau_d > 0 \\[2mm] 0 & g(x_j)\tau_d \le 0 \end{cases} \qquad (3.109)$$

and

$$\frac{\partial S_T(\cdot)}{\partial x_j} = \sum_{j=1}^{m} \left(2g(x_j)S(g(x_j)\tau_d)\frac{\partial g(x_j)}{\partial x_j} + g^2(x_j)\frac{\partial S(\cdot)}{\partial x_j} \right). \qquad (3.110)$$

It is not possible to explicitly calculate $\dfrac{\partial I_{dj}(\cdot)}{\partial \gamma_{dj}}$ *and* $\dfrac{\partial I_{dj}(\cdot)}{\partial q}$ *since the structure of these partial derivative terms will be dependent on the specific flux linkage model selected. From the partial derivative calculations given in (3.105–3.110), the only possible singularities in (3.105) and (3.106) occur when* $S_T(\cdot) = 0$ *which corresponds to* $\tau_d(t) = 0$. *While it is difficult to show for any number of phases that the singularity is avoided in these terms, symbolic calculations similar to those given in [12] can be used to state the following relationships*

$$\lim_{\tau_d \to 0} \frac{\partial \gamma_{dj}}{\partial \tau_d} = \left(\frac{g^3(x_j)}{\sum_{j=1}^{m} g^4(x_j)} \right) \qquad \lim_{\tau_d \to 0} \frac{\partial \gamma_{dj}}{\partial q} = 0 \quad \forall m = 3, 4, ..., 10.$$

$$(3.111)$$

Hence the above commutation strategy has been constructed such that these partial derivatives exist as required by the condition in Remark 3.6. If the structure of $\tau_j(\cdot)$ *in (3.60) facilitates the design of* $I_{dj}\left(q, \gamma_{dj}(q, \tau_d)\right)$ *such that* $\dfrac{\partial I_{dj}(\cdot)}{\partial \gamma_{dj}}, \dfrac{\partial I_{dj}(\cdot)}{\partial q} \in \mathcal{L}_\infty$ *given that* $q(t), \tau_d(t) \in \mathcal{L}_\infty$, *(3.83) and (3.104) can be used to state that* $\dot{I}_{dj}(\cdot) \in \mathcal{L}_\infty \; \forall m = 3, 4, ..., 10$ *given that* $q(t), \dot{q}(t), \tau_d(t),$ *and* $\dot{\tau}_d(t) \in \mathcal{L}_\infty$ *(i.e., provided condition (iii) in Remark 3.6 is satisfied).*

Saturated Flux Linkage Model

As stated previously, the SRM is typically operated "high" on the saturation curve, and hence, many electric machine control experts believe that flux linkage saturation effects should be incorporated in the overall model. In lieu of the linear flux linkage model proposed by [12], in this section the previous commutation strategy is redesigned under the assumption that the flux linkage model is saturated as follows

$$\lambda_j(q, I_j) = \lambda_s \arctan(\beta L I_j). \qquad (3.112)$$

In (3.112), λ_s, $\beta \in \mathbb{R}$ denote positive constant model parameters, and $L(x_j) \in \mathbb{R}$ is defined as follows

$$L = L_0 - L_1 \cos(x_j) > 0 \qquad (3.113)$$

where L_0, $L_1 \in \mathbb{R}$ are positive constant parameters. Based on the form of (3.112), the expressions given in (3.61) and (3.63) can be used to formulate the following electrical functions

$$\tau_j = \frac{\lambda_s N_r L_1}{2\beta L^2} \sin(x_j) \ln\left(1 + \beta^2 L^2 I_j^2\right),$$

$$L_j = \frac{\lambda_s \beta L}{1 + \beta^2 L^2 I_j^2}, \quad \text{and} \quad B_j = \frac{\lambda_s \beta N_r L_1 \sin(x_j) I_j}{1 + \beta^2 L^2 I_j^2}.$$

$$(3.114)$$

The torque transmission equation (3.114) can now be used to reconstruct the desired current trajectories to satisfy the commutation design equation given by (3.67) based on the saturated flux linkage model. Specifically, $I_{dj}(\cdot)$ must be designed to satisfy the following expression

$$\sum_{j=1}^{m} \frac{\lambda_s N_r L_1 \sin(x_j)}{2\beta L^2} \ln\left(1 + \beta^2 L^2 I_{dj}^2\right) = \tau_d. \tag{3.115}$$

Based on (3.115), $I_{dj}(\cdot)$ is redesigned as follows

$$I_{dj} = \frac{\sqrt{\exp\left(\dfrac{2\beta L^2}{\lambda_s N_r L_1}\gamma_{dj}\right) - 1}}{\beta L} \tag{3.116}$$

where $\gamma_{dj}(\cdot)$ was defined in (3.97). After substituting (3.116) for $I_{dj}(\cdot)$ in (3.115), the expression given in (3.95) can be obtained; hence, the rest of the general design procedure follows. In addition, the remaining partial derivative terms in (3.104) can now be calculated as follows

$$\frac{\partial I_{dj}}{\partial \gamma_{dj}} = \frac{1}{L}\frac{\dfrac{2L^2}{\lambda_s N_r L_1}\exp\left(\dfrac{2\beta L^2}{\lambda_s N_r L_1}\gamma_{dj}\right)}{2\sqrt{\exp\left(\dfrac{2\beta L^2}{\lambda_s N_r L_1}\gamma_{dj}\right) - 1}} \tag{3.117}$$

$$\frac{\partial I_{dj}}{\partial q} = \frac{\exp\left(\dfrac{2\beta L^2}{\lambda_s N_r L_1}\gamma_{dj}\right)}{\sqrt{\exp\left(\dfrac{2\beta L^2}{\lambda_s N_r L_1}\gamma_{dj}\right) - 1}}\left(\frac{2\gamma_{dj}}{\lambda_s}\sin(x_j) + \frac{L}{\lambda_s N_r L_1}\frac{\partial \gamma_{dj}}{\partial q}\right)$$

$$-\frac{N_r L_1 \sin(x_j)\sqrt{\exp\left(\dfrac{2\beta L^2}{\lambda_s N_r L_1}\gamma_{dj}\right) - 1}}{\beta L^2}.$$

$$(3.118)$$

As illustrated by (3.117) and (3.118), the structure of $\tau_j\left(\cdot\right)$ given in (3.114) ensures that $\dfrac{\partial I_{dj}\left(\cdot\right)}{\partial\gamma_{dj}}$, $\dfrac{\partial I_{dj}\left(\cdot\right)}{\partial q}\in\mathcal{L}_\infty$ given that $q\left(t\right),\tau_d\left(t\right)\in\mathcal{L}_\infty$. Hence, for the saturated flux linkage model given by (3.112), the controller exhibits no singularities; furthermore, all signals can be shown to be bounded following the argument outlined in Theorem 3.3 and Remark 3.9.

Remark 3.10 *The flux linkage model of (3.112) is motivated by the model given in [41] and the experimental work presented in [10]. Specifically, [41] used an exponentially saturated version of the linear flux linkage model to develop a feedback linearizing control for an SRM while [10] used experimental results to validate the use of the arc-tangent function for modeling saturation in a separately excited DC machine. Since the parameter λ_s appears linearly in the nonlinear terms in (3.114), an additional update law could be designed to compensate for any uncertainty associated with λ_s; however, it is not obvious how update laws can be designed to compensate for any uncertainty associated with β, L_0, and L_1.*

Remark 3.11 *In addition to the current dynamics and saturation effects in the flux linkage model, torque ripple is another phenomenon that can affect high-performance control of the SRM, and hence, the elimination of torque ripple in the SRM has been a subject of intense interest for several decades. This ripple has been attributed to three major sources: (i) imprecise torque transmission modeling, (ii) inability to track the desired current signal due to input voltage saturation or utilization of current-fed models that ignore the stator current dynamics, and (iii) improper commutation of the motor. For a more detailed discussion of these sources, the reader is referred to [15].*

Remark 3.12 *The primary advantage of the preceding full-order model based controller is that the saturation effects of the flux are incorporated. Since the control designs for other multi-phase machines (e.g., induction motors, brushless DC motors) are often based on a transformed model (e.g., the induction motor control design given in the previous section), it is not obvious how saturation effects can be incorporated into a general control design procedure.*

3.3.5 Experimental Setup and Results

An experiment was conducted to illustrate the performance of the SRM voltage controller. To implement the controller, an experimental testbed was constructed that consists of the following components: (i) a 486-66

MHz PC, (ii) a TMS320C30 DSP system board (Spectrum Signal Processing) operating with a sample time of 1 [ms], (iii) a DS-2 Motion Control Card to read the quadrature signals from the shaft-mounted encoder (Integrated Motions Inc.), (iv) a three-phase SRM (NSK Corp. Model RS-0810), (v) six linear amplifiers (Techron, Model 7570-60) connected in a master-slave configuration to power the motor with a possible of 4 [kW/pair], (vi) Hall-effect sensors (Microswitch, Model CSLB1AD) to measure the phase currents, and (vii) assorted electronics and hardware for signal conditioning (see Figure 3.5).

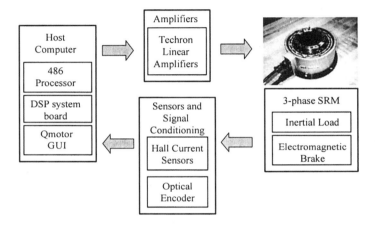

FIGURE 3.5. Block diagram configuration of the SRM experimental setup.

A large inertial load was connected to the SRM. The load can be modeled as follows

$$J = J_r + \frac{1}{2}m_1 r_1^2 + 2\left(\frac{1}{2}m_2 r_2^2 + m_2 l_1^2\right) + 2\left(\frac{1}{2}m_3 r_3^2 + m_3 l_2^2\right)$$

where $J_r \in \mathbb{R}$ represents the rotor inertia, m_1, m_2, $m_3 \in \mathbb{R}$ denote the cylindrical weights, r_1, r_2, $r_3 \in \mathbb{R}$ are the radii of the cylinders, and l_1, $l_2 \in \mathbb{R}$ are the distances of the weights from the axis of rotation. In addition to the inertial load, the motor was also connected to an electromagnetic brake. A constant voltage of 20 [V] was applied to the brake to maintain uniform braking on the rotor. Since the load used in the experiment is only dependent on the rotor velocity, the linear parameterization $W_m(\cdot)\theta_m$ defined in (3.60) is given by the following expression

$$W_m\theta_m = B\dot{q} + \sigma_s \arctan(100\dot{q}) \qquad (3.119)$$

where $B \in \mathbb{R}$ is the coefficient of viscous friction, and $\sigma_s \in \mathbb{R}$ is the coefficient of Coulomb friction. The modeling term $\sigma_s \arctan(100\dot{q})$ is used to account for the Coulomb friction in the motor and the torque applied by the brake. Through experimental trials, it was determined that the value of 100 associated with the argument of the arc-tangent function resulted in a close approximation of the signum function and did not introduce significant spikes in the voltage control inputs when the rotor velocity alternated signs. The rotor velocity signal was obtained by applying the backwards difference algorithm to the position signal. The resulting signal was then passed through a second-order digital filter. Based on previous experience, this method provides an adequate surrogate for the velocity signal in the absence of an output feedback control strategy. The various other parameters of the SRM were determined experimentally from standard test procedures or obtained from the manufacturer data sheets as follows

$$J_r = 0.02 \ [\text{kg} \cdot \text{m}^2] \qquad m_1 = 4.756 \ [\text{kg}] \quad m_2 = 22.68 \ [\text{kg}]$$

$$m_3 = 12.43 \ [\text{kg}] \qquad r_1 = 0.1803 \ [\text{m}] \quad r_2 = 0.2479 \ [\text{m}]$$

$$r_3 = 0.2254 \ [\text{m}] \qquad l_1 = 0.293 \ [\text{m}] \quad l_2 = 0.3606 \ [\text{m}]$$

$$B = 0.02 \ [\text{Nm} \cdot \text{s} \cdot \text{rad}^{-1}] \qquad N_r = 150 \qquad L_o = 0.189 \ [\text{H}]$$

$$\sigma_s = 1.3 \ [\text{Nm}] \qquad L_1 = 0.044 \ [\text{H}] \qquad R = 8.2 \ [\Omega].$$

The flux linkage model for the experiment was calculated by aligning the rotor with one of the phases and then energizing that phase with a sinusoidal voltage of known amplitude and frequency. The resulting current through the winding was then recorded and the flux linkage was calculated using the following relationship (see (3.62) and (3.63))

$$\lambda_j = \int_0^t (v_j - RI_j) \, dt. \tag{3.120}$$

As described by (3.112), an arc-tangent function was used to approximate the saturated flux linkage as follows

$$\lambda_j = 0.5 \arctan(1.8 L(x_j) I_j) \tag{3.121}$$

where $L(x_j)$ was introduced in (3.112). Figure 3.6 shows a comparison of the linear flux model defined by $\lambda_j = L(x_j)I_j$, the experimentally calculated flux given by (3.120), and the arc-tangent approximation defined in (3.121).

To illustrate the performance of the control strategy and to examine the effects of saturation in the flux model, experimental results were obtained

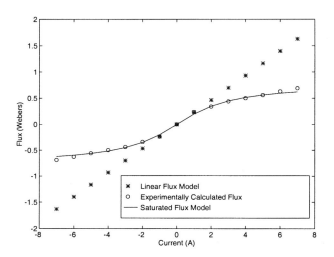

FIGURE 3.6. Flux model development.

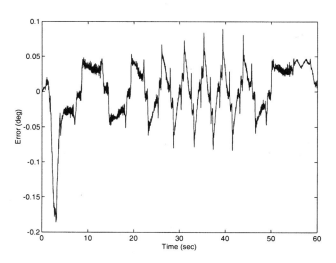

FIGURE 3.7. Position tracking error for the linear flux controller.

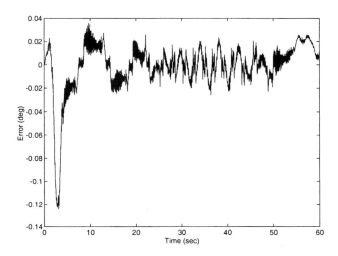

FIGURE 3.8. Position tracking error for the saturated flux controller.

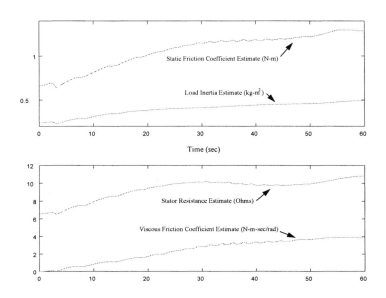

FIGURE 3.9. Adaptive estimates of unknown parameters.

by implementing a controller based on a linear flux model and a controller based on a saturated flux model. The desired rotor position trajectory for each experiment was selected as follows

$$q_d = 0.8\sin(0.8(-0.25\sin(0.1t) + 1.0)t)(1 - \exp(-0.05t^3)) \text{ [rad]}$$

where the initial condition for the desired rotor position and its first three time derivatives were selected as zero. Each controller was also implemented with the unknown parameters initialized to 80% or lower of their nominal values. To implement the controller based on the linear flux model, the following control gains were selected

$$\alpha = 160 \qquad K_s = 20 \qquad K_e = 750$$

$$\Gamma_1 = 0.001 \quad \Gamma_{21} = 0.002 \quad \Gamma_{22} = 0.005 \tag{3.122}$$

$$\Gamma_3 = 0.5 \qquad \gamma_{co} = 0.2 \qquad \epsilon_o = 15.0.$$

As illustrated in Figure 3.7, the steady state position tracking error for this controller was determined to be approximately 0.09 [deg]. To implement the controller based on the saturated flux model, the following control gains were selected

$$\alpha = 140 \qquad K_s = 80 \qquad K_e = 600$$

$$\Gamma_1 = 7 \times 10^{-5} \quad \Gamma_{21} = 0.02 \quad \Gamma_{22} = 0.001 \tag{3.123}$$

$$\Gamma_3 = 0.03 \qquad \gamma_{co} = 0.2 \qquad \epsilon_o = 15.0.$$

As illustrated in Figure 3.8, the steady state position tracking error for this controller was determined to be approximately 0.03 [deg] after a 10 [s] transient. For the nonlinear controller utilizing the saturated flux model, the dynamic estimates of the unknown parameters are depicted in Figure 3.9.

3.4   Active Magnetic Bearings

The electric machines described in the previous sections are modeled by coupled nonlinear ODEs that represent the mechanical and electrical dynamics. In a similar manner, the differential equations that describe a magnetic bearing can be subdivided into the mechanical subsystem dynamics, an algebraic force transmission relationship, and the electrical subsystem dynamics. Standard calculations can be applied to the flux linkage model to complete the description of the electrical subsystem dynamics and the algebraic force transmission relationship in a similar manner as in the model development for the SRM.

In this section, the rotor position of a 6-DOF vertical rotor shaft AMB system (see Figures 3.10 and 3.11) is forced to track a desired trajectory. To this end, a desired force trajectory signal is first designed to ensure that the rotor position tracks a desired trajectory. Then, the structure of the algebraic force transmission relationship is used to develop the desired current trajectory, which ensures that the desired force is delivered to the mechanical subsystem. The desired current trajectory signals are then used to motivate the design of a voltage control input for the electrical subsystem dynamics. An example flux linkage model is also provided to illustrate how the desired current trajectory signals can be constructed[8] to ensure that the desired force is delivered to the mechanical subsystem (i.e., the desired current trajectory signals satisfy the static equation associated with the algebraic force transmission relationship).

Remark 3.13 *For the AMB system depicted in Figure 3.11, the (x_t, y_t) and (x_b, y_b, z_b) axes are fixed at the center of the top and bottom radial magnetic bearing systems, respectively. The (x_o, y_o, z_o) frame is a moving coordinate system attached to the rotor center of mass that coincides with its principal axes such that the z_o axis is aligned with the rotor longitudinal axis.*

3.4.1   System Model

As illustrated in Figures 3.10 and 3.11, the AMB system consists of a rigid circular cylinder rotor suspended by several AMBs. A brushed DC (BDC) motor (not shown in Figures 3.10 and 3.11) is used to provide rotation of the rotor about its longitudinal axis. Specifically, a set of the AMBs depicted in Figure 3.10 are used to regulate the radial motion of the vertical rotor [31], while another set of AMBs are used to regulate the rotor axial motion as shown in Figure 3.11.

---

[8] This procedure is equivalent of the commutation strategy procedure developed in the previous section for the SRM.

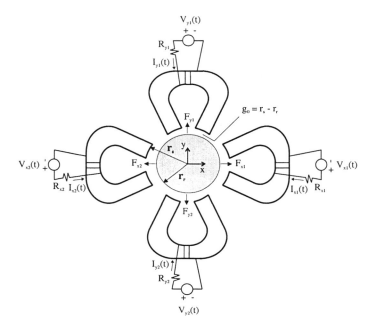

FIGURE 3.10. Top view of a vertical shaft AMB system.

A mathematical model of the mechanical subsystem can be written as follows[9]

$$M(q)\ddot{q} + V_m(q,\dot{q})\dot{q} + G(q) = \bar{F}(q, I_{ij}). \quad (3.124)$$

The matrix coefficients $M(q) \in \mathbb{R}^{6\times 6}$, $V_m(q,\dot{q}) \in \mathbb{R}^{6\times 6}$, $G(q) \in \mathbb{R}^6$ of (3.124) are explicitly defined in Section B.2.3 of Appendix B, and $\bar{F}(q, I_{ij}) \in \mathbb{R}^6$ represents the following transmitted force/torque vector

$$\bar{F}(q, I_{ij}) = \left[\sum_{j=1}^{2} (-1)^{j+1} F_{1j}(q_1, I_{1j}), \quad \sum_{j=1}^{2} (-1)^{j+1} F_{2j}(q_2, I_{2j}), \right.$$

$$\sum_{j=1}^{2} (-1)^{j+1} F_{3j}(q_3, I_{3j}), \quad \sum_{j=1}^{2} (-1)^{j+1} F_{4j}(q_4, I_{4j}),$$

$$\left. \sum_{j=1}^{2} (-1)^{j+1} F_{5j}(q_5, I_{5j}), \quad F_{61}(q_6, I_{61}) \right]^T.$$

$$(3.125)$$

---

[9] The mechanical dynamics associated with the rotor flexibility and unbalance, and the physical connection between the BDC motor and the rotor (i.e., the drive train) have been neglected to simplify the control design.

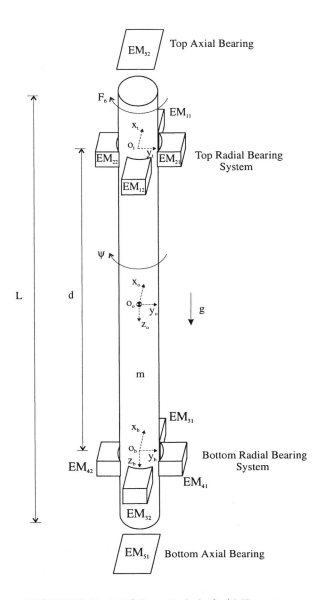

FIGURE 3.11. 6-DOF vertical shaft AMB system.

In (3.125), $F_{ij}(\cdot) \in \mathbb{R}$ $\forall i = 1, ..., 5$ and $j = 1, 2$ denote the absolute values of the forces produced by each electromagnetic circuit, $I_{ij}(t) \in \mathbb{R}$ $\forall i = 1, ..., 5$ and $j = 1, 2$ denote the currents in the AMB windings, $F_{61}(q_6, I_{61}) \in \mathbb{R}$ represents the torque produced by the BDC motor, and $I_{61}(t) \in \mathbb{R}$ represents the current in the windings of the BDC motor. In (3.124) and (3.125), $q(t) \in \mathbb{R}^6$ represents the rotor position vector where the components $q_i(t)$, $\forall i = 1, ..., 4$ represent the rotor position along the directions x_t, y_t, x_b, and y_b, respectively (i.e., the position of the point of intersection of the rotor longitudinal axis with the planes formed by (x_t, y_t) and (x_b, y_b)), the component $q_5(t)$ represents the position of the bottom end of the rotor longitudinal axis along the z_b axis, and $q_6(t)$ denotes the rotation of the rotor about the z_o axis (see Figure 3.11). Kinematic functions relating the position vector $q(t)$ to sensor measurements must be determined to facilitate the control implementation. In small gap applications, an approximation of this relationship is relatively straightforward to determine. However, in applications where the gaps cannot be considered "small" with respect to the radii r_r and r_s (see Figure 3.10), the kinematic functions are more difficult to derive. Specifically, the intersection of the circular cylinder rotor with the planes formed by (x_t, y_t) and (x_b, y_b) will result in two ellipses whose center and major and minor axes will vary in time as the rotor moves. If four position sensors are located at the top and bottom bearings (see Figure 3.11) with measurement beams aligned with the x_t, y_t, x_b, and y_b axes, respectively, the kinematic functions relating the center of the two ellipses (i.e., the points (q_1, q_2) and (q_3, q_4)) with the sensor measurements can be determined.

While the definition for the magnetic forces given in (3.125) are presented in a very general form, the common simplifying assumption is made that the applied magnetic forces are only dependent on the direction of major motion and the measured current in the coil winding (i.e., $F_{ij}(\cdot)$ only depends on $q_i(t)$ and $I_{ij}(t)$). The flux linkage model allows for the computation of the model for the force produced by each electromagnetic circuit. Specifically, the model for the magnetic bearing forces (i.e., $\forall\, i = 1, 2, ..., 5$ and $j = 1, 2$) can be computed as follows [49]

$$F_{ij}(q_i, I_{ij}) = \frac{\partial}{\partial q_i} \int_0^{I_{ij}} \lambda_{ij}(q_i, I_{ij}) \, dI_{ij} \tag{3.126}$$

where $\lambda_{ij}(q_i, I_{ij}) \in \mathbb{R}$ represents the flux linkage model. In addition, the flux linkage model allows for the computation of the electrical subsystem dynamics of the magnetic bearings (i.e., for $\forall\, i = 1, 2, ..., 5$ and $j = 1, 2$) as follows [49]

$$L_{ij}(q_i, I_{ij})\dot{I}_{ij} + R_{ij}I_{ij} + B_{ij}(q_i, I_{ij})\dot{q}_{ij} = v_{ij}. \qquad (3.127)$$

In (3.127), $v_{ij}(t) \in \mathbb{R}$ denotes the input voltage, $R_{ij} \in \mathbb{R}$ denotes the electrical resistance, $L_{ij}(q_i, I_{ij}) \in \mathbb{R}$ denotes an incremental inductance quantity (which is assumed to be positive), and $B_{ij}(q_i, I_{ij}) \in \mathbb{R}$ denotes the back-emf. The inductance and back-emf terms introduced in (3.127) can be calculated as follows [49]

$$L_{ij}(q_i, I_{ij}) = \frac{\partial \lambda_{ij}(q_i, I_{ij})}{\partial I_{ij}}, \qquad B_{ij}(q_i, I_{ij}) = \frac{\partial \lambda_{ij}(q_i, I_{ij})}{\partial q_i}. \qquad (3.128)$$

The dynamic system given in (3.124) exhibits the following properties that are used in the subsequent control development and stability analysis.

Property 3.1: Symmetric and Positive-Definite Inertia Matrix

The inertia matrix is symmetric positive-definite and satisfies the following inequalities

$$m_1 \|x\|^2 \le x^T M(q)x \le m_2(q) \|x\|^2 \qquad \forall x \in \mathbb{R}^6 \qquad (3.129)$$

where m_1, $m_2 \in \mathbb{R}$ are known positive constants, and $\|\cdot\|$ denotes the standard Euclidean norm.

Property 3.2: Skew-Symmetry

The inertia and centripetal-Coriolis matrices satisfy the following skew-symmetric relationship

$$x^T \left(\frac{1}{2}\dot{M}(q) - V_m(q, \dot{q})\right) x = 0 \qquad \forall x \in \mathbb{R}^6 \qquad (3.130)$$

where $\dot{M}(q)$ denotes the time derivative of the inertia matrix.

Remark 3.14 *For a simple BDC motor, a linear magnetic circuit model can be used to complete the description of the dynamical model (i.e., $i = 6$ and $j = 1$ in (3.125) and (3.127)) as follows [40]*

$$F_{61}(q_6, I_{61}) = K_\tau I_{61} \qquad (3.131)$$

and

$$L_{61}(q_6, I_{61}) = L_{61} \qquad B_{61}(q_6, I_{61}) = B_{61} \qquad (3.132)$$

where L_{61}, B_{61}, and $K_\tau \in \mathbb{R}$ denote positive constants.

Remark 3.15 *For the subsequent analysis, it is assumed that $F_{ij}(q_i, I_{ij})$, $L_{ij}(q_i, I_{ij})$, $B_{ij}(q_i, I_{ij})$, $\dfrac{\partial F_{ij}(q_i, I_{ij})}{\partial q_i}$, $\dfrac{\partial F_{ij}(q_i, I_{ij})}{\partial I_{ij}} \in \mathcal{L}_\infty$ provided $q_i(t)$ and $I_{ij}(t) \in \mathcal{L}_\infty$.*

3.4.2   AMB Tracking Control

The typical objective for an AMB problem is to regulate the rotor to a constant setpoint. To provide increased flexibility, the objective in this section is to develop a voltage control input to force the rotor position to track a desired trajectory (i.e., a "soft" desired position trajectory signal could be used when the bearing is initially energized). To quantify this objective a rotor position tracking error, denoted by $e(t) \in \mathbb{R}^6$, and a filtered tracking error, denoted by $r(t) \in \mathbb{R}^6$, are defined as follows

$$e = q_d - q \qquad r = \dot{e} + \alpha e \tag{3.133}$$

where $q_d(t) \in \mathbb{R}^6$ represents the desired position trajectory, and $\alpha \in \mathbb{R}$ is a positive control gain. As a means to develop the rotor position tracking controller, a desired current trajectory input is first constructed to ensure that the magnetic bearings produce the desired forces on the rotor shaft to achieve the desired trajectory. Integrator backstepping techniques are then used to construct a voltage control input that forces the actual current trajectory to follow the developed desired current trajectory. A current tracking error, denoted by $\eta_{ij}(t) \in \mathbb{R}$, is defined as follows to quantify the current tracking objective

$$\eta_{ij} = I_{dij} - I_{ij} \tag{3.134}$$

where $I_{dij}(q, f_d) \in \mathbb{R}$ denotes the subsequently designed desired current trajectory signal, and $f_d(q, \dot{q}, t) \in \mathbb{R}^6$ denotes the subsequently designed desired AMB force trajectory. For the subsequent development, it is assumed that $q(t)$, $\dot{q}(t)$, and $I_{ij}(t)$ are measurable, that $q_d(t)$, $\dot{q}_d(t)$, $\ddot{q}_d(t)$, $\dddot{q}_d(t) \in \mathcal{L}_\infty$, and that the electromechanical model is exactly known. In addition, it is assumed that $I_{dij}(t) \in \mathcal{L}_\infty$ provided that $q(t)$, $f_d(\cdot) \in \mathcal{L}_\infty$, and that $\dot{I}_{dij}(\cdot) \in \mathcal{L}_\infty$ provided that $q(t)$, $\dot{q}(t)$, $f_d(\cdot)$, $\dot{f}_d(\cdot) \in \mathcal{L}_\infty$.

A force/torque tracking error, denoted by $\eta_f(t) \in \mathbb{R}^6$, is defined as follows

$$\eta_f = \bar{F}(q, I_{dij}) - \bar{F}(q, I_{ij}) \tag{3.135}$$

to quantify the mismatch between the desired and actual transmitted force/torque vector defined in (3.125). In (3.135), $\bar{F}(q, I_{dij})$ denotes the desired force/torque profile that is defined later in this analysis.

Rotor Position Tracking

After using the tracking error definitions of (3.133) and (3.135), the mechanical subsystem dynamics of (3.124) can be written as follows

$$M(q)\dot{r} = -V_m(q, \dot{q})r + w(q, \dot{q}, t) - \bar{F}(q, I_{dij}) + \eta_f \tag{3.136}$$

where the force/torque tracking error $\eta_f(t)$ is defined (3.135), the desired force/torque profile $\bar{F}(q, I_{dij})$ is defined later in this analysis, and the auxiliary term $w(q, \dot{q}, t) \in \mathbb{R}^6$ is defined as follows

$$w(q, \dot{q}, t) = M(q)(\ddot{q}_d + \alpha\dot{e}) + V_m(q, \dot{q})(\dot{q}_d + \alpha e) + G(q). \qquad (3.137)$$

Based on the structure of (3.136), the desired current trajectory signal $I_{dij}(q, f_d)$ is designed to satisfy the following equation

$$\bar{F}(q, I_{dij}) = f_d \qquad (3.138)$$

where the desired AMB force trajectory is defined as

$$f_d = w(q, \dot{q}, t) + k_s\lambda_2(q)r. \qquad (3.139)$$

In (3.139), $k_s \in \mathbb{R}$ denotes a positive control gain, and $\lambda_2(q) \in \mathbb{R}$ denotes the following positive function

$$\lambda_2(q) = \max\{m_2(q), 1\}. \qquad (3.140)$$

After substituting (3.138) and (3.139) into (3.136) and then simplifying the resulting expression, the following closed-loop dynamics can be obtained

$$M(q)\dot{r} = -k_s\lambda_2(q)r - V_m(q, \dot{q})r + \eta_f. \qquad (3.141)$$

AMB Force and Current Tracking

To formulate a voltage control input that ensures that the actual AMB forces track the desired forces (i.e., to ensure that the actual current tracks the desired current), we take the time derivative of (3.134) and then substitute the electrical subsystem dynamics of (3.127) for $\dot{I}_{ij}(t)$ to obtain the following expression

$$\dot{\eta}_{ij} = \frac{\partial I_{dij}}{\partial f_d}\dot{f}_d + \frac{\partial I_{dij}}{\partial q}\dot{q} + \frac{1}{L_{ij}(q_i, I_{ij})}\left(R_{ij}I_{ij} + B_{ij}(q_i, I_{ij})\dot{q}_{ij} - v_{ij}\right). \qquad (3.142)$$

Given the definition of the desired force signal defined in (3.139), $\dot{f}_d(t)$ can be expressed as follows

$$\dot{f}_d = \frac{\partial f_d}{\partial \ddot{q}_d}\dddot{q}_d + \frac{\partial f_d}{\partial \dot{q}_d}\ddot{q}_d + \frac{\partial f_d}{\partial q_d}\dot{q}_d + \frac{\partial f_d}{\partial q}\dot{q} \qquad (3.143)$$

$$+ \frac{\partial f_d}{\partial \dot{q}}\left[M^{-1}(q)\left(\bar{F}(q, I_{ij}) - V_m(q, \dot{q})\dot{q} - G(q)\right)\right]$$

where (3.137) and (3.139) can be used to compute the partial derivatives $\dfrac{\partial f_d(\cdot)}{\partial \ddot{q}_d}, \dfrac{\partial f_d(\cdot)}{\partial \dot{q}_d}, \dfrac{\partial f_d(\cdot)}{\partial q_d}, \dfrac{\partial f_d(\cdot)}{\partial q}$, and $\dfrac{\partial f_d(\cdot)}{\partial \dot{q}}$ (see Definition B.2 of Appendix

B) and (3.124) was utilized. Based on the partial derivative expressions given in Section B.2 of Appendix B and the assumption of exact model knowledge of the mechanical parameters, the expression for $\dot{f}_d(t)$ given in (3.143) is measurable. After substituting (3.143) into (3.142) for $\dot{f}_d(\cdot)$, the following open-loop current tracking dynamics can be obtained

$$\dot{\eta}_{ij} = \Pi_{ij} - \frac{v_{ij}}{L_{ij}(q_i, I_{ij})} \tag{3.144}$$

where $\Pi_{ij}(t) \in \mathbb{R}$ is a measurable auxiliary function defined as follows

$$\Pi_{ij} = \frac{\partial I_{dij}}{\partial f_d}\left[\frac{\partial f_d}{\partial t} + \frac{\partial f_d}{\partial q}\dot{q} + \frac{\partial f_d}{\partial \dot{q}}M^{-1}(q)\left(\bar{F}(q, I_{ij}) - V_m(q, \dot{q})\dot{q} - G(q)\right)\right]$$

$$+ \frac{\partial I_{dij}}{\partial q}\dot{q} + \frac{1}{L_{ij}(q_i, I_{ij})}\left(R_{ij}I_{ij} + B_{ij}(q_i, I_{ij})\dot{q}_{ij}\right). \tag{3.145}$$

Based on the open-loop dynamics of (3.144) and the subsequent analysis, the voltage control input $v_{ij}(t)$ is designed as

$$v_{ij} = L_{ij}(q_i, I_{ij})\left(k_e\lambda_2(q)\eta_{ij} + \Pi_{ij} + (-1)^{j+1}u_{ij}\right) \tag{3.146}$$

where $k_e \in \mathbb{R}$ is a positive control gain, $\lambda_2(q)$ was defined in (3.140), and $u_{ij}(t) \in \mathbb{R}$ is an auxiliary control input utilized to sever the interconnections between the mechanical and electrical subsystem, defined as follows [39]

$$u_{ij} = \begin{cases} \dfrac{r_i\left(F_{ij}(q_i, I_{dij}) - F_{ij}(q_i, I_{ij})\right)}{\eta_{ij}} & \text{if } \eta_{ij}(t) \neq 0 \\[3mm] r_i\dfrac{\partial F_{ij}(q_i, I_{dij})}{\partial I_{dij}} & \text{if } \eta_{ij}(t) = 0 \end{cases} \tag{3.147}$$

where $r_i(t)$ denotes the ith element of $r(t)$. After substituting the control input given by (3.146) into (3.144), the closed-loop dynamics for $\eta_{ij}(t)$ can be written as follows

$$\dot{\eta}_{ij} = -k_e\lambda_2(q)\eta_{ij} - (-1)^{j+1}u_{ij}. \tag{3.148}$$

Stability Analysis

The voltage control input introduced in (3.146) and (3.147) confines the transient and steady state response of the rotor position tracking error to an exponentially decaying envelope provided a commutation strategy is selected to ensure that (3.138) and (3.139) are satisfied. This fact is stated in mathematical terms and is proven through the following stability analysis.

Theorem 3.4 *Provided (3.138) and (3.139) are satisfied, the voltage control input given in (3.146) and (3.147) ensure global exponential rotor position tracking in the sense that*

$$\|r(t)\| \leq \|p(t)\| \leq \sqrt{\frac{\lambda_2(q(0))}{\lambda_1}} \, \|p(0)\| \exp\left(-\lambda_3 t\right) \tag{3.149}$$

where $p(t) \in \mathbb{R}^{17}$ is defined as follows

$$p = \begin{bmatrix} r^T & \eta_{11} & \eta_{12} & \eta_{21} & \eta_{22} & \eta_{31} & \eta_{32} \\ & \eta_{41} & \eta_{42} & \eta_{51} & \eta_{52} & \eta_{61} \end{bmatrix}^T, \tag{3.150}$$

$\lambda_2(q)$ was defined in (3.140), and λ_1, $\lambda_3 \in \mathbb{R}$ are defined as follows

$$\lambda_1 = min\{m_1, 1\} \quad \lambda_3 = min\{k_s, k_e\} \ . \tag{3.151}$$

Proof: To prove Theorem 3.4, we define the nonnegative function $V(t) \in \mathbb{R}$ as follows

$$V = \frac{1}{2} r^T M(q) r + \frac{1}{2} \sum_{i=1}^{5} \sum_{j=1}^{2} \eta_{ij}^2 + \frac{1}{2} \eta_{61}^2. \tag{3.152}$$

Based on the structure of (3.152), the following lower and upper bounds can be developed

$$\frac{1}{2} \lambda_1 \|p\|^2 \leq V \leq \frac{1}{2} \lambda_2(q) \|p\|^2 \tag{3.153}$$

where $\lambda_2(q)$, $p(t)$, and λ_1 are defined in (3.140), (3.150), and (3.151), respectively. After taking the time derivative of (3.152), substituting for the closed-loop error systems given in (3.141) and (3.148), and using the skew-symmetry property given by (3.130), the following expression can be obtained

$$\dot{V} = -k_s \lambda_2(q) \|r\|^2 - k_e \lambda_2(q) \sum_{i=1}^{5} \sum_{j=1}^{2} \eta_{ij}^2 - k_e \lambda_2(q) \eta_{61}^2 \tag{3.154}$$

$$+ r^T \eta_f - \eta_{61} u_{61} - \sum_{i=1}^{5} \sum_{j=1}^{2} (-1)^{j+1} \eta_{ij} u_{ij}.$$

After substituting (3.125) and (3.135) into (3.154), $\dot{V}(t)$ can be rewritten as follows

$$\dot{V} = -k_s\lambda_2(q)\,\|r\|^2 - k_e\lambda_2(q)\sum_{i=1}^{5}\sum_{j=1}^{2}\eta_{ij}^2 - k_e\lambda_2(q)\eta_{61}^2$$

$$+r_{i=6}\left(F_{61}(q_6, I_{d61}) - F_{61}(q_6, I_{61})\right) - \eta_{61}u_{61}$$

$$+\sum_{i=1}^{5}\sum_{j=1}^{2}(-1)^{j+1}\left[r_i\left(F_{ij}(q_i, I_{dij}) - F_{ij}(q_i, I_{ij})\right) - \eta_{ij}u_{ij}\right].$$

$$(3.155)$$

When $\eta_{ij}(t) = 0$ (i.e., $I_{dij}(t) = I_{ij}(\cdot)$), then the definition of $\eta_f(t)$ given in (3.135) can be used to prove that $\eta_f(t) = 0$; hence, (3.155) can be written as follows

$$\dot{V} = -k_s\lambda_2(q)\,\|r\|^2 - k_e\lambda_2(q)\sum_{i=1}^{5}\sum_{j=1}^{2}\eta_{ij}^2 - k_e\lambda_2(q)\eta_{61}^2. \qquad (3.156)$$

When $\eta_{ij}(t) \neq 0$, the definition of $u_{ij}(t)$ from (3.147) can be substituted into (3.155) to yield the same result given by (3.156). The expression given in (3.156) can be upper bounded as follows

$$\dot{V} \leq -\lambda_3\lambda_2(q)\,\|p\|^2 \qquad (3.157)$$

where (3.150) and (3.151) were utilized. After using (3.153), the inequality in (3.157) can be further upper bounded as follows

$$\dot{V} \leq -2\lambda_3 V. \qquad (3.158)$$

After solving the differential expression given in (3.158) (See Lemma A.10 of Appendix A), the inequality given in (3.149) can be obtained. Since the position tracking error is related to the filtered tracking error according to the linear differential equation given by (3.64), Lemma A.14 of Appendix A can be invoked to show that the position tracking error is also upper bounded by an exponentially decaying envelope. From the above information and the structure of both the voltage control input and the electromechanical system, all of the system signals remain bounded during closed-loop operation.

Remark 3.16 *To compensate for the uncertainty associated with electromechanical parameters that appear linearly in the model, the above controller could be redesigned as an adaptive tracking controller that achieves global asymptotic rotor position tracking. For details regarding this extension, see [8]. Under certain conditions on the structure of the electromechanical dynamics, the adaptive controller may also be further modified* *to eliminate the requirement for velocity measurements. The control design given in [11] for robot manipulators can be used to illustrate this extension.*

3.4.3   Commutation Strategy

If fringing effects and leakage currents are neglected and the magnetic circuit is assumed to be linear, the following flux linkage model is often used to complete the description of the AMB model of Section 3.4.1 [38]

$$\lambda_{ij}(q_i, I_{ij}) = L_{ij}(q_i)I_{ij} \qquad \forall i = 1, 2, ..., 5 \qquad (3.159)$$

where $L_{ij}(q_i) \in \mathbb{R}$ is a positive inductance term. The flux linkage model given by (3.159) can now be used to calculate the quantities given by (3.126) and (3.128). For example, $F_{ij}(q_i, I_{ij})$ of (3.126) is explicitly given by

$$F_{ij}(q_i, I_{ij}) = \frac{1}{2} \frac{\partial L_{ij}(q_i)}{\partial q_i} I_{ij}^2 \qquad \forall i = 1, 2, ..., 5. \qquad (3.160)$$

Based on this development, the flux linkage model given by (3.159) can be used to design the desired current trajectories such that the design equation given by (3.138) is satisfied. Specifically, after substituting (3.160) into (3.138), $I_{dij}(\cdot)$ can be constructed as follows

$$\frac{1}{2} \sum_{j=1}^{2} (-1)^{j+1} \frac{\partial L_{ij}(q_i)}{\partial q_i} I_{dij}^2 = f_{di} \qquad \forall i = 1, 2, ..., 5 \qquad (3.161)$$

where $f_{di}(\cdot)$ denotes the ith component of the vector $f_d(\cdot)$. To satisfy (3.161), $I_{dij}(\cdot)$ can be designed as follows

$$I_{dij} = \left(\frac{\partial L_{ij}(q_i)}{\partial q_i} \right)^{-\frac{1}{2}} \sqrt{\gamma_{dij}} \qquad \forall i = 1, 2, ..., 5 \qquad (3.162)$$

where it is assumed that $\dfrac{\partial L_{ij}(q_i)}{\partial q_i} > 0$ for all $q_i(t)$, and $\gamma_{dij}(\cdot) \in \mathbb{R}$ is an auxiliary function that must be designed to be nonnegative. After substituting (3.162) into (3.161) for $I_{dij}(\cdot)$, $\gamma_{dij}(\cdot)$ can be designed to satisfy the following relationship

$$\frac{1}{2} \sum_{j=1}^{2} (-1)^{j+1} \gamma_{dij} = f_{di}; \qquad (3.163)$$

therefore, $\gamma_{dij}(\cdot)$ can be designed as follows

$$\gamma_{dij} = (-1)^{j+1} f_{di} + \sqrt{(f_{di})^2 + \gamma_0^2} \qquad (3.164)$$

where the positive design parameter $\gamma_0 \in \mathbb{R}$ is used to set the desired threshold winding current. Motivation for the structure of (3.164) is obtained by substituting (3.164) into (3.163) for $\gamma_{dij}(\cdot)$ as follows

$$\frac{1}{2}f_{di} + \frac{1}{2}\sqrt{(f_{di})^2 + \gamma_0^2} + \frac{1}{2}f_{di} - \frac{1}{2}\sqrt{(f_{di})^2 + \gamma_0^2} = f_{di}. \tag{3.165}$$

The expression given in (3.165) illustrates that the desired current trajectory given by (3.162) and (3.164) ensures that the desired force trajectory is delivered to the mechanical subsystem as required by (3.138) and (3.139).

Remark 3.17 *Based on the structure of (3.164), the following result can be obtained*

$$\lim_{f_{di} \to 0} \gamma_{dij} = \frac{1}{2}\gamma_0 \qquad \forall i = 1, 2, ..., 5. \tag{3.166}$$

Based on (3.162), (3.164), and (3.166), it can be determined that $\gamma_{dij}(\cdot)$, $I_{dij}(\cdot) \in \mathcal{L}_\infty$ given that $q_i(t)$, $f_{di}(\cdot) \in \mathcal{L}_\infty \; \forall i = 1, 2, ..., 5.$

Remark 3.18 *The formulation of the current tracking error dynamics requires the calculation of two partial derivatives associated with $I_{dij}(q, f_d)$ (i.e., $\dfrac{\partial I_{dij}}{\partial f_d}$ and $\dfrac{\partial I_{dij}}{\partial q}$ in (3.142)). Based on (3.162) and (3.164), the ith component of these partial derivative terms can now be calculated as follows*

$$\left(\frac{\partial I_{dij}}{\partial f_d}\right)_i = \left(\frac{\partial L_{ij}(q_i)}{\partial q_i}\right)^{-\frac{1}{2}} \frac{1}{2\sqrt{\gamma_{dij}}} \left((-1)^{j+1} + \frac{f_{di}}{\sqrt{(f_{di})^2 + \gamma_0^2}}\right)$$

$$\forall i = 1, 2, ..., 5 \tag{3.167}$$

and

$$\left(\frac{\partial I_{dij}}{\partial q}\right)_i = -\frac{1}{2}\left(\frac{\partial L_{ij}(q_i)}{\partial q_i}\right)^{-\frac{3}{2}} \sqrt{\gamma_{dij}} \qquad \forall i = 1, 2, ..., 5. \tag{3.168}$$

Hence, the desired current trajectory signal has been constructed such that the above partial derivatives can be calculated. From the stability result given in (3.149) and the structure of (3.167) and (3.168), $\dot{I}_{dij}(\cdot) \in \mathcal{L}_\infty$ given that $q_i(t)$, $\dot{q}_i(t)$, $f_{di}(\cdot)$, and $\dot{f}_{di}(\cdot) \in \mathcal{L}_\infty \; \forall i = 1, 2, ..., 5.$ Motivation for including the desired threshold winding current design parameter γ_0 during the construction of (3.164) is due to the structure of $\dfrac{\partial I_{dij}(\cdot)}{\partial f_d}$ and $\dfrac{\partial I_{dij}(\cdot)}{\partial q}$. Specifically, as a result of (3.166), the use of γ_0 ensures that the partial derivatives of $I_{dij}(\cdot)$ remain bounded as $f_{di}(t) \to 0$.

Remark 3.19 *The design of the desired current trajectory for the BDC motor is more direct than the technique used for the AMB. After substituting (3.131) into (3.138), the desired current trajectory $I_{d61}(q, f_d)$ can be constructed to satisfy the following expression*

$$K_\tau I_{d61} = f_{di}. \tag{3.169}$$

From the structure of (3.169), it is straightforward to design the desired current trajectory and hence, the required partial derivative terms in (3.142).

3.5 Background and Further Reading

Over the last decade, many induction motor control schemes have been developed with an IFOC-like control strategy at the core of the controller as a means to examine this control problem from a nonlinear control perspective as opposed to a more classical motor control perspective. For example, Espinosa-Perez and Ortega [17] presented a singularity-free velocity tracking controller that did not require rotor flux measurements; however, the performance of the controller was limited because the convergence rate of the velocity tracking error was restricted by the natural damping of the motor. Ortega, Nicklasson, and Espinosa-Perez [32] improved upon [17] by using a linear filtering technique to remove the damping restriction in [17]. Furthermore, the authors of [32] illustrated that for a desired constant velocity, the control algorithm of [32] reduced to the IFOC scheme [7]. In [14], Dawson, Hu, and Vedagarbha modified the control structure and stability analysis presented in [32] to construct an adaptive rotor position tracking controller which can be analyzed using standard Lyapunov-type arguments. Ortega et al. illustrated global asymptotic stability of the IFOC control scheme for rotor velocity setpoint applications in [34]. In [33], Ortega and Taoutaou illustrated how previously designed passivity-based control algorithms such as [35] can be expressed in the IFOC notation. With this link between the notation, Ortega et al. illustrated global asymptotic speed regulation for current-fed induction motors with a constant load torque. However, this proposed control algorithm appears to contain a possible control singularity in the stator current control input. The observer/controller algorithm of [33] does not appear to recover exponential stability even if the load torque is assumed to be known (i.e., an adaptive algorithm is required to compensate for the unknown initial conditions of the rotor flux). The authors of [16] analyzed the effects of varying the rotor resistance parameter in the indirect field-oriented control scheme on system stability. An output feedback controller was designed in [27] which achieved global exponential rotor velocity/rotor flux for the reduced-order model of the induction motor (i.e., current-fed induction motor). In addition, Marino, Peresada, and Tomei illustrated how the controller in [27] could be modified to compensate for the parametric uncertainty associated with the load torque and the rotor resistance parameter; however, the controller exhibited a singularity at motor start-up (see [28] for the original adaptive observer design). In the subsequent paper given in [26], Marino, Peresada, and Tomei removed the restriction in [27] by designing the controller for the full-order model. In [5], Behal, Feemster, and Dawson demonstrated how an improved IFOC design could be formulated to achieve exponential rotor tracking. The result in [5] also demonstrated how to transform the improved IFOC design into a structure that facilitates Lyapunov analysis techniques. Other induction motor designs using a Lyapunov-based approach include [2, 4, 18]. Specifically, the induction motor control problem is solved using an adaptive algorithm for estimation of rotor resistance in [18]. In [2], Aquino et al. designed a control algorithm for induction motor position tracking applications where the rotor resistance parameter was estimated in the absence of rotor velocity measurements. In [4], Behal et al. designed a rotor position tracking controller for an induction motor model that included saturation effects in the magnetic flux.

Significant efforts have been directed toward developing controllers for the SRM since it is a popular electric machine for high-precision applications. For example, in [41], Ilic'-Spong et al. introduced a detailed nonlinear model and an electronic commutation strategy for the SRM and applied a feedback linearizing control algorithm that compensates for the nonlinearities of the system. Given the assumption of constant motor velocity, Ilic'-Spong et al. [42] also introduced an instantaneous torque control for a SRM driving an inertial load. A composite control, based on a singularly perturbed model, using reduced-order feedback linearization techniques is presented by Taylor in [43] to reduce torque ripple in the actuation of an experimental load. A torque ripple study of SRMs was pursued in [47] where Wallace and Taylor investigated the performance of four related motors using finite-element analysis. Wallace and Taylor also presented a new method of computing the reference currents by utilizing a balanced commutator for current tracking feedback control in [48]. A certainty equivalence argument was used by Taylor in [44] to develop an adaptive feedback linearizing control for a single link robot with SRM actuators. Carroll, Dawson, and Leviner used a backstepping technique in [12] to develop a singularity-free adaptive trajectory tracking controller for an SRM driving an inertial load.

In [1], an instantaneous torque measurement method based on flux observations was used to develop an adaptive feedback linearizing torque controller for a three-phase SRM. In [21], an adaptive control for SRMs was developed that used B-spline functions to model the torque transmission relationship. A commutation strategy was designed in [20] for a sophisticated model of the SRM. In [45], Vedagarbha, Dawson, and Rhodes designed an adaptive controller for SRMs which does not require rotor velocity measurements. In the subsequent paper [46], Vedagarbha et al. also developed an SRM controller that compensates for uncertainty in an electromechanical model that utilizes a flux linkage model that includes magnetic saturation effects.

As pointed out in [38], many of the previous AMB control techniques are based on the linearized electromechanical system. For example, Matsumura and Yashimoto [29] designed an optimal controller to regulate the rotor position. In [30], Mohamed and Emad used Q-parameterization theory to stabilize the rotor position and evaluate noise rejection and robustness to parametric uncertainty. Mohamed and Busch-Vishniac illustrated how Q-parameterization theory could be used to autobalance the rotor of a vertical shaft AMB system in [31]. An adaptive forced balancing controller was developed in [3] which exhibited negligible effects on the bandwidth and the stability margin. The possible advantages or disadvantages of nonlinear control techniques for AMB applications were addressed in [13] and [38]. Specifically, in [13], Charara and Caron compared the unbalance rejection and energy saving of linear and nonlinear controllers considering linear and nonlinear models for AMBs while in [38], Smith and Weldon utilized input-output feedback linearization and sliding-smode control techniques to center the rotor in a set of magnetic bearings. In [36], a nonlinear tracking controller was developed for a 6-DOF active magnetic bearing system. For a review of the evolution of magnetic bearing hardware, the reader is referred to [6].

References

[1] L. B. Amor, O. Akhrif, L. A. Dessaint, and G. Oliver, "Adaptive Nonlinear Torque Control of a Switched Reluctance Motor," *Proceedings of the American Control Conference*, June 1993, pp. 2831–2836.

[2] P. Aquino, M. Feemster, D. Dawson, and A. Behal, "Adaptive Partial State Feedback Control of the Induction Motor: Elimination of Rotor Flux and Rotor Velocity Measurements," *International Journal*

of Adaptive Control and Signal Processing, Vol. 14, No. 2-3, 2000, pp. 83–108.

[3] S. Beale, B. Shafai, P. LaRocca, and E. Cusson, "Adaptive Forced Balancing for Multivariable Systems," *ASME Journal of Dynamic Systems, Measurement, and Control*, Vol. 117, No. 4, Dec. 1995, pp. 496–502.

[4] A. Behal, M. Feemster, D. M. Dawson, and A. Mangal, "Partial State Feedback Control of Induction Motors with Magnetic Saturation: Elimination of Flux Measurements," *Automatica*, Vol. 38, No. 2, Jan. 2002, pp. 191–203.

[5] A. Behal, M. Feemster, and D.M. Dawson, "An Improved Indirect Field Oriented Controller for the Induction Motor," *IEEE Transactions on Control Systems Technology*, accepted, to appear.

[6] H. Bleuler, D. Vischer, G. Schweitzer, A. Traxler, and D. Zlatnik, "New Concepts for Cost-effective Magnetic Bearing Control," *Automatica*, Vol. 30, No. 5, May 1994, pp. 871–876.

[7] B. Bose, *Power Electronics and AC Drives*, Englewood Cliffs, NJ: Prentice-Hall, 1986.

[8] M. M. Bridges, D. M. Dawson, and J. Hu, "Adaptive Control for a Class of Direct Drive Robot Manipulators," *International Journal of Adaptive Control and Signal Processing*, Vol. 10, No. 4–5, pp. 417–441.

[9] M. M. Bridges, D. M. Dawson, and C. T. Abdallah, "Control of Rigid-Link Flexible-Joint Robots: A Survey of Backstepping Approaches," *Journal of Robotic Systems*, Vol. 12, No. 3, Mar. 1995, pp. 199–216.

[10] T. Burg, D. Dawson, and J. Hu, "Velocity Tracking Control for a Separately Excited DC Motor Without Velocity Measurements," *Proceedings of the American Control Conference*, Baltimore, MD, June 1994, pp. 1051–1056.

[11] T. Burg, D. Dawson, J. Hu, and S. Lim, "An Adaptive Partial State Feedback Controller for RLED Robot Manipulators Actuated by BLDC Motors," *Proceedings of the IEEE Conference on Robotics and Automation*, Nagoya, Japan, May 1995, pp. 300–305.

[12] J. J. Carroll, D. M. Dawson, and M. D. Leviner, "Adaptive Tracking Control of a Switched Reluctance Motor Turning an Inertial Load," *Proceedings of the American Control Conference*, June 1993, pp. 670–674.

[13] A. Charara and B. Caron, "Magnetic Bearing: Comparison Between Linear and Nonlinear Functioning," *Proceedings of the 3rd International Symposium on Magnetic Bearings*, Alexandria, VA, July 1992, pp. 451–460.

[14] D. M. Dawson, J. Hu, and P. Vedagarbha, "An Adaptive Controller for a Class of Induction Motor Systems," *Proceedings of the IEEE Conference on Decision and Control*, New Orleans, LA, Dec. 1995, pp. 1567–1572.

[15] D. M. Dawson, J. Hu, and T. C. Burg, *Nonlinear Control of Electric Machinery*, New York, NY: Marcel Dekker, 1998.

[16] C. Canudas De Wit, R. Ortega, and I. Mareels, "Indirect Field Oriented Control of Induction Motors is Robustly Globally Stable," *Automatica*, Vol. 32. No. 10, Oct. 1996, pp. 1393–1402.

[17] G. Espinoza-Perez and R. Ortega, "State Observers are Unnecessary for Induction Motor Control," *Systems and Control Letters*, Vol. 23, No. 5, 1994, pp. 315–323.

[18] J. Hu and D. Dawson, "Adaptive Control of Induction Motor Systems Despite Rotor Resistance Uncertainty," *Automatica*, Vol. 32, No. 8, 1996, pp. 1127–1143.

[19] T. Kailath, *Linear Systems*, Englewood Cliffs, NJ: Prentice-Hall, 1980.

[20] C. Kim, I. Ha, H. Huh, and M. Ko, "A New Approach to Feedback Linearizing Control of Variable Reluctance Motors for Direct-Drive Applications," *Proceedings of the IECON*, 1994.

[21] R. Kohan and S. Bortoff, "Adaptive Control of Variable Reluctance Motors Using Spline Functions," *Proceedings of the IEEE Conference on Decision and Control*, Lake Buena Vista, FL, 1994, pp. 1694–1699.

[22] P. Krause, *Analysis of Electric Machinery*, New York, NY: McGraw-Hill, 1986.

[23] F. L. Lewis, C. T. Abdallah, and D. M. Dawson, *Control of Robot Manipulators*, New York, NY: Macmillan Publishing Co., 1993.

[24] M. S. Loffler, N. P. Costescu, and D. M. Dawson, "QMotor 3.0 and the QMotor Robotic Toolkit: A PC-Based Control Platform," *Control Systems Magazine*, Vol. 22, No. 3, June 2002, pp. 12–26.

[25] R. Lozano and B. Brogliato, "Adaptive Control of Robot Manipulators with Flexible Joints," *IEEE Transactions on Automatic Control*, Vol. 37, No. 2, Feb. 1992, pp. 174–181.

[26] R. Marino, S. Peresada, and P. Tomei, "Global Adaptive Output Feedback Control of Induction Motors with Uncertain Rotor Resistance," *IEEE Transactions on Automatic Control*, Vol. 44, No. 5, 1999, pp. 967–983.

[27] R. Marino, S. Peresada, and P. Tomei, "Output Feedback Control of Current-Fed Induction Motors with Unknown Rotor Resistance," *IEEE Transactions on Control Systems Technology*, Vol. 4, No. 4, July 1996, pp. 336–346.

[28] R. Marino, S. Peresada, and P. Valigi, "Exponentially Convergent Rotor Resistance Estimation for Induction Motors," *IEEE Transactions on Industrial Electronics*, Vol. 42, No. 5, Oct. 1995, pp. 508–515.

[29] F. Matsumura and T. Yoshimoto, "System Modeling and Control Design of a Horizontal-Shaft Magnetic-Bearing System," *IEEE Transactions on Magnetics*, Vol. MAG-22, No. 3, May 1986, pp. 196–203.

[30] A. M. Mohamed and F. P. Emad, "Conical Magnetic Bearings with Radial and Thrust Control," *IEEE Transactions on Automatic Control*, Vol. 37, No. 12, Dec. 1992, pp. 1859–1868.

[31] A. M. Mohamed and I. Busch-Vishniac, "Imbalance Compensation and Automatic Balancing in Magnetic Bearing Systems Using the Q-Parameterization Theory," *Proceedings of the American Control Conference*, Baltimore, MD, June 1994, pp. 2952–2957.

[32] R. Ortega, P. Nicklasson, and G. Espinosa-Perez, "On Speed Control of Induction Motors," *Automatica*, Vol. 32, No. 3, Mar. 1996, pp. 455–460.

[33] R. Ortega, D. Taoutaou, R. Rabinovici, and J. Vilain, "On Field Oriented and Passivity-based Control of Induction Motors: Downward Compatibility," *Proceedings of the IFAC NOLCOS Conference*, Tahoe City, CA., 1995, pp. 672–677.

[34] R. Ortega and D. Taoutaou, "Indirect Field Oriented Speed Regulation for Induction Motors is Globally Stable," *IEEE Transactions on Industrial Electronics*, Vol. 43, No. 2, Apr. 1996, pp. 340–341.

[35] R. Ortega and G. Espinosa, "Torque Regulation of Induction Motors," *Automatica*, Vol. 29, No. 3, May 1993, pp. 621–633.

[36] M. de Queiroz and D. M. Dawson, "Nonlinear Control of Active Magnetic Bearings: A Backstepping Approach," *IEEE Transactions on Control Systems Technology*, Vol. 4, No. 5, Sept. 1996, pp. 545–552.

[37] S. Sastry and M. Bodson, *Adaptive Control: Stability, Convergence, and Robustness*, Englewood Cliffs, NJ: Prentice-Hall, 1989.

[38] R. D. Smith and W. F. Weldon, "Nonlinear Control of a Rigid Rotor Magnetic Bearing System: Modeling and Simulation with Full State Feedback," *IEEE Transactions on Magnetics*, Vol. 31, No. 2, Mar. 1995, pp. 973–980.

[39] E. Sontag and H. Sussmann, "Further Comments on the Stabilization of the Angular Velocity of a Rigid Body," *Systems and Control Letters*, Vol. 12, 1988, pp. 213–217.

[40] M. W. Spong and M. Vidyasagar, *Robot Dynamics and Control*, New York: John Wiley, 1989.

[41] M. Ilic'-Spong, R. Marino, S. M. Peresada, and D. G. Taylor, "Feedback Linearizing Control of Switched Reluctance Motors," *IEEE Transactions on Automatic Control*, Vol. AC-32, No. 5, 1987, pp. 371–379.

[42] M. Ilic'-Spong, T. J. Miller, S. R. Macminn, and J. S. Thorp, "Instantaneous Torque Control of Electric Motor Drives," *IEEE Transactions on Power Electronics*, Vol. PE-2, No. 1, 1987, pp. 55–61.

[43] D. G. Taylor, "An Experimental Study on Composite Control of Switched Reluctance Motors," *IEEE Control Systems Magazine*, 1991, pp. 31–36.

[44] D. G. Taylor, "Adaptive Control Design for a Class of Doubly-Salient Motors," *Proceedings of the IEEE Conference on Decision and Control*, 1991, pp. 2903–2908.

[45] P. Vedagarbha, T. Burg, J. Hu, and D. Dawson, "A Systematic Procedure for the Design of Adaptive Partial State Feedback Position Tracking Controllers for Electric Machines," *Proceedings of the IEEE Conference on Decision and Control*, New Orleans, LA, Dec. 1995, pp. 2133–2138.

[46] P. Vedagarbha, D. M. Dawson, and W. Rhodes, "An Adaptive Control for General Class of Switch Reluctance Motor Models," *Automatica*, Vol. 33, No. 9, Sept. 1997, pp. 1647–1655.

[47] R. S. Wallace and D. G. Taylor, "Low-Torque-Ripple Switched Reluctance Motors for Direct-Drive Robotics," *IEEE Transactions on Robotics and Automation*, Vol. 7, No. 6, 1991, pp. 733–742.

[48] R. S. Wallace and D. G. Taylor, "A Balanced Commutator for Switched Reluctance Motors to Reduce Torque Ripple," *IEEE Transactions on Power Electronics*, Vol. 7, No. 4, 1992, pp. 617–626.

[49] H. H. Woodson and J. R. Melcher, *Electromechanical Dynamics-Part I: Discrete Systems*, New York, NY: John Wiley, 1968.

4
Robotic Systems

4.1 Introduction

In this chapter, several different robotic applications are examined. Given that a myriad of industrial applications require robots to perform repetitious tasks (e.g., assembly, manipulation, inspection), the first robotic control application examined in this chapter the development of learning control methods that exploit the periodic nature of the robot dynamics to improve link position tracking performance. Some of the advantages of a learning-based controller over some other approaches include the ability to compensate for disturbances without high-frequency or high-gain feedback terms and the ability to compensate for time-varying disturbances that can include time-varying parametric effects. In the first section of this chapter, we illustrate how a saturated learning-based estimate can be used to achieve asymptotic tracking in the presence of periodic nonlinear disturbances. Since the learning-based controller estimate is generated from a Lyapunov-based stability analysis, we also illustrate how additional control terms can be integrated to compensate for nonperiodic components of the unknown dynamics. Specifically, a hybrid adaptive/learning controller is designed for the robot manipulator dynamics. Experimental results are provided to illustrate that the link position tracking performance of a robot manipulator improves with each repetitive motion due to the mitigating action of the learning estimate.

In many industrial/manufacturing applications, a robot manipulator is required to make contact with the environment (e.g., contour following, grinding, scribing, deburring, and assembly-related tasks). In these applications, interaction forces between the robot manipulator end-effector and the environment are generated that constrain motion. Motivated by the desire to precisely control both the motion and the interaction forces of a robot, in the second section of this chapter, an adaptive tracking controller is developed for constrained robot manipulators that ensures global asymptotic position/force tracking performance despite parametric uncertainty in the robot manipulator. The control strategy is primarily based on the fact that the total degrees of freedom (denoted by n) for the position/force control problem can be partitioned into m position control objectives and k force control objectives (i.e., $n = m + k$). This decoupling is accomplished by employing the reduced-order transformation of [42] to develop n dynamic equations which are used to develop the position and force controllers separately. Specifically, the open-loop dynamics for the filtered tracking error system are developed independently of the contact force. The filtered tracking error dynamics are then used to formulate an adaptive position controller that achieves asymptotic position/velocity tracking despite parametric uncertainty. Next, the n dynamic equations developed from the reduced-order transformation are combined to form k dynamic equations that allow the contact force to be written in terms of nonlinear, dynamic quantities that are functions of only position and velocity (i.e., the dependence on acceleration has been eliminated). The k dynamic equations are then rewritten in terms of an open-loop tracking error system which describes the trajectory of the *integral* of the force tracking error. A judicious arrangement of the dynamic terms and the multiplication by a matrix determinant term facilitates the linear parameterization[1] of these k open-loop tracking error dynamic equations. The k dynamic equations are then used to design an adaptive force controller that achieves asymptotic tracking of the integral of the force tracking error despite parametric uncertainty. Further analysis is then presented to illustrate that the actual force trajectory tracks the desired force trajectory. An extension is provided that also describes how the full-state feedback controller can be modified to eliminate the need for velocity measurements. Specifically, based on the additional constraint of output feedback, high-gain control terms are injected to damp out bounded state dependant disturbances. The use of high-gain

---

[1] We note that this technique for obtaining a linear parametrizable dynamic equation was first used in [40] for the control of flexible joint robots.

control terms coupled with higher-order nonlinearities of the constrained robot manipulator model results in a semi-global asymptotic tracking result for the output feedback case. Experimental results of the output feedback controller are presented to illustrate the controller's performance.

Based on advances in control and sensor technologies, new robotic applications are being developed in which the robot is required to operate in unstructured environments. Based on the ability of camera systems to provide a passive noncontact sense of perception in unstructured environments, many of these robotic applications have exploited camera-based vision systems to enable trajectory planning and control. In the third section of this chapter, a visual servoing control design is developed, in which visual information is embedded directly in the feedback loop of the control algorithm. By closing the feedback loop in the image space, the vision and robotic motion problems are fused together, resulting in a cohesive control problem. Following this line of reasoning, in the third section of this chapter the design of visual servoing controllers for a nonredundant planar robot manipulator with a fixed camera [23, 32] configuration is developed. Given different levels of modeling uncertainty for the system, two position tracking controllers are formulated which account for the nonlinear robot dynamics and compensate for parametric uncertainty associated with the robot parameters and/or the camera calibration parameters. Provided that the initial orientation of the camera is in the interval $(-90°, 90°)$, a global adaptive tracking controller is first developed that compensates for unknown camera calibration parameters given that the mechanical parameters are exactly known. Assuming the camera orientation is restricted to the same interval, an adaptive controller is then developed that compensates for parametric uncertainties throughout the entire robot-camera system while producing global asymptotic position tracking. The design of the example controllers is facilitated by the construction of a novel design matrix that premultiplies the uncalibrated camera matrix into a more suitable positive-definite symmetric structure for use in the stability proofs. An extension is provided to illustrate how the fixed camera visual servoing controllers can be extended to incorporate redundant robot manipulators. Moreover, motivated by applications that are not well suited to the fixed camera configuration, an adaptive controller is also developed for the camera-in-hand problem.

4.2 Learning Control Applications

Driven by the desire to exploit the repetitive nature of many robotic applications, researchers have investigated a variety of learning-based controllers. Typically, these controllers are characterized by the use of a standard repetitive update law as the core part of the controller. However, to ensure that the stability analysis validates the proposed results, some researchers have investigated the use of additional rules or complex modifications to the standard repetitive update law that inject additional complexity in the control design. As demonstrated in this section, these additional rules and additional complexity injected into the stability analysis are not necessary for the development of learning controllers. For example, consider the following simple closed-loop system

$$\dot{x} = -x + \varphi(t) - \hat{\varphi}(t) \tag{4.1}$$

where $x(t) \in \mathbb{R}$ is a tracking error signal, $\varphi(t) \in \mathbb{R}$ is an unknown nonlinear function that is periodic with a known period T (i.e., $\varphi(t-T) = \varphi(t)$), and $\hat{\varphi}(t) \in \mathbb{R}$ is a learning-based estimate of $\varphi(t)$. For the system given in (4.1), the standard repetitive update law is given by the following expression

$$\hat{\varphi}(t) = \hat{\varphi}(t - T) + x. \tag{4.2}$$

With regard to the error system given in (4.1) and (4.2), Messner et al. noted that the techniques used in [43] could not be used to show that $\hat{\varphi}(t) \in \mathcal{L}_\infty$ if $\hat{\varphi}(t)$ is generated by (4.2). To address the boundedness problem associated with the standard repetitive update law, a heuristic approach would be to saturate the entire right-hand side of (4.2) as follows [48]

$$\hat{\varphi}(t) = \text{sat}\left(\hat{\varphi}(t - T) + x\right), \tag{4.3}$$

and hence, guarantee that $\hat{\varphi}(t) \in \mathcal{L}_\infty$ (the function sat($\cdot$) in 4.3 is the standard linear piecewise bounded saturation function). Unfortunately, although it is well known how to apply a projection algorithm to the adaptive estimates of a gradient adaptive update law and still accommodate a Lyapunov-based stability analysis, similar arguments cannot be made for the saturated update law given in (4.3), and hence, it is not clear how the stability of the system is affected. Another approach to ensure the boundedness of the update law is to saturate the standard repetitive update law as follows [20]

$$\hat{\varphi}(t) = \text{sat}\left(\hat{\varphi}(t - T)\right) + x. \tag{4.4}$$

Based on the structure of the repetitive update law given in (4.4), a Lyapunov-based approach is used in this section to (i) incorporate the saturation function in (4.4) in the stability analysis, (ii) prove that $x(t)$ is forced asymptotically to zero, and (iii) show that $\hat{\varphi}(t) \in \mathcal{L}_\infty$. To illustrate the generality of the learning-based update law given by (4.4), we illustrate how the update law can be used to force the origin of a general error system with a nonlinear periodic disturbance to achieve global asymptotic tracking. Moreover, to illustrate the fact that other Lyapunov-based techniques can be exploited in tandem with the repetitive update law to compensate for additional disturbances that are not periodic, a hybrid adaptive/repetitive learning scheme is designed to achieve global asymptotic link position tracking for a robot manipulator. In comparison with several other results found in the literature, the learning-based controllers developed in this section (i) utilize standard Lyapunov-based techniques (hence, other Lyapunov-based tools can be easily incorporated), (ii) can be analyzed through a straightforward stability analysis, (iii) utilize a simple modification of the standard repetitive update law as opposed to the use of a multiple-step process or menu, and (iv) are continuously updated during the transient response (versus during the steady state), and hence, facilitate improved transient response.

4.2.1   General Problem

To illustrate the generality of the learning control scheme developed in this section, we first examine the following error dynamics [43]

$$\dot{e} = f\left(t, e\right) + B\left(t, e\right)\left[w(t) - \hat{w}\left(t\right)\right] \tag{4.5}$$

where $e(t) \in \mathbb{R}^n$ is an error vector, $w(t) \in \mathbb{R}^m$ is an unknown nonlinear function, $\hat{w}(t) \in \mathbb{R}^m$ is a subsequently designed learning-based estimate of $w(t)$, and the auxiliary functions $f(t,e) \in \mathbb{R}^n$ and $B\left(t, e\right) \in \mathbb{R}^{n \times m}$ are bounded provided $e(t)$ is bounded. To facilitate the subsequent development, the following assumptions are made regarding the error dynamics of (4.5).

Assumption 4.1: Asymptotic Stability

The origin of the error system $e(t) = 0$ is uniformly asymptotically stable for

$$\dot{e} = f\left(t, e\right); \tag{4.6}$$

furthermore, there exists a first-order differentiable positive-definite function $V_1(e, t) \in \mathbb{R}$, a positive-definite symmetric matrix $Q(t) \in \mathbb{R}^{n \times n}$, and a known matrix $R(t) \in \mathbb{R}^{n \times m}$ such that

$$\dot{V}_1 \leq -e^T Q e + e^T R\left[w - \hat{w}\right]. \tag{4.7}$$

Assumption 4.2: Periodicity and Boundedness

The unknown nonlinear function $w(t)$ is periodic with a known period T; hence,

$$w(t - T) = w(t). \tag{4.8}$$

Furthermore, we assume that an upper bound for the unknown function $w(t)$ can be formulated as follows

$$|w_i| \leq \beta_i \qquad \text{for } i = 1, 2, ..., m \tag{4.9}$$

where $\beta = \begin{bmatrix} \beta_1 & \beta_2 & ... & \beta_m \end{bmatrix} \in \mathbb{R}^m$ is a vector of known positive bounding constants.

Control Objective

The control objective for the general problem given in (4.5) is to design a learning-based estimate $\hat{w}(t)$ such that

$$\lim_{t \to \infty} e(t) = 0 \tag{4.10}$$

for any bounded initial condition denoted by $e(0)$. To quantify the mismatch between the learning-based estimate and $w(t)$, we define an estimation error term, denoted by $\tilde{w}(t) \in \mathbb{R}^m$, as follows

$$\tilde{w}(t) = w(t) - \hat{w}(t). \tag{4.11}$$

Learning-Based Control

Based on the error system given in (4.5) and the subsequent stability analysis, we design the learning-based estimate $\hat{w}(t)$ as follows[2]

$$\hat{w}(t) = \text{sat}_\beta \left(\hat{w}(t - T) \right) + k_e R^T e \tag{4.12}$$

where $k_e \in \mathbb{R}$ is a positive constant control gain, and $\text{sat}_\beta (\cdot) \in \mathbb{R}^m$ is a vector function whose elements are defined as follows

$$\text{sat}_{\beta i} (\xi_i) = \begin{cases} \xi_i & \text{for } |\xi_i| \leq \beta_i \\ \text{sgn} (\xi_i) \beta_i & \text{for } |\xi_i| > \beta_i \end{cases} \qquad \forall \xi_i \in \mathbb{R}, i = 1, 2, ..., m \tag{4.13}$$

where β_i represent the elements of β defined in (4.9), and $\text{sgn}(\cdot)$ denotes the standard signum function. From the definition of $\text{sat}_\beta (\cdot)$ given in (4.13), the following inequality can be obtained (see Lemma B.13 of Appendix B)

$$(\xi_{1i} - \xi_{2i})^2 \geq (\text{sat}_{\beta i} (\xi_{1i}) - \text{sat}_{\beta i} (\xi_{2i}))^2 \tag{4.14}$$
$$\forall |\xi_{1i}| \leq \beta_i, \xi_{2i} \in \mathbb{R}, i = 1, 2, ..., m.$$

---

[2]Note that the learning-based estimate given in (4.12) has the same structure as (4.4).

Based on (4.8), (4.9), and (4.13), the following relationship can be formulated

$$w(t) = \text{sat}_\beta \left(w(t) \right) = \text{sat}_\beta \left(w(t - T) \right); \tag{4.15}$$

hence, after substituting (4.12) and (4.15) into (4.11) for $\hat{w}(t)$ and $w(t)$, respectively, the following expression can be developed for $\tilde{w}(t)$

$$\tilde{w}(t) = \text{sat}_\beta \left(w(t - T) \right) - \text{sat}_\beta \left(\hat{w}(t - T) \right) - k_e R^T e. \tag{4.16}$$

Remark 4.1 *In the subsequent stability analysis, we exploit the fact that the learning-based feedforward term given in (4.12) is composed of a saturation function. That is, based on the structure of (4.12), we exploit the fact that $\hat{w}(t) \in \mathcal{L}_\infty$ provided $e(t) \in \mathcal{L}_\infty$.*

Stability Analysis

Given the learning-based estimate in (4.12), the stability of the error system given in (4.5) can now be examined through the following theorem.

Theorem 4.1 *The learning-based estimate defined in (4.12) ensures that*

$$\lim_{t \to \infty} e(t) = 0 \tag{4.17}$$

for any bounded initial condition denoted by $e(0)$.

Proof: To prove Theorem 4.1, a nonnegative function $V_2(t, e, \tilde{w}) \in \mathbb{R}$ is defined as follows

$$V_2 = V_1 + \frac{1}{2k_e} \int_{t-T}^{t} \left([\text{sat}_\beta \left(w(\tau) \right) - \text{sat}_\beta \left(\hat{w}(\tau) \right)]^T \right. \\ \left. \cdot [\text{sat}_\beta \left(w(\tau) \right) - \text{sat}_\beta \left(\hat{w}(\tau) \right)] \right) d\tau \tag{4.18}$$

where $V_1(t, e)$ is described in Assumption 4.1. After taking the time derivative of (4.18), the following expression is obtained

$$\dot{V}_2 \leq -e^T Q e + e^T R \tilde{w} \\ + \frac{1}{2k_e} [\text{sat}_\beta \left(w(t) \right) - \text{sat}_\beta \left(\hat{w}(t) \right)]^T [\text{sat}_\beta \left(w(t) \right) - \text{sat}_\beta \left(\hat{w}(t) \right)] \\ - \frac{1}{2k_e} ([\text{sat}_\beta \left(w(t - T) \right) - \text{sat}_\beta \left(\hat{w}(t - T) \right)]^T \\ \cdot [\text{sat}_\beta \left(w(t - T) \right) - \text{sat}_\beta \left(\hat{w}(t - T) \right)]) \tag{4.19}$$

where (4.7) and (4.11) were utilized. After using (4.16), the expression given in (4.19) can be rewritten as follows

$$\dot{V}_2 \leq -e^T Q e + e^T R \tilde{w} - \frac{1}{2k_e} \left(\tilde{w} + k_e R^T e \right)^T \left(\tilde{w} + k_e R^T e \right) \\ + \frac{1}{2k_e} [\text{sat}_\beta \left(w(t) \right) - \text{sat}_\beta \left(\hat{w}(t) \right)]^T [\text{sat}_\beta \left(w(t) \right) - \text{sat}_\beta \left(\hat{w}(t) \right)]. \tag{4.20}$$

After performing some simple algebraic operations, (4.20) can be further simplified as follows

$$\dot{V}_2 \leq -e^T \left(Q + \frac{k_e}{2} RR^T \right) e$$

$$-\frac{1}{2k_e} \left(\tilde{w}^T \tilde{w} - [\text{sat}_\beta(w(t)) - \text{sat}_\beta(\hat{w}(t))]^T \right. \tag{4.21}$$

$$\left. \cdot [\text{sat}_\beta(w(t)) - \text{sat}_\beta(\hat{w}(t))] \right)$$

Finally, (4.9), (4.11), and (4.14) can be utilized to simplify (4.21) as shown below

$$\dot{V}_2 \leq -e^T Q e. \tag{4.22}$$

Based on (4.18), (4.22), and the fact that $Q(t)$ is a positive-definite symmetric matrix, it is clear that $e(t) \in \mathcal{L}_2 \cap \mathcal{L}_\infty$. Since $e(t) \in \mathcal{L}_\infty$, we can use (4.5), (4.12), (4.13), and (4.16) to prove that $\hat{w}(t)$, $\tilde{w}(t)$, $f(t,e)$, $B(t,e) \in \mathcal{L}_\infty$. Given that $\hat{w}(t)$, $\tilde{w}(t)$, $f(t,e)$, $B(t,e) \in \mathcal{L}_\infty$, it is clear from (4.5) that $\dot{e}(t) \in \mathcal{L}_\infty$, and hence, $e(t)$ is uniformly continuous. Since $e(t) \in \mathcal{L}_2 \cap \mathcal{L}_\infty$ and is uniformly continuous, Barbalat's Lemma (see Lemma A.16 of Appendix A) can now be invoked to prove (4.17). □

4.2.2   Robot Control Problem

The control design in the previous section exploited the fact that the unknown nonlinear dynamics were periodic with a known period. Unfortunately, some robotic systems may not adhere to the ideal assumption that all of the unknown nonlinear dynamics are entirely periodic. Since the learning-based feedforward term developed in the previous section is generated from a straightforward Lyapunov-based stability analysis, other Lyapunov-based control design techniques can be used to develop hybrid control schemes that used learning-based feedforward terms to compensate for periodic dynamics, and other Lyapunov-based approaches (e.g., adaptive-based feedforward terms) can be used to compensate for nonperiodic dynamics. To illustrate this point, a hybrid adaptive/learning control scheme is developed for an n-rigid link revolute direct-drive robot manipulator in the following sections.

Dynamic Model

The dynamic model for a n-rigid link revolute direct-drive robot is assumed to have the following form [50]

$$M(q)\ddot{q} + V_m(q,\dot{q})\dot{q} + G(q) + F_d\dot{q} + F_s\text{sgn}(\dot{q}) = \tau \tag{4.23}$$

where $q(t)$, $\dot{q}(t)$, $\ddot{q}(t) \in \mathbb{R}^n$ denote the link position, velocity, and acceleration vectors, respectively, $M(q) \in \mathbb{R}^{n \times n}$ represents the link inertia matrix, $V_m(q, \dot{q}) \in \mathbb{R}^{n \times n}$ represents centripetal-Coriolis matrix, $G(q) \in \mathbb{R}^n$ represents the gravity effects, $F_d \in \mathbb{R}^{n \times n}$ is the constant diagonal positive-definite viscous friction coefficient matrix, $F_s \in \mathbb{R}^{n \times n}$ is a constant diagonal positive-definite matrix composed of static friction coefficients, $\text{sgn}(\dot{q})$ denotes the vector composed of the standard signum function applied to each link position, and $\tau(t) \in \mathbb{R}^n$ represents the torque input vector. With regard to the dynamics given in (4.23), the standard assumption, that all of the terms on the left-hand side of (4.23) are bounded provided $q(t)$, $\dot{q}(t)$, and $\ddot{q}(t)$ are bounded, is made.

The dynamic system given in (4.23) exhibits the following properties that are utilized in the subsequent control development and stability analysis.

Property 4.1: Symmetric and Positive-Definite Inertia Matrix

The inertia matrix is symmetric positive-definite and satisfies the following inequalities

$$m_1 \|\xi\|^2 \leq \xi^T M(q)\xi \leq m_2 \|\xi\|^2 \qquad \forall \xi \in \mathbb{R}^n \qquad (4.24)$$

where m_1, $m_2 \in \mathbb{R}$ are known positive constants, and $\|\cdot\|$ denotes the standard Euclidean norm.

Property 4.2: Skew-Symmetry

The inertia and centripetal-Coriolis matrices satisfy the following skew-symmetric relationship

$$\xi^T \left(\frac{1}{2}\dot{M}(q) - V_m(q, \dot{q}) \right) \xi = 0 \qquad \forall \xi \in \mathbb{R}^n \qquad (4.25)$$

where $\dot{M}(q)$ denotes the time derivative of the inertia matrix.

Property 4.3: Bounding Inequalities

The norm of the centripetal-Coriolis, gravity, and viscous friction terms of (4.23) can be upper bounded as follows

$$\|V_m(q, \dot{q})\|_{i\infty} \leq \zeta_{c1} \|\dot{q}\| \qquad \|G(q)\| \leq \zeta_g \qquad \|F_d\|_{i\infty} \leq \zeta_{fd} \qquad (4.26)$$

where ζ_{c1}, ζ_g, $\zeta_{fd} \in \mathbb{R}$ denote known positive bounding constants, and $\|\cdot\|_{i\infty}$ denotes the induced infinity-norm of a matrix.

Property 4.4: Linearity in the Static Friction Parameters

The static friction terms given in (4.23) can be linear parameterized as follows

$$Y_s(\dot{q})\theta_s = F_s\text{sgn}(\dot{q}) \tag{4.27}$$

where $\theta_s \in \mathbb{R}^n$ contains the unknown, constant static friction coefficients, and the regression matrix $Y_s(\dot{q}) \in \mathbb{R}^{n \times n}$ contains known functions of the link velocity.

Control Objective

The objective of this section is to design a global link position tracking controller despite parametric uncertainty in the dynamic model given in (4.23). The control objective is based on the assumption that $q(t)$ and $\dot{q}(t)$ are measurable. To quantify this objective, the link position tracking error, denoted by $e(t) \in \mathbb{R}^n$, is defined as follows

$$e = q_d - q \tag{4.28}$$

where the desired link position $q_d(t) \in \mathbb{R}^n$ and its first two time derivatives are assumed to be bounded periodic functions of time with a known period T such that

$$q_d(t) = q_d(t - T) \qquad \dot{q}_d(t) = \dot{q}_d(t - T) \qquad \ddot{q}_d(t) = \ddot{q}_d(t - T). \tag{4.29}$$

Control Development

To facilitate the subsequent control development and stability analysis, the order of the dynamic expression given in (4.23) is reduced by defining a filtered tracking error variable, denoted by $r(t) \in \mathbb{R}^n$, as follows

$$r = \dot{e} + \alpha e \tag{4.30}$$

where $\alpha \in \mathbb{R}$ is a positive constant control gain. After taking the time derivative of (4.30), premultiplying the resulting expression by $M(q)$, using (4.23) and (4.28), and then performing some algebraic manipulation, the following expression can be obtained

$$M\dot{r} = -V_m r + w_r + \chi + Y_s\theta_s - \tau \tag{4.31}$$

where the auxiliary expressions $w_r(t), \chi(t) \in \mathbb{R}^n$ are defined as follows

$$w_r = M(q_d)\ddot{q}_d + V_m(q_d, \dot{q}_d)\dot{q}_d + G(q_d) + F_d\dot{q}_d \tag{4.32}$$

$$\chi = M(q)(\ddot{q}_d + \alpha\dot{e}) + V_m(q, \dot{q})(\dot{q}_d + \alpha e) + G(q) + F_d\dot{q} - w_r. \tag{4.33}$$

By exploiting Properties 4.1 and 4.3 and then using (4.28) and (4.30), the results given in [47] can be used to prove that

$$\|\chi\| \leq \rho\left(\|z\|\right)\|z\| \tag{4.34}$$

where the auxiliary signal $z(t) \in \mathbb{R}^{2n}$ is defined as follows

$$z(t) = \left[\begin{array}{cc} e^T(t) & r^T(t) \end{array}\right]^T \tag{4.35}$$

and $\rho\left(\cdot\right) \in \mathbb{R}$ is a known positive bounding function. Furthermore, based on the expression given in (4.32) and the boundedness assumptions with regard to the robot dynamics and the desired trajectory, it is clear that

$$|w_{ri}| \leq \beta_{ri} \qquad \text{for } i = 1, 2, ..., n \tag{4.36}$$

where $\beta_r = \left[\begin{array}{ccc} \beta_{r1}, & ..., & \beta_{rn} \end{array}\right] \in \mathbb{R}^n$ is a vector of known positive bounding constants.

Motivated by the open-loop error system in (4.31) and the subsequent stability analysis, we design the following hybrid adaptive/learning controller

$$\tau = kr + k_n\rho^2\left(\|z\|\right)r + e + \hat{w}_r + Y_s\hat{\theta}_s. \tag{4.37}$$

In the controller given in (4.37), k, $k_n \in \mathbb{R}$ are positive constant control gains, $\hat{w}_r(t) \in \mathbb{R}^n$ is generated according to the following learning-based algorithm

$$\hat{w}_r(t) = \text{sat}_{\beta_r}(\hat{w}_r(t-T)) + k_L r, \tag{4.38}$$

$\hat{\theta}_s(t) \in \mathbb{R}^n$ is generated according to the following gradient update law

$$\dot{\hat{\theta}}_s(t) = \Gamma_s Y_s^T r, \tag{4.39}$$

and $Y_s(\dot{q})$, $\rho(\cdot)$ were defined in (4.27) and (4.34), respectively. For the learning-based update law given in (4.38), $k_L \in \mathbb{R}$ denotes a positive constant control gain, and $\text{sat}_{\beta_r}(\cdot)$ is defined in the same manner as in (4.13). For the adaptive gradient update law given in (4.39), $\Gamma_s \in \mathbb{R}^{n\times n}$ is a constant diagonal positive-definite adaptation gain matrix.

To develop the closed-loop error system for $r(t)$, we substitute (4.37) into (4.31) to obtain the following expression

$$M\dot{r} = -V_m r - kr - e + Y_s\tilde{\theta}_s + \tilde{w}_r + \chi - k_n\rho^2\left(\|z\|\right)r \tag{4.40}$$

where the parameter estimation error vector, denoted by $\tilde{\theta}_s(t) \in \mathbb{R}^n$, is defined as follows

$$\tilde{\theta}_s = \theta_s - \hat{\theta}_s, \tag{4.41}$$

and $\tilde{w}_r(t) \in \mathbb{R}^n$ is a learning estimation error signal defined as follows

$$\tilde{w}_r = w_r - \hat{w}_r. \tag{4.42}$$

In a similar manner as in (4.15), the expressions introduced in (4.29), (4.32), (4.36), and the fact that $\mathrm{sat}_{\beta r}(\cdot)$ is defined in the same manner as in (4.13), can be used to develop the following relationship

$$w_r(t) = \mathrm{sat}_{\beta r}\left(w_r(t)\right) = \mathrm{sat}_{\beta r}\left(w_r(t-T)\right). \tag{4.43}$$

Thus, after substituting (4.38) and (4.43) into (4.42) for $\hat{w}_r(t)$ and $w_r(t)$, respectively, the following expression can be developed for $\tilde{w}_r(t)$

$$\tilde{w}_r = \mathrm{sat}_{\beta r}\left(w_r(t-T)\right) - \mathrm{sat}_{\beta_r}\left(\hat{w}_r(t-T)\right) - k_L r. \tag{4.44}$$

Remark 4.2 *As in the stability analysis for Theorem 4.1, the fact that the learning-based feedforward term given in (4.38) is composed of a saturation function is exploited. That is, from the structure of (4.38), it is evident that $\hat{w}_r(t) \in \mathcal{L}_\infty$ provided $r(t) \in \mathcal{L}_\infty$.*

Stability Analysis

The stability of the hybrid adaptive/learning controller given in (4.37), (4.38), and (4.39) can now be examined through the following theorem.

Theorem 4.2 *The proposed hybrid adaptive/learning controller given in (4.37–4.39) ensures global asymptotic link position tracking in the sense that*

$$\lim_{t \to \infty} e(t) = 0 \tag{4.45}$$

where the control gains α, k, k_n, and k_L introduced in (4.30), (4.37), and (4.38) are selected to satisfy the following sufficient condition

$$\min\left(\alpha, k + \frac{k_L}{2}\right) > \frac{1}{4k_n}. \tag{4.46}$$

Proof: To prove Theorem 4.2, we define a nonnegative function $V(t) \in \mathbb{R}$ as follows

$$V = \frac{1}{2}e^T e + \frac{1}{2}r^T M r + \frac{1}{2}\tilde{\theta}_s^T \Gamma_s^{-1} \tilde{\theta}_s$$

$$+ \frac{1}{2k_L}\int_{t-T}^{t} \left[\mathrm{sat}_{\beta r}\left(w_r(\tau)\right) - \mathrm{sat}_{\beta r}\left(\hat{w}_r(\tau)\right)\right]^T \tag{4.47}$$

$$\cdot \left[\mathrm{sat}_{\beta r}\left(w_r(\tau)\right) - \mathrm{sat}_{\beta r}\left(\hat{w}_r(\tau)\right)\right] d\tau.$$

After taking the time derivative of (4.47) and then utilizing (4.25), (4.30), (4.39), and (4.40), the following expression can be obtained

$$
\dot{V} = e^T (r - \alpha e) + r^T \left(-kr - e - k_n \rho^2 (\|z\|) r + Y_s \tilde{\theta}_s + \tilde{w}_r + \chi \right)
$$

$$
- \tilde{\theta}_s^T Y_s^T r + \frac{1}{2k_L} [\mathrm{sat}_{\beta r} (w_r(t)) - \mathrm{sat}_{\beta r} (\hat{w}_r (t))]^T
$$

$$
\cdot [\mathrm{sat}_{\beta r} (w_r(t)) - \mathrm{sat}_{\beta r} (\hat{w}_r (t))]
$$

$$
- \frac{1}{2k_L} [\mathrm{sat}_{\beta r} (w_r(t - T)) - \mathrm{sat}_{\beta r} (\hat{w}_r (t - T))]^T
$$

$$
\cdot [\mathrm{sat}_{\beta r} (w_r(t - T)) - \mathrm{sat}_{\beta r} (\hat{w}_r (t - T))].
$$

(4.48)

After utilizing (4.34), (4.36), (4.44), and then simplifying the resulting expression, we can rewrite (4.48) as follows

$$
\dot{V} \leq -\alpha e^T e - kr^T r + r^T \tilde{w}_r + \left[\rho (\|z\|) \|z\| \|r\| - k_n \rho^2 (\|z\|) \|r\|^2 \right]
$$

$$
- \frac{1}{2k_L} (\tilde{w}_r + k_L r)^T (\tilde{w}_r + k_L r)
$$

$$
+ \frac{1}{2k_L} [\mathrm{sat}_{\beta r} (w_r(t)) - \mathrm{sat}_{\beta r} (\hat{w}_r (t))]^T
$$

$$
\cdot [\mathrm{sat}_{\beta r} (w_r(t)) - \mathrm{sat}_{\beta r} (\hat{w}_r (t))].
$$

(4.49)

After expanding the second line of (4.49) and then cancelling common terms, the following expression can be obtained

$$
\dot{V} \leq -\alpha e^T e - \left(k + \frac{k_L}{2} \right) r^T r + \left[\rho (\|z\|) \|z\| \|r\| - k_n \rho^2 (\|z\|) \|r\|^2 \right]
$$

$$
- \frac{1}{2k_L} \left[\tilde{w}_r^T \tilde{w}_r - [\mathrm{sat}_{\beta r} (w_r(t)) - \mathrm{sat}_{\beta r} (\hat{w}_r (t))]^T \right.
$$

$$
\left. \cdot [\mathrm{sat}_{\beta r} (w_r(t)) - \mathrm{sat}_{\beta r} (\hat{w}_r (t))] \right].
$$

(4.50)

By exploiting the property given in (4.14), completing the square on the bracketed term in the first line of (4.50) (or by invoking the nonlinear damping tool given in Lemma A.17 of Appendix A), and then utilizing the definition of $z(t)$ given in (4.35), the expression given in (4.50) can be simplified as follows

$$\dot{V} \leq - \left(\min \left(\alpha, k + \frac{k_L}{2} \right) - \frac{1}{4k_n} \right) \|z\|^2 . \qquad (4.51)$$

Based on (4.35), (4.46), (4.47), and (4.51), we can now prove that $e(t)$, $r(t) \in \mathcal{L}_2 \cap \mathcal{L}_\infty$. Based on the fact that $r(t) \in \mathcal{L}_\infty$, we can utilize (4.30), (4.38), and (4.44) to prove that $\hat{w}_r(t)$, $\tilde{w}_r(t)$, $\dot{e}(t) \in \mathcal{L}_\infty$, and hence, $e(t)$ is uniformly continuous. Since $e(t) \in \mathcal{L}_2 \cap \mathcal{L}_\infty$ and is uniformly continuous, Lemma A.16 of Appendix A can now be invoked to prove (4.45).   □

Remark 4.3 *One of the advantages of the saturated learning-based feedforward term is that it is developed through Lyapunov-based techniques. By utilizing Lyapunov-based design and analysis techniques, the boundedness of the feedforward term can be proven in a straightforward manner (see Remark 4.2), and the ability to utilize additional Lyapunov-based techniques to augment the control design (as in the example of the hybrid adaptive/learning controller) is facilitated. These traits are in contrast to learning-based designs in which additional analysis is required to examine the boundedness of the feedforward terms and the structure is less amenable to the incorporation of additional control elements (e.g., an adaptive control component).*

4.2.3   Experimental Setup and Results

To illustrate the performance of the previous learning-based controller, the following controller[3] was implemented on an Integrated Motion Inc. (IMI) two-link direct-drive planar robot manipulator

$$\tau = kr + \hat{w}_r \qquad (4.52)$$

where $r(t)$ was defined in (4.30), k is given in (4.37), and $\hat{w}_r(t)$ is generated according to (4.38). The dynamic model for the IMI robot manipulator can be formulated as follows [19]

$$\begin{bmatrix} \tau_1 \\ \tau_2 \end{bmatrix} = \begin{bmatrix} p_1 + 2p_3c_2 & p_2 + p_3c_2 \\ p_2 + p_3c_2 & p_2 \end{bmatrix} \begin{bmatrix} \ddot{q}_1 \\ \ddot{q}_2 \end{bmatrix}$$
$$+ \begin{bmatrix} -p_3s_2\dot{q}_2 & -p_3s_2(\dot{q}_1 + \dot{q}_2) \\ p_3s_2\dot{q}_1 & 0 \end{bmatrix} \begin{bmatrix} \dot{q}_1 \\ \dot{q}_2 \end{bmatrix} \qquad (4.53)$$
$$+ \begin{bmatrix} f_{d1} & 0 \\ 0 & f_{d2} \end{bmatrix} \begin{bmatrix} \dot{q}_1 \\ \dot{q}_2 \end{bmatrix} + \begin{bmatrix} f_{s1} & 0 \\ 0 & f_{s2} \end{bmatrix} \begin{bmatrix} \mathrm{sgn}\,(\dot{q}_1) \\ \mathrm{sgn}\,(\dot{q}_2) \end{bmatrix}$$

---

[3] To emphasize the effects of the learning-based feedforward control term, the nonlinear damping term and the adaptive feedforward term utilized in (4.37) were omitted from the experimental trials.

where $p_1 = 3.473$ [kg-m^2], $p_2 = 0.193$ [kg-m^2], $p_3 = 0.242$ [kg-m^2], $f_{d1} = 5.3$ [Nm-sec], $f_{d2} = 1.1$ [Nm-sec], $f_{s1} = 8.45$ [Nm], $f_{s2} = 2.35$ [Nm], sgn($\cdot$) denotes the standard signum function, $c_2 \triangleq \cos(q_2)$, and $s_2 \triangleq \sin(q_2)$. A Pentium 266 [MHz] PC with the real-time extension of the Linux operating system (RT-Linux) hosted the control algorithm. A Matlab/Simulink environment with the Real-Time Linux Target [62] for RT-Linux was used to implement the controller. The Servo-To-Go I/O board provided for data transfer between the computer subsystem and the robot at a rate of 1 [kHz].

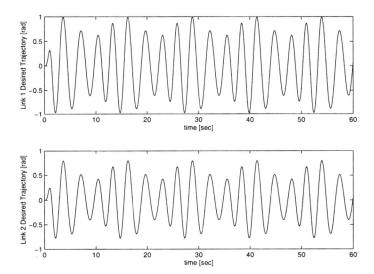

FIGURE 4.1. Periodic desired trajectory.

The experiment was performed using the following periodic desired position trajectory (see Figure 4.1)

$$\begin{bmatrix} q_{d1}(t) \\ q_{d2}(t) \end{bmatrix} = \begin{bmatrix} (0.8 + 0.2\sin(0.5t))\sin\left(0.5\sin(0.5t)\right)\left(1 - \exp\left(-0.6t^3\right)\right) \\ (0.6 + 0.2\sin(0.5t))\sin\left(0.5\sin(0.5t)\right)\left(1 - \exp\left(-0.6t^3\right)\right) \end{bmatrix}$$
$$[\text{rad}]$$
$$(4.54)$$

where the exponential term was included to provide a "smooth-start" to the system. After a tuning process, the control gains were selected as follows

$$k = \text{diag}\{40, 12\} \qquad \alpha = \text{diag}\{20, 14\} \qquad k_L = \text{diag}\{30, 10\} \qquad (4.55)$$

where diag$\{\cdot\}$ denotes the diagonal elements of a matrix. Note that in previous sections, the control gains k, α, and k_L are defined as scalars for simplicity, whereas in (4.55), the control gains are selected as diagonal

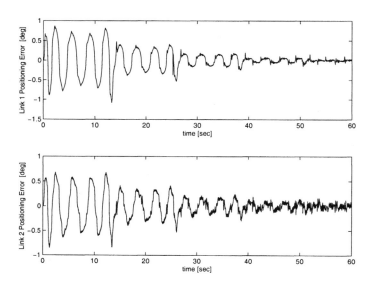

FIGURE 4.2. Link position tracking error.

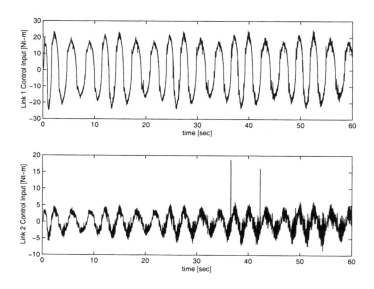

FIGURE 4.3. Control torque input.

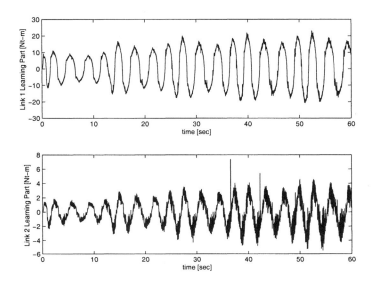

FIGURE 4.4. Learning-based feedforward component of the control input.

matrices to facilitate the "tuning" process. The link position tracking errors
are depicted in Figure 4.2. The results given in Figure 4.2 illustrate that
the tracking error reduces after each period of the desired trajectory. The
control torque input for each link motor is shown in Figure 4.3, and the
feedforward learning component of the controller is given in Figure 4.4.

4.3   Position and Force Control Applications

As described in the introduction, many industrial applications often require
that a robot apply forces to the environment. When a robot is in contact
with the environment, rigid constraints are imposed on the robot motion.
Since robots are typically designed with the capability of exerting large
forces that have the potential to be destructive to the robot and/or the
environment, precise control of both the position and the forces exerted by
the robot is well motivated. Adaptive position/force tracking controllers are
developed in this section for robot manipulators during constrained motion.
Specifically, we prove that global asymptotic position/force tracking can be
obtained despite parametric uncertainty in the mechanical dynamics. An
extension is also provided that illustrates how a high-pass filter can be constructed to facilitate the development of an adaptive position/force tracking
controller when link velocity measurements are not available (output feed-

back). Experimental results are provided to demonstrate the performance of the developed output feedback controller.

4.3.1 System Model

In this section, the kinematic and dynamic models of a rigid-link robot are developed. Based on the fact that the control objective is focused on position/force tracking when the robot is in contact with the environment, motion constraint equations are also developed.

Kinematic Model

Given an open set $O_1 \subset \mathbb{R}^n$ and a function[4] $H \in C^3(O_1)$, $H : O_1 \to \mathbb{R}^n$, the forward kinematic model that relates the task space coordinates, denoted by $x(t) \in \mathbb{R}^n$, of a nonredundant robot manipulator (i.e., $n \leq 6$) to the joint space variables, denoted by $q(t) \in O_1$, can be developed as follows

$$x = H(q). \tag{4.56}$$

By taking the time derivative of (4.56), the time derivative of the task space coordinates can be related to the time derivative of the joint space variables as follows

$$\dot{x} = J\dot{q} \tag{4.57}$$

where $J(q) \in \mathbb{R}^{n \times n}$ represents the manipulator Jacobian and is defined as follows

$$J = \frac{\partial H}{\partial q}. \tag{4.58}$$

By defining a function $h \in C^3(H(O_1))$, $h : H(O_1) \to O_1$, the joint space variables and the task space coordinates can also be related through the following inverse kinematic model

$$q = h(H(q)) = h(x) \quad \forall x \in H(O_1). \tag{4.59}$$

As described previously, motivation exists to develop a reduced-order dynamic model. To facilitate the development of the reduced-order model, the task space variables are partitioned as follows

$$x(t) = [x_1^T \quad x_2^T]^T \tag{4.60}$$

where $x_1(t) \in \mathbb{R}^m$, $x_2(t) \in \mathbb{R}^k$, and $m + k = n$. With regard to the manipulator Jacobian, it is assumed that $J(x) \in C^2(O_1)$ and $J^{-1}(x) \in C^2(H(O_1))$.

---

[4] The notation $f \in C^\varsigma$ is utilized to indicate that f is ς^{th} order differentiable.

Moreover, the inverse of the manipulator Jacobian matrix is assumed to exist (i.e., all kinematic singularities can be avoided) and is given by the following expression

$$J^{-1} = \frac{\partial h}{\partial x}. \tag{4.61}$$

Constraint Equations

As described previously, this section is focused on the control of a rigid robot manipulator when it comes in contact with a rigid environment. When in contact with the environment, the motion of the robot manipulator is constrained by interaction forces. Specifically, the interaction forces (also referred to as constraint forces) between the robot manipulator and the environment reduce the degrees of freedom of the manipulator. As a means of incorporating the effects of the interaction forces in the overall control design, a constraint model is formulated. To facilitate the subsequent control development and stability analysis, we assume that the robot manipulator end-effector is always on the constraint surface during closed-loop operation and that the known constraints are frictionless[5] and holonomic. Specifically, an open set $O_2 \subset H(O_1)$ and a known function $\Theta \in C^3(H(O_1))$, $\Theta : H(O_1) \to \mathbb{R}^k$, are defined where $\Theta(x)$ is assumed to satisfy the following holonomic constraint

$$\Theta = 0_k \qquad \forall x \in O_2 \tag{4.62}$$

where the notation 0_ζ denotes a $\zeta \times 1$ vector of zeros for all $\zeta \in \mathbb{R}$. The time derivative of $\Theta(x)$ can be expressed by the following relationship

$$\dot{\Theta} = A\dot{x} \tag{4.63}$$

where the Jacobian matrix $A(x) \in \mathbb{R}^{k \times n}$ is a C^2 function that has full rank (i.e., the k columns of $A(x)$ are linearly independent) and is defined as follows

$$A = \frac{\partial \Theta}{\partial x}. \tag{4.64}$$

To facilitate the development of the subsequent reduced-order dynamic model, the Jacobian matrix $A(x)$ of (4.64) is partitioned as follows

$$A = \begin{bmatrix} \Sigma & \Pi \end{bmatrix} \tag{4.65}$$

---

[5] From a practical standpoint, the constraint forces will be subject to surface friction effects; however, these effects can simply be treated as a disturbance since friction is a function of the applied normal force [39].

where the partitioned matrices $\Sigma(x) \in \mathbb{R}^{k \times m}$ and $\Pi(x) \in \mathbb{R}^{k \times k}$ are defined as follows

$$\Sigma = \frac{\partial \Theta}{\partial x_1} \qquad \Pi = \frac{\partial \Theta}{\partial x_2} \tag{4.66}$$

where $x_2(t)$ is assumed to be selected such that $\Pi(x)$ has full rank, and hence, is invertible. Based on the previous development, the Implicit Function Theorem (see Lemma A.2 of Appendix A) can be used to define an open set $O_3 \subset \mathbb{R}^{m \times 1}$ and a unique function $\Omega \in C^3(O_3)$, $\Omega : O_3 \to \mathbb{R}^k$, such that

$$\Omega(x_1) = x_2 \qquad \Theta(x_1, \Omega) = 0_k \quad \forall x_1 \in O_3. \tag{4.67}$$

Since $A(x)$ is partitioned as in (4.65), the expression given in (4.63) can be rewritten as follows

$$\dot{\Theta} = \left(\frac{\partial \Theta}{\partial x_1} + \frac{\partial \Theta}{\partial x_2} \frac{\partial \Omega}{\partial x_1} \right) \dot{x}_1$$
$$= (\Sigma + \Pi \Delta) \, \dot{x}_1 = 0_{k \times m} \tag{4.68}$$

where $\Sigma(x)$ and $\Pi(x)$ are defined in (4.66) and $\Delta(x) \in \mathbb{R}^{k \times m}$ is a C^2 function that is defined as follows

$$\Delta = \frac{\partial \Omega}{\partial x_1}. \tag{4.69}$$

Remark 4.4 *From a heuristic standpoint, the assumption that the interaction forces are frictionless allows the number of forces that must be controlled to be reduced. By assuming that the interaction forces impose a holonomic constraint, the manipulator motion can be shown to belong to an invariant manifold $C \in \mathbb{R}^{2n}$ as follows [39, 42]*

$$C = \{(x, \dot{x}) : \Theta(x) = 0_k, \qquad A(x)\dot{x} = 0\}. \tag{4.70}$$

As stated in [42], C is singular; therefore, the order of the motion dynamics can also be reduced (see [39] for a further heuristic discussion on how the position and force variables have been reduced in dimension).

Reduced-Order Dynamic Model

The full-order dynamic model for an n-rigid link direct-drive revolute nonredundant robot is assumed to have the following form [50]

$$M\ddot{q} + V_m\dot{q} + N + f = \tau \tag{4.71}$$

where $q(t)$, $\dot{q}(t)$, $\ddot{q}(t) \in \mathbb{R}^n$ denote the joint position, velocity, and acceleration vectors, respectively, $M(q) \in \mathbb{R}^{n \times n}$ is the inertia matrix, $V_m(q, \dot{q})$

$\in \mathbb{R}^{n \times n}$ is the centripetal-Coriolis matrix, $N(q, \dot{q}) \in \mathbb{R}^n$ contains the gravity and friction vectors, $f(t) \in \mathbb{R}^n$ describes the joint space end-effector forces exerted on the environment by the robot manipulator, and $\tau(t) \in \mathbb{R}^n$ represents the input torque vector.

To facilitate the subsequent position/force tracking control objective, it is common to transform the robot joint space dynamics, given in (4.71), into the task space [50]. This coordinate transformation allows the robot end-effector forces of (4.71) to be written as task space forces, which is the most common way of specifying a force control problem. Specifically, after premultiplying (4.71) by the transpose of the inverse Jacobian given in (4.61) and utilizing the kinematic relationship given in (4.57), the following task space dynamic model can be developed

$$M^* \ddot{x} + V_m^* \dot{x} + N^* + J^{-T} f = \tau^* \tag{4.72}$$

where $x(t)$ is given in (4.56), and $M^*(x) \in \mathbb{R}^{n \times n}$, $V_m^*(x, \dot{x}) \in \mathbb{R}^{n \times n}$, $N^*(x, \dot{x}) \in \mathbb{R}^n$, and $\tau^*(t) \in \mathbb{R}^n$ are defined as follows

$$M^* = J^{-T} M J^{-1} \qquad V_m^* = J^{-T} \left(V_m - M J^{-1} \dot{J} \right) J^{-1} \tag{4.73}$$

$$N^* = J^{-T} N \qquad \tau^* = J^{-T} \tau.$$

The dynamic equation given in (4.72) has n degrees of freedom in the absence of motion constraints. However, given the k frictionless and holonomic motion constraints in (4.62), the resulting degrees of freedom are reduced to $n - k$. Moreover, although there are only n control inputs, there are $n + k$ variables to control (i.e., n position control variables and k force control variables). To reduce the number of states of the constrained robot system, a reduced-order transformation [42] variable $u(x) = \begin{bmatrix} u_1^T(x) & u_2^T(x) \end{bmatrix}^T \in \mathbb{R}^n$ can be defined as follows

$$\begin{bmatrix} u_1 \\ u_2 \end{bmatrix} = \begin{bmatrix} x_1 \\ x_2 - \Omega(x_1) \end{bmatrix}, \tag{4.74}$$

and a transformation matrix $T(x) \in \mathbb{R}^{n \times n}$ can be defined as follows

$$T = \begin{bmatrix} I_{m \times m} & 0_{m \times k} \\ \Delta & I_{k \times k} \end{bmatrix} \in \mathbb{R}^{n \times n} \tag{4.75}$$

where $x_1(t)$, $x_2(t)$ were defined in (4.60), $\Omega(x)$ is defined in (4.67), and $\Delta(x)$ is defined in (4.69). Based on (4.69), (4.74), and (4.75), the following relationship can be developed

$$\dot{x} = Tu. \tag{4.76}$$

After premultiplying (4.72) by $T^T(x)$ and utilizing (4.76), the constrained robot manipulator dynamics can written as follows

$$\bar{M}\ddot{u} + \bar{V}_m\dot{u} + \bar{N} + \bar{A}^T\lambda = \bar{\tau} \tag{4.77}$$

where $\bar{M}(u) \in \mathbb{R}^{n \times n}$, $\bar{V}_m(u, \dot{u}) \in \mathbb{R}^{n \times n}$, $\bar{N}(u, \dot{u}) \in \mathbb{R}^n$, $\bar{A}(u) \in \mathbb{R}^{k \times n}$, and $\bar{\tau}(t) \in \mathbb{R}^n$ are defined as follows[6]

$$\bar{M} = T^T M^* T \qquad \bar{V}_m = T^T \left(M^* \dot{T} + V_m^* T \right) \tag{4.78}$$

$$\bar{N} = T^T N^* \qquad \bar{A} = AT \qquad \bar{\tau} = T^T \tau^*$$

where $A(x)$ was defined in (4.64), and the generalized force multipliers $\lambda(t) \in \mathbb{R}^k$ satisfy the following relationship [42]

$$J^{-T}f = A^T\lambda \tag{4.79}$$

where $J(q)$ was defined in (4.58), and $f(t)$ is given in (4.71). Based on (4.65), (4.75), and (4.78), $\bar{A}(u)$ can be written in the following partitioned form

$$\bar{A} = \begin{bmatrix} 0_{k \times m} & \Pi \end{bmatrix}. \tag{4.80}$$

Furthermore, based on the definition of $\Omega(x)$ given in (4.67) and the definition of $u_2(t)$ given in (4.74), the dynamics for $u_2(t)$ can be reduced as follows

$$u_2 = \dot{u}_2 = \ddot{u}_2 = 0_k \tag{4.81}$$

on the constraint surface. Given the partition of (4.74) and the relationships in (4.80) and (4.81), the dynamic model given in (4.77) can be expressed in the following reduced-order partitioned form

$$
\begin{bmatrix} \bar{\tau}_1 \\ \bar{\tau}_2 \end{bmatrix} = \begin{bmatrix} \bar{M}_{11}(u_1) & \bar{M}_{12}(u_1) \\ \bar{M}_{21}(u_1) & \bar{M}_{22}(u_1) \end{bmatrix} \begin{bmatrix} \ddot{u}_1 \\ 0_k \end{bmatrix}
$$
$$
+ \begin{bmatrix} \bar{V}_{m11}(u_1, \dot{u}_1) & \bar{V}_{m12}(u_1, \dot{u}_1) \\ \bar{V}_{m21}(u_1, \dot{u}_1) & \bar{V}_{m22}(u_1, \dot{u}_1) \end{bmatrix} \begin{bmatrix} \dot{u}_1 \\ 0_k \end{bmatrix} \tag{4.82}
$$
$$
+ \begin{bmatrix} \bar{N}_1(u_1, \dot{u}_1) \\ \bar{N}_2(u_1, \dot{u}_1) \end{bmatrix} + \begin{bmatrix} 0_{m \times k} \\ \Pi^T \end{bmatrix} \lambda.
$$

---

[6] We make the standard assumption that $\bar{V}_m(u, \dot{u})$, $\bar{N}(u, \dot{u})$, and $\bar{A}^T(u)$ of (4.77) and the first time derivatives of these quantities are bounded provided that $u(t)$, $\dot{u}(t)$, $\ddot{u}(t) \in L_\infty$. In addition, we assume that $\bar{M}(u)$ of (4.77) and its first and second time derivatives are bounded provided that $u(t)$, $\dot{u}(t)$, $\ddot{u}(t) \in L_\infty$.

Property 4.5: Symmetric and Positive-Definite Inertia Matrix

The inertia matrix is symmetric positive-definite and satisfies the following inequalities

$$m_1 \left\| \xi \right\|^2 \leq \xi^T \bar{M}(u) \xi \leq m_2 \left\| \xi \right\|^2 \qquad \forall \xi \in \mathbb{R}^n \qquad (4.83)$$

where m_1, $m_2 \in \mathbb{R}$ are known positive constants, and $\left\| \cdot \right\|$ denotes the standard Euclidean norm.

Property 4.6: Skew-Symmetry

The inertia and centripetal-Coriolis matrices satisfy the following skew-symmetric relationship

$$\xi^T \left(\frac{1}{2} \dot{\bar{M}}(u) - \bar{V}_m(u, \dot{u}) \right) \xi = 0 \qquad \forall \xi \in \mathbb{R}^n \qquad (4.84)$$

where $\dot{M}(q)$ denotes the time derivative of the inertia matrix.

Property 4.7: Linearity in the Parameters

The first m rows of the reduced dynamic model given in (4.82) can be linear parameterized as follows

$$Y_d \theta_1 = \bar{M}_{11}(u_{d1}) \ddot{u}_{d1} + \bar{V}_{m11}(u_{d1}, \dot{u}_{d1}) \dot{u}_{d1} + \bar{N}_1(u_{d1}, \dot{u}_{d1}) \qquad (4.85)$$

where $\theta \in \mathbb{R}^p$ contains the unknown constant mechanical parameters (e.g., friction coefficients and inertia-related quantities), $u_{d1}(t), \dot{u}_{d1}(t), \ddot{u}_{d1}(t) \in \mathbb{R}^m$ represent the desired position, velocity, and acceleration trajectories, respectively, and $Y_d(t) \in \mathbb{R}^{m \times p}$ represents a known regression matrix which is only a function of the desired motion trajectory and its time derivatives. It is assumed that upper bounds for $u_{d1}(t)$ and its first and second time derivatives can be formulated as follows

$$\left\| u_{d1}(t) \right\| \leq \zeta_{dp} \qquad \left\| \dot{u}_{d1}(t) \right\| \leq \zeta_{dv} \qquad \left\| \ddot{u}_{d1}(t) \right\| \leq \zeta_{da} \qquad (4.86)$$

where ζ_{dp}, ζ_{dv}, and $\zeta_{da} \in \mathbb{R}$ are known positive bounding constants. The determinant of the submatrix $\bar{M}_{11}(u_1)$ defined in (4.82) can also be parameterized as follows

$$\det \left(\bar{M}_{11}(u_1) \right) = h^T(u_1) \theta_m \qquad (4.87)$$

where $\theta_m \in \mathbb{R}^{p_0}$ contains the unknown constant parameters, $h(u_1) \in \mathbb{R}^{p_0}$ is a known vector function, and $\det(\cdot)$ denotes the determinant of a matrix.

Property 4.8: Convex Region

To avoid singularities in the subsequent control law, a convex region can be defined [40] for the parameter vector θ_m introduced in (4.87). Specifically, the structure of (4.83) and (4.87) can be used to define the following relationship

$$\underline{m}_d \leq h^T(u_1)\theta_m \leq \overline{m}_d(u_1) \tag{4.88}$$

where $\underline{m}_d \in \mathbb{R}$ is a known positive bounding constant, and $\overline{m}_d(\cdot) \in \mathbb{R}$ is a known positive bounding function. Based on the parameterization of (4.87), a space H spanned by the vector function $h(u_1)$ can be defined as follows

$$H = \{v : v = h(u_1)\}. \tag{4.89}$$

A convex region Λ (see [40] for the proof of convexity of Λ) is also defined as follows

$$\Lambda = \{v : v^T h \geq \underline{m}_d \quad \forall h \in H\} \tag{4.90}$$

where $\underline{m}_d$ was given in (4.88). In addition, the following definitions concerning the region Λ and the subsequently designed parameter estimate vector $\hat{\theta}_m(t) \in \mathbb{R}^{p0}$ (i.e., the dynamic estimate of $\theta_m \in \Lambda$ given in (4.87)) are provided as follows: int(Λ) denotes the interior of the region Λ, $\partial(\Lambda)$ denotes the boundary for the region Λ, $\hat{\theta}_m^\perp(t) \in \mathbb{R}^{p0}$ is a unit vector normal to $\partial(\Lambda)$ at the point of intersection of the boundary surface $\partial(\Lambda)$ and $\hat{\theta}_m(t)$ where the positive direction for $\hat{\theta}_m^\perp(t)$ is defined as pointing away from int(Λ) (note that $\hat{\theta}_m^\perp(t)$ is only defined for $\hat{\theta}_m(t) \in \partial(\Lambda)$), $P_r^t(\mu)$ is the component of the vector $\mu \in \mathbb{R}^{p0}$ that is tangent to $\partial(\Lambda)$ at the point of intersection of the boundary surface $\partial(\Lambda)$ and the vector $\hat{\theta}_m(t)$, and

$$P_r^\perp(\mu) = \mu - P_r^t(\mu) \in \mathbb{R}^{p0} \tag{4.91}$$

is the component of the vector $\mu \in \mathbb{R}^{p0}$ that is perpendicular to $\partial(\Lambda)$ at the point of intersection of the boundary surface $\partial(\Lambda)$ and the vector $\hat{\theta}_m(t)$.

4.3.2 Control Objective

The objective of this section is to enable the end-effector of a robot manipulator to simultaneously track a desired task space position and force trajectory despite parametric uncertainty associated with the manipulator dynamic model given in (4.71). The control objective is based on the assumption that $x(t)$, $\dot{x}(t)$, and $\lambda(t)$ are measurable. To quantify the task space position control objective, the position tracking error $e(t) \in \mathbb{R}^m$ is defined as follows

$$e = u_{d1} - u_1 \tag{4.92}$$

where $u_1(t)$ and $u_{d1}(t)$ are defined in (4.74) and (4.85), respectively. To reduce the order of the subsequent error dynamics, a filtered tracking error $r(t) \in \mathbb{R}^m$ is defined as follows

$$r = \dot{e} + \alpha e \qquad (4.93)$$

where $\alpha \in \mathbb{R}$ is a positive constant control gain. To quantify the force control objective, a force tracking error-like term $e_\lambda(t) \in \mathbb{R}^k$ is defined as follows

$$e_\lambda = \int_0^t \left(\lambda_d(\sigma) - \lambda(\sigma)\right) d\sigma \qquad (4.94)$$

where $\lambda(t)$ was defined in (4.77) and $\lambda_d(t) \in \mathbb{R}^k$ represents the desired force trajectory. It is assumed that $\lambda_d(t)$ and its first time derivative are bounded. Based on (4.94), the time derivative of $e_\lambda(t)$ is given by the following expression

$$\dot{e}_\lambda = \lambda_d - \lambda. \qquad (4.95)$$

In the subsequent stability analysis, $e_\lambda(t)$ and $\dot{e}_\lambda(t)$ are proved to be asymptotically driven to zero; hence, if $\dot{e}_\lambda$ goes to zero, (4.95) can be used to prove that the mismatch between the actual and desired forces is driven to zero.

4.3.3   Full-State Feedback Control

Position Control Development

To develop the open-loop error system for $r(t)$, we take the time derivative of (4.93), premultiply the resulting expression by $\bar{M}_{11}(u_1)$, and then substitute (4.82) for $\bar{M}_{11}(u_1)\ddot{u}_1(t)$ to obtain the following expression

$$\bar{M}_{11}\dot{r} = -\bar{V}_{m11}r + Y_1\theta_1 - \bar{\tau}_1 \qquad (4.96)$$

where the time derivative of (4.92) and (4.93) were utilized and the linear parameterization $Y_1(e, \dot{e}, t)\theta_1$ is defined as follows

$$Y_1\theta_1 = \bar{M}_{11}\ddot{u}_{d1} + \bar{V}_{m11}\dot{u}_{d1} + \alpha\left(\bar{M}_{11}\dot{e} + \bar{V}_{m11}e\right) + \bar{N}_1 \qquad (4.97)$$

where $Y_1(\cdot) \in \mathbb{R}^{k \times p}$ represents a known regression matrix, and θ_1 is given in (4.85). Based on the structure of (4.96) and the subsequent stability analysis, the following adaptive position tracking controller can be developed

$$\bar{\tau}_1 = Y_1\hat{\theta}_1 + kr \qquad (4.98)$$

where $k \in \mathbb{R}$ is a positive constant control gain, and $\hat{\theta}_1(t) \in \mathbb{R}^p$ is a dynamic estimate for the unknown parameter vector θ_1 that is generated according to the following gradient update law

$$\dot{\hat{\theta}}_1 = \Gamma Y_1^T r \qquad (4.99)$$

where $\Gamma \in \mathbb{R}^{p \times p}$ is a constant diagonal positive definite gain matrix. After substituting (4.98) into (4.96) for $\bar{\tau}_1(t)$, the following closed-loop dynamics can be obtained

$$\bar{M}_{11}\dot{r} = -\bar{V}_{m11}r + Y_1\tilde{\theta}_1 - kr \qquad (4.100)$$

where the parameter estimation error $\tilde{\theta}_1(t) \in \mathbb{R}^p$ is defined as follows

$$\tilde{\theta}_1 = \theta_1 - \hat{\theta}_1. \qquad (4.101)$$

Since the unknown parameter vector θ_1 is constant, the closed-loop error system for $\tilde{\theta}_1(t)$ is obtained by taking the time derivative of (4.101) and substituting (4.99) for $\dot{\hat{\theta}}_1(t)$ as follows

$$\dot{\tilde{\theta}}_1 = -\Gamma Y_1^T r. \qquad (4.102)$$

Force Control Development

To develop the open-loop error system for $\dot{e}_\lambda(t)$, the first m rows of (4.82) can be used to solve for $\ddot{u}_1(t)$ and the resulting expression substituted into the last k rows of (4.82) to obtain the following expression

$$\bar{\tau}_2 = \bar{M}_{21}\bar{M}_{11}^{-1}\left(\bar{\tau}_1 - \bar{V}_{m11}\dot{u}_1 - \bar{N}_1\right) + \bar{V}_{m21}\dot{u}_1 + \bar{N}_2 + \Pi^T\lambda. \qquad (4.103)$$

After premultiplying (4.103) by $\det(\bar{M}_{11})\Pi^{-T}(x)$, the following expression is obtained

$$\det(\bar{M}_{11})\lambda = -\Pi^{-T}\bar{M}_{21}\text{adj}(\bar{M}_{11})\left(\bar{\tau}_1 - \bar{V}_{m11}\dot{u}_1 - \bar{N}_1\right)$$

$$- \det(\bar{M}_{11})\Pi^{-T}\left(\bar{V}_{m21}\dot{u}_1 + \bar{N}_2\right) + \det(\bar{M}_{11})\Pi^{-T}\bar{\tau}_2 \qquad (4.104)$$

where $\text{adj}(\bar{M}_{11})$ denotes the adjoint of $\bar{M}_{11}(u_1)$. To obtain the open-loop force tracking error dynamics, (4.95) can be used to rewrite (4.104) as follows

$$\det(\bar{M}_{11})\dot{e}_\lambda = -\left[\frac{1}{2}\frac{d\left(\det\left(\bar{M}_{11}\right)\right)}{dt}e_\lambda\right] + Y_2\theta_2 - \det(\bar{M}_{11})\Pi^{-T}\bar{\tau}_2 \qquad (4.105)$$

where the linear parameterization $Y_2(e_\lambda, u_1, \dot{u}_1, r, \hat{\theta}_1, t)\theta_2$ is defined as follows

$$Y_2\theta_2 = \left[\frac{1}{2}\frac{d\left(\det\left(\bar{M}_{11}\right)\right)}{dt}e_\lambda\right] + \Pi^{-T}\bar{M}_{21}\text{adj}(\bar{M}_{11})\left(\bar{\tau}_1 - \bar{V}_{m11}\dot{u}_1 - \bar{N}_1\right)$$

$$+ \det(\bar{M}_{11})\Pi^{-T}\left(\bar{V}_{m21}\dot{u}_1 + \bar{N}_2\right) + \det(\bar{M}_{11})\lambda_d$$

$$(4.106)$$

where $\theta_2 \in \mathbb{R}^{p2}$ denotes a vector of unknown constant parameters, $Y_2(\cdot) \in \mathbb{R}^{k \times p2}$ is a known regression matrix, and the bracketed term in (4.105) and (4.106) was added and subtracted to facilitate the subsequent control design and stability analysis.

Given the open-loop force tracking error dynamics of (4.105), the following adaptive force controller can be designed

$$\bar{\tau}_2 = \frac{1}{h^T\hat{\theta}_m}\Pi^T\left(Y_2\hat{\theta}_2\right) + \Pi^T\frac{k}{m_d}e_\lambda \qquad (4.107)$$

where k is the same positive control gain given in (4.98), $h\left(u_1\right)$ is defined in (4.87), $\hat{\theta}_m(t)$ is given in Property 4.8, $\hat{\theta}_2(t) \in \mathbb{R}^{p2}$ is subsequently designed dynamic estimate for θ_2, and m_d is the positive constant defined in (4.88). After substituting $\bar{\tau}_2(t)$ of (4.107) into (4.105) and then adding and subtracting $Y_2(\cdot)\hat{\theta}_2(t)$ to the resulting expression, the following closed-loop force tracking error dynamics are obtained

$$\det(\bar{M}_{11})\dot{e}_\lambda = -\frac{1}{2}\frac{d}{dt}\left(\det\left(\bar{M}_{11}\right)\right)e_\lambda + Y_2\tilde{\theta}_2 - \frac{h^T\tilde{\theta}_m}{h^T\hat{\theta}_m}Y_2\hat{\theta}_2 - \frac{k\det(\bar{M}_{11})}{m_d}e_\lambda$$

$$(4.108)$$

where the parameter estimation error terms $\tilde{\theta}_2(t) \in \mathbb{R}^{p2}$ and $\tilde{\theta}_m(t) \in \mathbb{R}^{p0}$ are defined as follows

$$\tilde{\theta}_2 = \theta_2 - \hat{\theta}_2 \qquad \tilde{\theta}_m = \theta_m - \hat{\theta}_m. \qquad (4.109)$$

Based on the subsequent stability analysis, the structure of (4.108), and Property 4.8, the parameter update laws for the unknown vectors θ_2 and θ_m are designed as follows

$$\dot{\hat{\theta}}_2 = \Gamma_1 Y_2^T e_\lambda \qquad (4.110)$$

and

$$\dot{\hat{\theta}}_m = \begin{cases} \mu_1 & \text{if } \hat{\theta}_m \in \text{int}(\Lambda) \\ \mu_1 & \text{if } \hat{\theta}_m \in \partial(\Lambda) \text{ and } \mu_1^T\hat{\theta}_m^\perp \leq 0 \\ P_r^t(\mu_1) & \text{if } \hat{\theta}_m \in \partial(\Lambda) \text{ and } \mu_1^T\hat{\theta}_m^\perp > 0 \end{cases} \qquad (4.111)$$

$$\hat{\theta}_m(0) \in \text{int}(\Lambda)$$

where $\Gamma_1 \in \mathbb{R}^{p2 \times p2}$ is a positive definite diagonal matrix, and $\mu_1 \in \mathbb{R}^{p0}$ is defined as follows

$$\mu_1 = \frac{-h}{h^T \hat{\theta}_m} \left[Y_2 \hat{\theta}_2 \right]^T e_\lambda. \tag{4.112}$$

Because θ_2 and θ_m are constant vectors, the closed-loop dynamics for $\tilde{\theta}_2(t)$ and $\tilde{\theta}_m(t)$ can be obtained by taking the time derivative of (4.109) and substituting (4.110) and (4.111) into the resulting expressions for $\tilde{\theta}_2(t)$ and $\tilde{\theta}_m(t)$, respectively, as follows

$$\dot{\tilde{\theta}}_2(t) = -\Gamma_1 Y_2^T e_\lambda \tag{4.113}$$

$$\dot{\tilde{\theta}}_m = \begin{cases} -\mu_1 & \text{if } \hat{\theta}_m \in \text{int}(\Lambda) \\ -\mu_1 & \text{if } \hat{\theta}_m \in \partial(\Lambda) \text{ and } \mu_1^T \hat{\theta}_m^\perp \le 0 \\ -P_r^t(\mu_1) & \text{if } \hat{\theta}_m \in \partial(\Lambda) \text{ and } \mu_1^T \hat{\theta}_m^\perp > 0 \end{cases} \tag{4.114}$$

Stability Analysis

The stability of the adaptive position and force tracking controller given in (4.98) and (4.107), with the adaptive update laws given in (4.99), (4.110), and (4.111) can be examined through the following theorem.

Theorem 4.3 *The position controller of (4.98) and (4.99) and the force controller of (4.107), (4.110), and (4.111) ensure global asymptotic position and force tracking in the sense that*

$$\lim_{t \to \infty} e(t), e_\lambda(t), \dot{e}_\lambda(t) = 0. \tag{4.115}$$

where $e(t)$, $e_\lambda(t)$, and $\dot{e}_\lambda(t)$ were defined in (4.92), (4.94), and (4.95), respectively.

Proof: To prove Theorem 4.3, we define a nonnegative function $V(t) \in \mathbb{R}$ as follows

$$V = \frac{1}{2} r^T \bar{M}_{11} r + \frac{1}{2} \det(\bar{M}_{11}) e_\lambda^T e_\lambda + \frac{1}{2} \tilde{\theta}_1^T \Gamma^{-1} \tilde{\theta}_1$$

$$+ \frac{1}{2} \tilde{\theta}_2^T \Gamma_1^{-1} \tilde{\theta}_2 + \frac{1}{2} \tilde{\theta}_m^T \tilde{\theta}_m. \tag{4.116}$$

After taking the time derivative of (4.116), substituting (4.100) and (4.108) for the closed-loop tracking error dynamics for $r(t)$ and $e_\lambda^T(t)$, respectively, utilizing (4.83) and (4.88), the following simplified expression is obtained

$$\dot{V} \le r^T \left(Y_1 \tilde{\theta}_1 - kr \right) + e_\lambda^T \left[+Y_2 \tilde{\theta}_2 - \frac{h^T \tilde{\theta}_m}{h^T \hat{\theta}_m} Y_2 \hat{\theta}_2 - ke_\lambda \right]$$

$$+ \tilde{\theta}_1^T \Gamma^{-1} \dot{\tilde{\theta}}_1 + \tilde{\theta}_2^T \Gamma_1^{-1} \dot{\tilde{\theta}}_2 + \tilde{\theta}_m^T \dot{\tilde{\theta}}_m. \tag{4.117}$$

After substituting the closed-loop dynamics given in (4.102), (4.113), and (4.114) into (4.117) and cancelling common terms, the following expression can be obtained (see Lemma B.14 of Appendix B)

$$\dot{V} \leq -kr^T r - k e_\lambda^T e_\lambda \qquad (4.118)$$

where (4.87) and (4.112) were utilized.

The direct implication of (4.116) and (4.118) is that $r(t)$, $e_\lambda(t)$, $\tilde{\theta}_1(t)$, $\tilde{\theta}_2(t)$, and $\tilde{\theta}_m(t) \in \mathcal{L}_\infty$ and that $r(t)$, $e_\lambda(t) \in \mathcal{L}_2$. Hence, by invoking Lemma A.13 of Appendix A, we can prove that $e(t)$, $\dot{e}(t) \in \mathcal{L}_\infty$. Based on the fact that θ, θ_1, and θ_m are composed of constant bounded parameters and that $\tilde{\theta}_1(t)$, $\tilde{\theta}_2(t)$, and $\tilde{\theta}_m(t) \in \mathcal{L}_\infty$, (4.101) and (4.109) can be used to prove that $\hat{\theta}_1(t)$, $\hat{\theta}_2(t)$, and $\hat{\theta}_m(t) \in \mathcal{L}_\infty$. Since $e(t)$, $\dot{e}(t)$, $\hat{\theta}_1(t)$, $r(t) \in \mathcal{L}_\infty$ and $u_{d1}(t)$, $\dot{u}_{d1}(t)$, $\ddot{u}_{d1}(t) \in \mathcal{L}_\infty$, the expressions in (4.92), (4.97), and (4.98) can be used to prove that $u_1(t)$, $\dot{u}_1(t)$, $Y_1(\cdot)$, $\bar{\tau}_1(t) \in \mathcal{L}_\infty$. Given that $Y_1(\cdot) \in \mathcal{L}_\infty$, all the terms on the right-hand side of (4.97) are bounded, and hence, (4.93), (4.99), and (4.100) can be utilized to prove that $\dot{r}(t)$, $\ddot{e}(t)$, $\dot{\hat{\theta}}_1(t) \in \mathcal{L}_\infty$. From the fact that $\dot{r}(t) \in \mathcal{L}_\infty$, (4.92) and (4.93) can be used to prove that $\ddot{u}_1(t) \in \mathcal{L}_\infty$. The expression in (4.80) and the fact that $u_1(t)$, $\dot{u}_1(t)$, $\ddot{u}_1(t) \in \mathcal{L}_\infty$ can now be used to prove that $\bar{V}_m(u, \dot{u})$, $\bar{N}(u, \dot{u})$, $\bar{A}(u)$, $\Pi(x)$, $\bar{M}(u) \in \mathcal{L}_\infty$. The fact that $u_1(t)$, $\dot{u}_1(t)$, $e_\lambda(t)$, $\Pi(x)$, $Y_1(\cdot)$, $\bar{\tau}_1(t) \in \mathcal{L}_\infty$ and $\lambda_d(t)$ is assumed to be bounded, (4.106) can be used to prove that $Y_2(\cdot) \in \mathcal{L}_\infty$. From the previous boundedness arguments, the expressions in (4.103), (4.107), (4.108), (4.110) and the development in Lemma B.14 of Appendix B can be used to prove that $\dot{e}_\lambda(t)$, $\dot{\hat{\theta}}_2(t)$, $\bar{\tau}_2(t)$, $\lambda(t) \in \mathcal{L}_\infty$.[7] Furthermore, from (4.112), it can be shown that $\mu_1(t)$, $P_r^t(\mu_1)$, $P_r^\perp(\mu_1) \in \mathcal{L}_\infty$; hence, (4.111) can be used to prove that $\dot{\hat{\theta}}_m(t) \in \mathcal{L}_\infty$. Having proved that $r(t)$, $e_\lambda(t) \in \mathcal{L}_\infty \cap \mathcal{L}_2$ and $\dot{r}(t)$, $\dot{e}_\lambda(t) \in \mathcal{L}_\infty$, Lemmas A.15 and A.16 of Appendix A can be invoked to prove that

$$\lim_{t \to \infty} r(t), \, e(t), \, e_\lambda(t) = 0. \qquad (4.119)$$

From (4.119), it is clear that the task space position control objective is satisfied. However, to prove that the force control objective is satisfied (i.e., to prove that $\lambda(t)$ tracks $\lambda_d(t)$), we must now prove that $\dot{e}_\lambda(t)$ goes

---

[7]Note that the only appearances of $\int \lambda_d(t)dt$ and $\int \lambda(t)dt$ in (4.107), (4.110), and (4.111) are through the variable $e_\lambda(t)$ which was shown to be bounded. Hence, it is not necessary to assume that $\int \lambda_d(t)dt$ is bounded to ensure the boundedness of $\bar{\tau}_2(t)$, $\dot{\hat{\theta}}_2(t)$, and $\dot{\hat{\theta}}_m(t)$.

to zero as indicated by (4.95). Note that $e_\lambda(t)$ can be rewritten as

$$e_\lambda = \int_0^t \frac{de_\lambda(\sigma)}{d\sigma} d\sigma + C \qquad (4.120)$$

where $C \in \mathbb{R}^k$ is a vector of integration constants. Based on (4.119), the expression given in (4.120) can be used to prove that

$$\lim_{t \to \infty} \int_0^t \frac{de_\lambda(\sigma)}{d\sigma} d\sigma \qquad (4.121)$$

exists and is finite. Moreover, based on the previous development, it can be shown that $\ddot{e}_\lambda(t) \in \mathcal{L}_\infty$ (see Lemma B.15 in Appendix B), and hence, $\dot{e}_\lambda(t)$ is uniformly continuous. Given (4.121) and the fact that $\dot{e}_\lambda(t)$ is uniformly continuous, the integral form of Barbalat's Lemma (see Lemma A.20 of Appendix A) can be invoked to prove that

$$\lim_{t \to \infty} \dot{e}_\lambda = 0.$$

Hence, the position and force control objectives are satisfied according to the result given in (4.115). $\square$

4.3.4   Output Feedback Control

The development given in Section 4.3.3 is based on the assumption that link position and velocity measurements are available. However, link velocity measurements are usually not available in robotic applications due to the additional sensor cost. One popular method for obtaining position and velocity measurements without incorporating an additional sensor is to use the backwards difference algorithm in which the link velocity is numerically calculated from the link position measurements. As described in [18], this approach is not theoretically satisfying and introduces noise into the system; hence, motivated by the desire to eliminate the need for velocity measurements, an output feedback extension of the previous position/force tracking result is developed in this section.

Position Control Modifications

Since link velocity (and hence $\dot{e}(t)$ and $\dot{u}_1(t)$) is now assumed to be unmeasurable, a filter can be constructed to generate a surrogate velocity tracking error signal. This filter can be thought of as a high-pass filter having as an input the position tracking error of (4.92) and producing as an output a pseudo-velocity tracking error signal. The filter is given by the following dynamic relationship [7]

$$\dot{p} = -(k_p + 1)p + (k_p^2 + 1)e \qquad p(0) = k_p e(0) \qquad (4.122)$$

$$e_f = -k_p e + p \qquad (4.123)$$

where $e_f(t) \in \mathbb{R}^m$ is the output of the filter that will be used as a link velocity substitute, $p(t) \in \mathbb{R}^m$ is an auxiliary filter variable, and $k_p \in \mathbb{R}$ is a filter gain selected as follows

$$k_p = \frac{1}{m_1}(1 + 2k_n) \qquad (4.124)$$

where $m_1 \in \mathbb{R}$ is a known bounding constant for $\bar{M}_{11}(u)$ of (4.82) such that

$$m_1 \|\xi\|^2 \leq \xi^T \bar{M}_{11}(u)\xi \qquad \forall \xi \in \mathbb{R}^n \qquad (4.125)$$

and $k_n \in \mathbb{R}$ is a positive nonlinear damping gain [36].

To further facilitate the output feedback design, the open-loop position tracking error system is developed in terms of the high-pass filter dynamics. Specifically, after taking the time derivative of (4.123) and substituting the right-hand side of (4.122) for $\dot{p}(t)$ the following expression is obtained

$$\dot{e}_f = -k_p \dot{e} - (k_p + 1) p + (k_p^2 + 1) e. \qquad (4.126)$$

After substituting (4.123) into (4.126) for $p(t)$, the following open-loop position tracking error dynamics for $e_f(t)$ can be obtained

$$\dot{e}_f = -k_p \eta - e_f + e \qquad (4.127)$$

where the filtered tracking error variable $\eta(t) \in \mathbb{R}^m$ is defined as follows

$$\eta = \dot{e} + e_f + e. \qquad (4.128)$$

The open-loop dynamics for $\eta(t)$ can be developed as follows [17]

$$\bar{M}_{11}\dot{\eta} = \chi - \bar{\tau}_1 + Y_d \theta_1 - k_p \bar{M}_{11}\eta - \bar{V}_{m11}\eta \qquad (4.129)$$

where the state-dependent disturbance variable $\chi(u_1, \dot{u}_1, e_f, e, \eta) \in \mathbb{R}^m$ is defined as follows

$$\chi = \bar{M}_{11}\ddot{u}_{d1} + \bar{V}_{m11}\dot{u}_{d1} + \bar{N}_1 - Y_d \theta_1 - 2\bar{M}_{11}e_f + \bar{M}_{11}\eta + \bar{V}_{m11}(e_f + e), \qquad (4.130)$$

and the linear parameterization $Y_d(u_{d1}, \dot{u}_{d1}, \ddot{u}_{d1})\theta_1$ is given in (4.85).

Based on the structure of (4.129), the following adaptive output feedback position tracking controller can be designed

$$\bar{\tau}_1 = Y_d \hat{\theta}_1 - k_p e_f + e \qquad (4.131)$$

where k_p was defined in (4.124), and $\hat{\theta}_1(t) \in \mathbb{R}^p$ of (4.99) is redesigned according to the following velocity independent gradient update law

$$\hat{\theta}_1 = \Gamma_1 \int_0^t Y_d^T(\sigma)(e(\sigma) + e_f(\sigma))\, d\sigma + \Gamma_1 Y_d^T e - \Gamma_1 \int_0^t \dot{Y}_d^T(\sigma)e(\sigma)d\sigma. \qquad (4.132)$$

Force Control Modifications

To rewrite the open-loop force tracking error dynamics of (4.105) in a form that facilitates the development of a velocity independent controller, (4.95) can be used to rewrite (4.104) in the following form

$$\det(\bar{M}_{11})\dot{e}_\lambda = -\frac{1}{2}\frac{d\left(\det\left(\bar{M}_{11}\right)\right)}{dt}e_\lambda + \tilde{Y}_v + Y_v\theta_v - \det(\bar{M}_{11})\Pi^{-T}\bar{\tau}_2 \quad (4.133)$$

where $\tilde{Y}_v(u_1, \dot{u}_1, e_f, e_\lambda) \in \mathbb{R}^k$ is a state-dependent disturbance variable defined as follows

$$\tilde{Y}_v = \frac{1}{2}\frac{d\left(\det\left(\bar{M}_{11}\right)\right)}{dt}e_\lambda + \det(\bar{M}_{11})\lambda_d - Y_v\theta_v$$

$$+\Pi^{-T}\bar{M}_{21}\mathrm{adj}(\bar{M}_{11})\left(\bar{\tau}_1 - \bar{V}_{m11}\dot{u}_1 - \bar{N}_1\right) \quad (4.134)$$

$$+\det(\bar{M}_{11})\Pi^{-T}\left(\bar{V}_{m21}\dot{u}_1 + \bar{N}_2\right),$$

and the linear parameterization $Y_v(u_1, e_f, e_\lambda, \hat{\theta}, t)\theta_v$ is defined as follows

$$Y_v\theta_v = \frac{1}{2}\frac{\partial\left(\det\left(\bar{M}_{11}\right)\right)}{\partial u_1}\dot{u}_{d1}e_\lambda + \det(\bar{M}_{11})\lambda_d$$

$$+\Pi^{-T}\bar{M}_{21}\mathrm{adj}(\bar{M}_{11})\left(\bar{\tau}_1 - \bar{V}_{m11}\left(u_1, \dot{u}_{d1}\right)\dot{u}_{d1} - \bar{N}_1(u_1, \dot{u}_{d1})\right)$$

$$\det(\bar{M}_{11})\Pi^{-T}\left(\bar{V}_{m21}\left(u_1, \dot{u}_{d1}\right)\dot{u}_{d1} + \bar{N}_2(u_1, \dot{u}_{d1})\right)$$

$$(4.135)$$

where $\theta_v \in \mathbb{R}^{p4}$ denotes a vector of unknown constant parameters, and $Y_v(u_1, e_f, e_\lambda, \hat{\theta}) \in \mathbb{R}^{k \times p4}$ is a known regression matrix. It is important to note that the parameterization given in (4.135) has been constructed to avoid the use of velocity measurements by replacing the link velocity signal $\dot{u}_1(t)$ with the desired link velocity $\dot{u}_{d1}(t)$.

Given the open-loop force tracking error dynamics of (4.133), the adaptive force controller given in (4.107) is redesigned as follows

$$\bar{\tau}_2 = \frac{1}{h^T\hat{\theta}_m}\Pi^T\left(Y_v\hat{\theta}_v\right) + \Pi^T\frac{k_p m_1}{m_d}e_\lambda \quad (4.136)$$

where m_1 is the positive bounding constant defined in (4.125) and $\underline{m}_d$ is the positive bounding constant defined in (4.88). In (4.136), $\hat{\theta}_v(t) \in \mathbb{R}^{p4}$ denotes a dynamic estimate for the unknown parameter vector θ_3 that is generated according to the following gradient update law

$$\dot{\hat{\theta}}_v = \Gamma_v Y_v^T e_\lambda \quad (4.137)$$

where $\Gamma_v \in \mathbb{R}^{p4 \times p4}$ is a constant diagonal positive definite gain matrix and $\hat{\theta}_m(t)$ is designed in the same manner as in (4.111), where $\mu(t)$ of (4.112) is redefined as follows

$$\mu = \frac{-h}{h^T \hat{\theta}_m} \left[Y_v \hat{\theta}_v \right]^T e_\lambda. \tag{4.138}$$

After substituting $\bar{\tau}_2(t)$ of (4.136) into (4.133), the following closed-loop force tracking error dynamics can be obtained

$$\det(\bar{M}_{11})\dot{e}_\lambda = \quad -\frac{1}{2}\frac{d}{dt}\left(\det\left(\bar{M}_{11}\right)\right) e_\lambda + Y_v \hat{\theta}_v - \frac{h^T \tilde{\theta}_m}{h^T \hat{\theta}_m} Y_v \hat{\theta}_v$$

$$+\tilde{Y}_v - \frac{k_p m_1 \det(\bar{M}_{11})}{m_d} e_\lambda \tag{4.139}$$

where the parameterization of $\det(\bar{M}_{11})$ in (4.87) has been utilized, the parameter estimation error term $\tilde{\theta}_v(t) \in \mathbb{R}^{p4}$ is defined as follows

$$\tilde{\theta}_v = \theta_v - \hat{\theta}_v, \tag{4.140}$$

and $\tilde{\theta}_m(t)$ is defined in (4.109).

Remark 4.5 *By adding and subtracting suitable terms to the right-hand side of (4.130) and (4.134), Lemma B.16 and Lemma B.17 of Appendix B can be invoked to construct the following inequality*

$$\rho(\zeta_{dp}, \zeta_{dv}, \zeta_{da}, \|y\|) \|y\| \geq \max\left\{ \|\chi\|, \left\|\tilde{Y}_v\right\| \right\} \tag{4.141}$$

where ζ_{dp}, ζ_{dv}, and ζ_{da} were defined in (4.86), $y(t) \in \Re^{3m+k}$ is defined as

$$y = \left[\begin{array}{cccc} e^T & e_f^T & \eta^T & e_\lambda^T \end{array} \right]^T, \tag{4.142}$$

and $\rho(\cdot) \in \mathbb{R}$ is a positive nondecreasing function.

Stability Analysis

The stability of the adaptive output feedback position/force tracking controller given in (4.131), (4.132), (4.136), (4.137), and (4.111) with (4.138) can now be examined through the following theorem.

Theorem 4.4 *Given the position controller of (4.131), the force controller of (4.136), and the parameter update laws of (4.132), (4.137), and (4.111) with (4.138), if k_n of (4.124) is selected as follows*

$$k_n \geq \rho^2 \left(\zeta_{dp}, \zeta_{dv}, \zeta_{da}, \sqrt{\frac{\lambda_2(u_1(0))}{\lambda_1}} \|z(0)\| \right) \tag{4.143}$$

then

$$\lim_{t \to \infty} \|y(t)\| = 0 \qquad (4.144)$$

where $\rho(\cdot)$ and $y(t)$ were defined in (4.141) and (4.142), respectively, $z(t) \in \mathbb{R}^{3m+k+p+p3+p4}$ is defined as

$$z = \begin{bmatrix} e^T & e_f^T & \eta^T & e_\lambda^T & \tilde{\theta}_1^T & \tilde{\theta}_m^T & \tilde{\theta}_v^T \end{bmatrix}^T, \qquad (4.145)$$

$\lambda_1 \in \mathbb{R}$ *is a positive bounding constant defined as*

$$\lambda_1 = \frac{1}{2} \min \left\{ 1, m_1, \underline{m}_d, \lambda_{\min} \left\{ \Gamma_1^{-1} \right\}, \lambda_{\min} \left\{ \Gamma_v^{-1} \right\} \right\}, \qquad (4.146)$$

and $\lambda_2(u_1) \in \mathbb{R}$ is a positive bounding function defined as follows

$$\lambda_2(u_1) = \frac{1}{2} \max \left\{ 1, m_2(u_1), \overline{m}_d(u_1), \lambda_{\max} \left\{ \Gamma_1^{-1} \right\}, \lambda_{\max} \left\{ \Gamma_v^{-1} \right\} \right\}. \qquad (4.147)$$

The notations $\lambda_{\min} \{\cdot\}$ and $\lambda_{\max} \{\cdot\}$ represent the minimum and maximum eigenvalues of a matrix, respectively.

Proof: To prove Theorem 4.5, we define a nonnegative function $V(t) \in \mathbb{R}$ as follows

$$V = \frac{1}{2} e^T e + \frac{1}{2} e_f^T e_f + \frac{1}{2} \eta^T \bar{M}_{11} \eta + \frac{1}{2} \det(\bar{M}_{11}) e_\lambda^T e_\lambda$$

$$+ \frac{1}{2} \tilde{\theta}_v^T \Gamma_v^{-1} \tilde{\theta}_v + \frac{1}{2} \tilde{\theta}_1^T \Gamma_1^{-1} \tilde{\theta}_1 + \frac{1}{2} \tilde{\theta}_m^T \tilde{\theta}_m \qquad (4.148)$$

where $V(t)$ can be bounded as follows

$$\lambda_1 \|y\|^2 \leq \lambda_1 \|z\|^2 \leq V \leq \lambda_2(u_1) \|z\|^2 \qquad (4.149)$$

and $y(t)$, $z(t)$, λ_1, and $\lambda_2(u_1)$ were defined in (4.142) and (4.145–4.147), respectively. After taking the time derivative of (4.148) and following a similar analysis as given in the proof for Theorem 4.3, the following upper bound can be obtained

$$\dot{V} \leq -\|e\|^2 - \|e_f\|^2 - k_p m_1 \|e_\lambda\|^2 - k_p m_1 \|\eta\|^2$$

$$+ \|\eta\| \|\chi\| + \|e_\lambda\| \left\| \tilde{Y}_v \right\|. \qquad (4.150)$$

After substituting (4.124) and (4.141) into (4.150), the following upper bound can be formulated

$$\dot{V} \leq -\|e\|^2 - \|e_f\|^2 - \|e_\lambda\|^2 - \|\eta\|^2$$

$$+ \left[\rho(\zeta_{dp}, \zeta_{dv}, \zeta_{da}, \|y\|) \|y\| \|\eta\| - 2k_n \|\eta\|^2 \right] \qquad (4.151)$$

$$+ \left[\rho(\zeta_{dp}, \zeta_{dv}, \zeta_{da}, \|y\|) \|y\| \|e_\lambda\| - 2k_n \|e_\lambda\|^2 \right]$$

where $y(t)$ was defined in (4.142). After invoking Lemma A.17 of Appendix A, and utilizing (4.149), (4.151) can be upper bounded as follows

$$\dot{V} \leq -\left[1 - \frac{1}{k_n}\rho^2\left(\zeta_{dp}, \zeta_{dv}, \zeta_{da}, \sqrt{\frac{V}{\lambda_1}}\right)\right]\|y\|^2. \qquad (4.152)$$

To ensure that $\dot{V}(t)$ is negative semi-definite, the bracketed term in (4.152) must be positive; hence, the analysis to this point can be summarized by the following inequality

$$\dot{V} \leq -\beta\|y\|^2 \qquad \text{for} \qquad k_n \geq \rho^2\left(\zeta_{dp}, \zeta_{dv}, \zeta_{da}, \sqrt{\frac{V}{\lambda_1}}\right) \qquad (4.153)$$

where $\beta \in \mathbb{R}$ is a positive bounding constant. Provided the gain condition given in (4.153) is satisfied, $\dot{V}(t)$ will be negative semi-definite, and hence

$$V(t) \leq V(z(0)) \quad \forall t \geq 0. \qquad (4.154)$$

Based on (4.149) and (4.154), the following result can be obtained

$$\dot{V} \leq -\beta\|y\|^2 \qquad \text{for} \qquad k_n \geq \rho^2\left(\zeta_{dp}, \zeta_{dv}, \zeta_{da}, \sqrt{\frac{\lambda_2(u_1(0))}{\lambda_1}}\|z(0)\|\right).$$
$$(4.155)$$

From (4.148) and (4.155), we can prove that $y(t) \in \mathcal{L}_\infty \cap \mathcal{L}_2$ provided the condition given in (4.143) is satisfied. Using similar techniques as in the proof for Theorem 4.3, all signals can be proved to be bounded during closed-loop operation, and hence, $y(t)$ is uniformly continuous. Since it can be proven that $y(t) \in \mathcal{L}_\infty \cap \mathcal{L}_2$ and $\dot{y}(t) \in \mathcal{L}_\infty$, Barbalat's Lemma (see Lemma A.16 of Appendix A) can be invoked to prove the result given in (4.144). □

4.3.5   Experimental Setup and Results

Although a considerable amount of theoretical work has been done regarding the design of advanced position/force controllers, few experimental implementations of these algorithms have been reported. Furthermore, many of these experiments do not forward results on simultaneous position/force tracking. Hence, in this section, the performance of the output feedback version of the adaptive position/force controller is experimentally demonstrated in response to both time-varying position and force trajectories, which constitutes the most general position/force control task. The objective of the experiment is to force the robot end-effector to move along a vertical wall, parallel to the x_1 axis at a known distance b while applying a force in the direction of the x_2 axis (see Figure 4.5). Hence, the constraint function given in (4.62) is calculated to be

$$\Theta(x_1, x_2) = x_2 - b = 0.$$

The desired position trajectory for the end-effector was selected as follows

$$u_{d1}(t) = 11 \sin(2.8t) \left(1 - \exp(-0.3t^3)\right) \text{ [cm]},$$

where the exponential term is included to ensure that $u_{d1}(0) = \dot{u}_{d1}(0) = \ddot{u}_{d1}(0) = 0$ (note from (4.74) that $u_1(t) = x_1(t)$), and the desired force trajectory was selected as follows

$$\lambda_d(t) = 0.35t \exp(-0.0005t^2) \text{ [N]}.$$

The controller was implemented on the 2-link IMI direct-drive manipulator described in Section 4.2.3 with the exception that a JR-3 force sensor was mounted between the end-effector and a wheel tool (see Figure 4.5) to measure the end-effector force applied to the wall along the x_2 axis. The control gains that resulted in the best performance (in terms of smallest steady state error) were determined as follows

$$k = \quad 60.0 \quad \Gamma = \text{diag}\{2.0, 1.0, 2.0, 7.0, 20.0\}$$

$$\Gamma_1 = \quad \text{diag}\{4.0, 1.0, 1.0, 3.0, 3.0, 4.0, 4.0,$$
$$70.0, 3.0, 1.0, 1.0, 1.0, 1.0, 1.0, 1.0\} \times 10^{-4}.$$

The initial values of the parameter estimate vectors of (4.110) and (4.137) were set to zero (i.e., no prior knowledge of the nominal parameter values is assumed) while the parameter estimate vector of (4.111) was directly set to the nominal values (i.e., perfect knowledge of the parameter values was assumed). A standard trapezoidal algorithm was used to compute the integrals for the parameter estimates while the control algorithm was executed at a sampling rate of 1 [msec].

The position and force tracking errors (i.e., $e(t)$ and $\dot{e}_\lambda(t)$) are depicted in Figure 4.6. For illustration purposes, Figure 4.7 depicts three of the parameter estimate trajectories while Figure 4.8 illustrates a sample input control torque (i.e., the torque input for link 2). From Figure 4.6, the maximum position tracking error is approximately ± 1.375 [cm], which is approximately 12.5% of the maximum peak value of 11.0 [cm], and the maximum force tracking error is approximately ± 8.0 [N], which is approximately 15.5% of the corresponding desired value at $t = 20$ [sec]. During

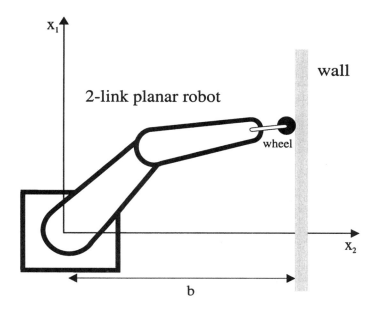

FIGURE 4.5. Experimental setup.

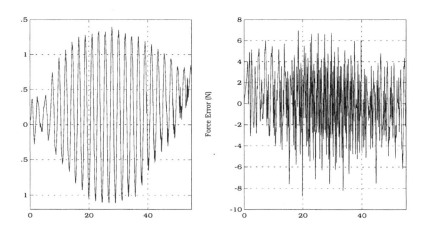

FIGURE 4.6. End-effector position and force tracking errors.

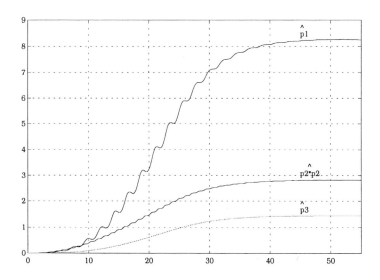

FIGURE 4.7. Sample of the parameter estimates.

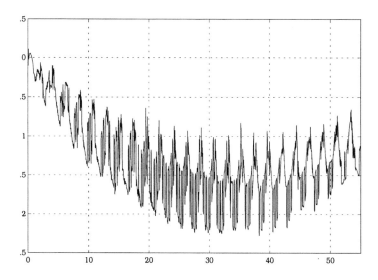

FIGURE 4.8. Link 2 input torque.

the experimental trials, we noticed that if the force controller was turned off, the maximum position tracking error could be tuned to approximately ± 0.1 [cm] (i.e., approximately 0.91% of the maximum peak value). Therefore, we believe that while both the direct-drive robot manipulator system and the experimental setup (i.e., the constraint wall) were both constructed to be very rigid, small flexibilities in the system can cause the position and force control loops to affect each other during operation degrading the position/force tracking performance. Furthermore, the force measurements contain significant low-frequency and high-frequency noise, degrading the force tracking performance.

4.4 Visual Servo Control Application

As described in the introduction to this chapter, many robotic applications are facilitated by the use of a camera-based vision system that can provide a sense of perception of the environment. Many current vision-based approaches assume that the camera is perfectly calibrated or that the unknown parameters can be numerically estimated. Unfortunately, these approaches may lead to errors in the calibration parameters that can result in unpredictable motion by the robot. In this section, a Lyapunov-based visual servoing approach is employed to construct adaptive controllers that compensate for the camera uncertainty and to incorporate the effects of the robot dynamics. Specifically, for the fixed camera configuration, an adaptive controller that achieves global asymptotic tracking despite camera calibration is developed. In addition, a second adaptive controller that compensates for camera calibration parameters and parametric uncertainty associated with the robot dynamics is designed. An extension to these controllers is presented to illustrate how redundant robot manipulators can be incorporated into the design. An extension is also presented to illustrate how Lyapunov-based methods can be used to construct an adaptive controller for the camera-in-hand regulation problem. Experimental results are used to demonstrate the performance of the controllers for the fixed camera configuration.

4.4.1 System Model

Camera Model

The development in this section assumes that a camera is fixed above the robot workspace (see Figure 4.9) such that the image plane of the camera is parallel to the robot's plane of motion (i.e., the degrees of motion of the robot manipulator is constrained to be $n = 2$ for this application), and the camera can capture images throughout the entire robot workspace. Moreover, the fixed camera system is assumed to be modeled by the pinhole-lens model [6]. Specifically, by using the pinhole-lens model the task space position of the end-effector of a robot manipulator, denoted by $x(t)$ (see Section 4.3.1 for the definition of $x(t)$ and for the forward kinematic model of the robot manipulator), can be written in terms of image space coordinates, denoted by $y(t) = \begin{bmatrix} y_1(t) & y_2(t) \end{bmatrix} \in \mathbb{R}^2$, as follows

$$y = BR \left(x - \begin{bmatrix} O_{o1} \\ O_{o2} \end{bmatrix} \right) + \begin{bmatrix} O_{i1} \\ O_{i2} \end{bmatrix}. \tag{4.156}$$

In (4.156), $[O_{i1}, O_{i2}]^T \in \mathbb{R}^2$ denotes the image center that is defined as the frame buffer coordinates of the intersection of the optical axis with the image plane (see [38] for more details), $[O_{o1}, O_{o2}]^T \in \mathbb{R}^2$ denotes the projection of the camera's optical center on the task space plane, $B \in \mathbb{R}^{2 \times 2}$ is a constant positive-definite diagonal scaling matrix defined as follows

$$B = \frac{\lambda}{z} \begin{bmatrix} \beta_1 & 0 \\ 0 & \beta_2 \end{bmatrix}, \tag{4.157}$$

and $R(\theta) \in SO(2)$ is a constant rotation matrix defined as follows

$$R = \begin{bmatrix} \cos(\theta) & -\sin(\theta) \\ \sin(\theta) & \cos(\theta) \end{bmatrix} \tag{4.158}$$

where $\theta \in \mathbb{R}$ represents the constant clockwise rotation angle of the camera coordinate system with respect to the task space coordinate system that is assumed to satisfy the following inequalities

$$-90° < \theta < 90°. \tag{4.159}$$

In (4.157), $\lambda \in \mathbb{R}$ represents the camera's constant focal length, $z \in \mathbb{R}$ is the positive constant distance from the optical center of the camera to the robot's task space plane along the normal (i.e., the robot motion is assumed to be confined to a plane that is parallel to the image plane), and $\beta_1, \beta_2 \in \mathbb{R}$ represent the camera scale factors (in [pixels/m]) along each image space axis, respectively.

To facilitate the subsequent development and stability analysis, (4.156) is written in the following compact form

$$y = BRx + p \tag{4.160}$$

where $p \in \mathbb{R}^2$ is a constant vector defined as follows

$$p = \begin{bmatrix} O_{i1} \\ O_{i2} \end{bmatrix} - B R \begin{bmatrix} O_{o1} \\ O_{o2} \end{bmatrix}. \tag{4.161}$$

Based on the definitions given in (4.157) and (4.158), respectively, B and $R(\theta)$ are invertible. Hence, the inverse of the image space to task space relationship given in (4.160) can be developed as follows

$$x(t) = R^{-1} B^{-1} (y(t) - p). \tag{4.162}$$

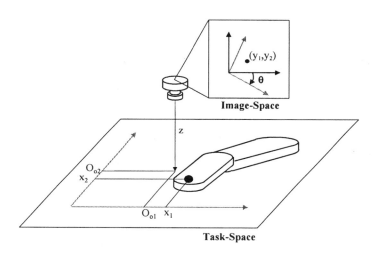

FIGURE 4.9. Robot and camera configuration.

Dynamic Model

The dynamic model for a two-rigid link revolute direct-drive planar robot manipulator is assumed to have the following form [50]

$$M\ddot{q} + V_m \dot{q} + G + F = \tau \tag{4.163}$$

where $q(t)$, $\dot{q}(t)$, $\ddot{q}(t) \in \mathbb{R}^2$ denote the link position, velocity, and acceleration vectors, respectively, $M(q) \in \mathbb{R}^{2 \times 2}$ represents the link inertia matrix, $V_m(q, \dot{q}) \in \mathbb{R}^{2 \times 2}$ represents centripetal-Coriolis matrix, $G(q) \in \mathbb{R}^2$ represents the gravity effects, $F(\dot{q}) \in \mathbb{R}^2$ represents the friction effects, and $\tau(t) \in \mathbb{R}^2$ represents the torque input vector. Based on the fact that the pinhole-lens model relates the image space to the task space, one is motivated to express the dynamic model given in (4.163) in terms of the task space. To this end, we premultiply (4.163) by the transpose of the inverse Jacobian[8] given in (4.61) and utilize the kinematic relationship given in (4.57) to develop the following task space dynamic model [50]

$$M^* \ddot{x} + V_m^* \dot{x} + G^* + F^* = \tau^* \qquad (4.164)$$

where $x(t)$ is defined in (4.56), $M^*(x) \in \mathbb{R}^{2 \times 2}$, $V_m^*(x, \dot{x}) \in \mathbb{R}^{2 \times 2}$, $G^*(x) \in \mathbb{R}^2$, $F^*(x, \dot{x}) \in \mathbb{R}^2$, and $\tau^*(t) \in \mathbb{R}^2$ are defined as follows

$$M^* = J^{-T} M \, J^{-1} \qquad V_m^* = J^{-T}(V_m - MJ^{-1}\dot{J}) \, J^{-1} \qquad (4.165)$$

$$G^* = J^{-T}G \qquad F^* = J^{-T}F, \qquad \tau^* = J^{-T}\tau,$$

and $J(q)$ was defined in (4.58) (where $n = 2$ for this application).

Property 4.9: Symmetric and Positive-Definite Inertia Matrix

The inertia matrix is symmetric positive-definite and satisfies the following inequalities

$$m_1 \|\xi\|^2 \leq \xi^T M^*(x) \xi \leq m_2 \|\xi\|^2 \qquad \forall \xi \in \mathbb{R}^n \qquad (4.166)$$

where m_1, $m_2 \in \mathbb{R}$ are known positive constants, and $\|\cdot\|$ denotes the standard Euclidean norm.

Property 4.10: Skew-Symmetry

The inertia and centripetal-Coriolis matrices satisfy the following skew-symmetric relationship

$$\xi^T \left(\frac{1}{2} \dot{M}^*(x) - V_m^*(x, \dot{x}) \right) \xi = 0 \qquad \forall \xi \in \mathbb{R}^n \qquad (4.167)$$

where $\dot{M}(q)$ denotes the time derivative of the inertia matrix.

Property 4.11: Linearity in the Parameters

The left-hand side of (4.164) can be linearly parameterized as shown below

$$M^* \ddot{x} + V_m^* \dot{x} + G^* + F^* = Y\phi \qquad (4.168)$$

where $\phi \in \mathbb{R}^m$ contains the constant system parameters, and the regression matrix $Y(x, \dot{x}, \ddot{x}) \in \mathbb{R}^{2 \times m}$ is assumed to be bounded provided that $x(t)$, $\dot{x}(t)$, $\ddot{x}(t) \in \mathcal{L}_\infty$.

---

[8] As in the previous sections, we assume that kinematic singularities of the robot manipulator are always avoided (i.e., $J^{-1}(q)$ always exists).

4.4.2 Control Objective

The control objective in this section is to enable a robot end-effector to track a desired time-varying image space trajectory despite the fact that the camera is uncalibrated (i.e., the constant parameters in (4.157) and (4.158) are assumed to be unknown). To quantify the tracking control objective, an image space end-effector position tracking error $e(t) = \begin{bmatrix} e_1(t) & e_2(t) \end{bmatrix} \in \mathbb{R}^2$ is defined as follows

$$e(t) = y_d(t) - y(t) \qquad (4.169)$$

where $y_d(t) \in \mathbb{R}^2$ denotes the desired image space trajectory[9] and $y(t)$ is defined in (4.160). By utilizing (4.156), the image space end-effector position tracking error can also be defined in terms of the mismatch between the actual and desired task space end-effector trajectory as follows

$$e(t) = BR\left(x_d(t) - x(t)\right) \qquad (4.170)$$

where the desired task space trajectory, denoted by $x_d(t) \in \mathbb{R}^2$, can be formulated from the desired image space trajectory by using the inverse camera model given in (4.162) as follows

$$x_d = R^{-1}B^{-1}(y_d - p) \qquad (4.171)$$

where $x_d(t)$ is assumed to be always located in the robot workspace and that $y(t)$ is always in the image plane of the camera. Based on the expressions given in (4.169) and (4.170), if the image space position of the robot manipulator is proven to track the desired image space trajectory (i.e., if $\lim_{t \to \infty} e(t) = 0$), then the task space position of the manipulator can be proven to track the desired task space trajectory (i.e., $\lim_{t \to \infty} x(t) = x_d(t)$). That is, convergence of the end-effector position tracking error in the image space will result in convergence of the end-effector position tracking error in the task space. Since the desired trajectory is being generated via an uncalibrated camera (i.e., the parameters of the matrices B and $R(\theta)$ are unknown), the expression in (4.171) indicates that the desired task space trajectory is unknown. Hence, the measurable image space tracking error expression given in (4.169) will be utilized in the subsequent control development. That is, the camera system will be used for the dual purpose of generating the desired trajectory and as a feedback signal in the subsequent closed-loop controller. As in previous sections, to facilitate the control development, a filtered tracking error, denoted by

---

[9]We make the standard assumption that the desired image-space trajectory of the robot end-effector is constructed such that $y_d(t), \dot{y}_d(t), \ddot{y}_d(t) \in \mathcal{L}_\infty$.

$r(t) = \begin{bmatrix} r_1(t) & r_2(t) \end{bmatrix} \in \mathbb{R}^2$, is defined as follows

$$r(t) = \dot{e}(t) + \alpha e(t) \tag{4.172}$$

where $\alpha \in \mathbb{R}^{2 \times 2}$ is a diagonal positive-definite control gain matrix.

4.4.3 Adaptive Control Development

In this section, two adaptive control laws are developed that ensure global asymptotic tracking control. Specifically, the first adaptive control design compensates for parametric uncertainty associated with the camera calibration (i.e., the constant parameters in (4.157) and (4.158)), while the parameters of the dynamic model are assumed to be exactly known. The second controller compensates for parametric uncertainty throughout the entire camera and robot system. Each of these controllers "close-the-loop" in the image space; hence, the actual control inputs to the robot actuators must be computed according to (4.165). Based on the fact that $x(t)$, $\dot{x}(t)$, $\ddot{x}(t)$ cannot be directly measured due to the parametric uncertainty in the camera calibration parameters, the link positions are directly measured from encoders at each joint. Specifically, the subsequent control laws require $y(t)$, $\dot{y}(t)$, $q(t)$, and $\dot{q}(t)$ to be measurable.

Adaptive Control Example 1

In this example, a solution to the so-called adaptive camera calibration problem [6] is developed. That is, our aim is to design a controller that provides link position tracking and compensates for all of the unknown camera calibration parameters. In this first example, the parameters associated with the robot dynamics (i.e., inertia, mass, and friction) are assumed to be exactly known. In the subsequent example, the control design will target end-effector position tracking despite parametric uncertainty in the camera model and the robot dynamic model.

Under the assumption that exact model knowledge of the robot dynamics is available, a feedback linearizing approach can be employed to directly cancel the nonlinear dynamic effects (i.e., dynamic inversion [49]). To this end, we take the second time derivative of (4.162) as follows

$$A\ddot{y} = \ddot{x} \tag{4.173}$$

where the constant matrix $A \in \mathbb{R}^{2 \times 2}$ is defined as follows

$$A = \begin{bmatrix} A_1 & A_2 \\ A_3 & A_4 \end{bmatrix} = R^{-1}B^{-1} = \begin{bmatrix} \dfrac{z\cos(\theta)}{\lambda\beta_1} & \dfrac{z\sin(\theta)}{\lambda\beta_2} \\ -\dfrac{z\sin(\theta)}{\lambda\beta_1} & \dfrac{z\cos(\theta)}{\lambda\beta_2} \end{bmatrix} \tag{4.174}$$

where B and $R(\theta)$ are defined in (4.157) and (4.158), respectively. After substituting (4.164) into (4.173) for $\ddot{x}(t)$ the following expression is obtained

$$A\ddot{y} = (M^*)^{-1}(\tau^* - V_m^*\dot{x} - G^* - F^*). \qquad (4.175)$$

Based on the form of (4.175), the following feedback linearizing control input can be designed

$$\tau^* = M^*u + V_m^*\dot{x} + G^* + F^* \qquad (4.176)$$

where $u(t) = \begin{bmatrix} u_1(t), & u_2(T) \end{bmatrix}^T \in \mathbb{R}^2$ is a subsequently designed control term. After substituting (4.176) into (4.175) for $\tau^*(t)$, the following simplified open-loop dynamic system is obtained

$$A\ddot{y}(t) = u. \qquad (4.177)$$

To develop the open-loop error system for $r(t)$ of (4.172), we take the time derivative of (4.172), premultiply the resulting expression by A, and then substitute (4.177) for $A\ddot{y}(t)$ to obtain the following expression

$$A\dot{r} = A(\ddot{y}_d + \alpha\dot{e}) - u. \qquad (4.178)$$

Motivated by the subsequent stability analysis to ensure that the matrix that premultiplies $\dot{r}(t)$ in (4.178) is positive-definite, a constant upper-triangular transformation matrix, denoted by $T \in \mathbb{R}^{2\times2}$, is defined as follows

$$T = \begin{bmatrix} \dfrac{1}{\det(A)} & \dfrac{A_3}{A_4} - \dfrac{A_2}{A_4\det(A)} \\ 0 & 1 \end{bmatrix}. \qquad (4.179)$$

After premultiplying (4.178) by T, the following expression is obtained

$$Z\dot{r} = Z(\ddot{y}_d + \alpha\dot{e}) - Tu \qquad (4.180)$$

where the constant symmetric matrix $Z \in \mathbb{R}^{2\times2}$ is defined as

$$Z = TA = \begin{bmatrix} \dfrac{1+A_3^2}{A_4} & A_3 \\ A_3 & A_4 \end{bmatrix}. \qquad (4.181)$$

Based on the structure of (4.181), the expressions in (4.159) and (4.174) can be used to prove that Z is positive-definite.

To facilitate the adaptive camera calibration design, (4.180) is linear parameterized as follows

$$Z\dot{r} = \begin{bmatrix} \phi_3^{-1}(Y_{c1}\phi_1 - u_1) \\ Y_{c2}\phi_2 - u_2 \end{bmatrix} \qquad (4.182)$$

where $Y_{c1}(\dot{e}, u_2, t) \in \mathbb{R}^{1 \times 3}$ and $Y_{c2}(\dot{e}, t) \in \mathbb{R}^{1 \times 2}$ are known regression matrices defined as

$$Y_{c1} = \begin{bmatrix} \ddot{y}_{d1} + \alpha_1 \dot{e}_1 & \ddot{y}_{d2} + \alpha_2 \dot{e}_2 & -u_2 \end{bmatrix}$$

$$Y_{c2} = \begin{bmatrix} \ddot{y}_{d1} + \alpha_1 \dot{e}_1 & \ddot{y}_{d2} + \alpha_2 \dot{e}_2 \end{bmatrix}$$

(4.183)

and $\phi_1 \in \mathbb{R}^3$, $\phi_2 \in \mathbb{R}^2$, and $\phi_3 \in \mathbb{R}$ represent unknown constant parameter vectors that are defined as follows

$$\phi_1 = \begin{bmatrix} \dfrac{(1 + A_3^2)\det(A)}{A_4} & A_3\det(A) & \dfrac{A_3\det(A) - A_2}{A_4} \end{bmatrix}^T$$

(4.184)

$$\phi_2 = \begin{bmatrix} A_3 & A_4 \end{bmatrix}^T \qquad \phi_3 = \det(A).$$

Based on the open-loop dynamics of (4.182) and the subsequent stability analysis, $u(t)$ is designed as follows

$$u_1 = Y_{c1}\hat{\phi}_1 + k_{c1}r_1 \qquad u_2 = Y_{c2}\hat{\phi}_2 + k_{c2}r_2 \qquad (4.185)$$

where k_{c1}, $k_{c2} \in \mathbb{R}$ are positive control gains. In (4.185), $\hat{\phi}_1(t) \in \mathbb{R}^3$ and $\hat{\phi}_2(t) \in \mathbb{R}^2$ denote dynamic parameter estimates that are defined by the following gradient update laws

$$\dot{\hat{\phi}}_1 = \Gamma_{c1} Y_{c1}^T r_1 \qquad \dot{\hat{\phi}}_2 = \Gamma_{c2} Y_{c2}^T r_2 \qquad (4.186)$$

where $\Gamma_{c1} \in \mathbb{R}^{3 \times 3}$ and $\Gamma_{c2} \in \mathbb{R}^{2 \times 2}$ are diagonal positive-definite gain matrices. After substituting (4.185) into (4.182), the following closed-loop dynamics for $r(t)$ can be obtained

$$Z\dot{r} = \begin{bmatrix} \phi_3^{-1} Y_{c1}\tilde{\phi}_1 - \phi_3^{-1} k_{c1} r_1 \\ Y_{c2}\tilde{\phi}_2 - k_{c2} r_2 \end{bmatrix}$$

(4.187)

where the parameter estimation error vectors $\tilde{\phi}_1(t) \in \mathbb{R}^3$ and $\tilde{\phi}_2(t) \in \mathbb{R}^2$ are defined as follows

$$\tilde{\phi}_i = \phi_i - \hat{\phi}_i \qquad \forall i = 1, 2. \qquad (4.188)$$

The stability of the adaptive camera calibration controller can now be examined by using the following theorem.

Theorem 4.5 *The adaptive camera calibration controller given in (4.176), (4.185), and (4.186) ensure global asymptotic end-effector position tracking in the sense that*

$$\lim_{t \to \infty} e(t) = 0 \qquad (4.189)$$

provided the condition given in (4.159) is satisfied.

Proof: To prove Theorem 4.5, we define a nonnegative function $V(t) \in \mathbb{R}$ as follows

$$V = \frac{1}{2}r^T Z r + \frac{1}{2}\phi_3^{-1}\tilde{\phi}_1^T \Gamma_{c1}^{-1}\tilde{\phi}_1 + \frac{1}{2}\tilde{\phi}_2^T \Gamma_{c2}^{-1}\tilde{\phi}_2. \tag{4.190}$$

After differentiating (4.190), and then substituting (4.187) and the time derivative of (4.188) into the resulting expression, the following expression can be obtained

$$\dot{V} = -\phi_3^{-1}k_{c1}r_1^2 - k_{c2}r_2^2 \leq -\min\left\{\phi_{c3}^{-1}k_{c1}, k_{c2}\right\}\|r\|^2 \tag{4.191}$$

where (4.186) was utilized. Due to the structure of (4.190) and (4.191), $r(t), \tilde{\phi}_1(t), \tilde{\phi}_2(t) \in \mathcal{L}_\infty$ and $r(t) \in \mathcal{L}_2$ can be proved. Based on the fact that $r(t), \tilde{\phi}_1(t), \tilde{\phi}_2(t) \in \mathcal{L}_\infty$, the closed-loop error systems (i.e., standard signal chasing arguments) can be used to show that all system and controller signals are bounded. Having proved that $r(t) \in \mathcal{L}_\infty \cap \mathcal{L}_2$ and $\dot{r}(t) \in \mathcal{L}_\infty$, Barbalat's Lemma (see Lemma A.16 of Appendix A) can be used along with Lemma A.15 of Appendix A to prove the result given in (4.189).   □

Adaptive Control Example 2

The adaptive calibration controller in the previous example required exact knowledge of the mechanical parameters to provide asymptotic position tracking. In this example, an adaptive backstepping control law is designed to take into account parametric uncertainty throughout the entire system. To this end, we take the time derivative of (4.169) and premultiply the resulting expression by A as follows

$$A\dot{e} = A\dot{y}_d - v + \eta \tag{4.192}$$

where (4.174) and the time derivative of (4.162) were utilized, and the auxiliary tracking error signal $\eta(t) \in \mathbb{R}^2$ is defined as follows

$$\eta = v - \dot{x} \tag{4.193}$$

where $v(t) = \begin{bmatrix} v_1(t) & v_2(t) \end{bmatrix}^T \in \mathbb{R}^2$ is a subsequently designed control signal. After premultiplying (4.192) by the transformation matrix T defined in (4.179), the following expression is obtained

$$Z\dot{e} = Z\dot{y}_d - Tv + T\eta \tag{4.194}$$

where the matrix Z was defined in (4.181). In a similar manner as in the previous section, the open-loop dynamics of (4.194) can be rewritten in the following advantageous form

$$Z\dot{e} = \begin{bmatrix} \phi_3^{-1}(Y_1\phi_1 - v_1) \\ Y_2\phi_2 - v_2 \end{bmatrix} + T\eta \tag{4.195}$$

where ϕ_1, ϕ_2, and ϕ_3 were defined in (4.184), and $Y_1(v_2, t) \in \mathbb{R}^{1\times3}$ and $Y_2(t) \in \mathbb{R}^{1\times2}$ are known regression matrices defined as follows

$$Y_1 = \begin{bmatrix} \dot{y}_{d1} & \dot{y}_{d2} & -v_2 \end{bmatrix} \qquad Y_2 = \begin{bmatrix} \dot{y}_{d1} & \dot{y}_{d2} \end{bmatrix}. \tag{4.196}$$

Based on the form of (4.195) and the subsequent stability analysis, the control input $v(t)$ is designed as follows

$$v_1 = Y_1 \hat{\phi}_1 + k_1 e_1 \qquad v_2 = Y_2 \hat{\phi}_2 + k_2 e_2 \tag{4.197}$$

where k_1, $k_2 \in \mathbb{R}$ are positive constant control gains, and $\hat{\phi}_1(t)$, $\hat{\phi}_2(t)$ of (4.186) are redesigned in this example as follows

$$\dot{\hat{\phi}}_1 = \Gamma_1 Y_1^T e_1 \qquad \dot{\hat{\phi}}_2 = \Gamma_2 Y_2^T e_2 \tag{4.198}$$

where $\Gamma_1 \in \mathbb{R}^{3\times3}$ and $\Gamma_2 \in \mathbb{R}^{2\times2}$ are diagonal positive-definite gain matrices. After substituting (4.197) into (4.195) for $v(t)$, the closed-loop dynamics for $e(t)$ are obtained as follows

$$Z\dot{e} = \begin{bmatrix} \phi_3^{-1}\left(Y_1 \tilde{\phi}_1 - k_1 e_1\right) \\ Y_2 \tilde{\phi}_2 - k_2 e_2 \end{bmatrix} + T\eta. \tag{4.199}$$

To develop the open-loop dynamics for $\eta(t)$, we take the time derivative of (4.193), premultiply the resulting equation by $M^*(x)$, and then make use of (4.164) and (4.193) to obtain the following expression

$$M^*\dot{\eta} = -V_m^*\eta + \Psi - \tau^* - T^T e \tag{4.200}$$

where $\Psi(x, \dot{x}, v, \dot{v}, e) \in \mathbb{R}^2$ is defined as follows

$$\Psi = M^*\dot{v} + V_m^* v + G^* + F^* + T^T e \tag{4.201}$$

and $T^T e$ was added and subtracted to (4.200) to cancel the interconnection term $T\eta(t)$ of (4.199) during the subsequent stability analysis. By exploiting Property 4.11 and the structure of T given in (4.179), the following expression can be developed

$$\Psi = Y_n \phi_n + \Lambda \tag{4.202}$$

where $\Lambda(e) \in \mathbb{R}^2$ and the linear parametrization $Y_n(x, \dot{x}, v, \dot{v}, e)\phi_n \in \mathbb{R}^2$ are defined as follows

$$\Lambda = \begin{bmatrix} 0 & e_2 \end{bmatrix}^T \tag{4.203}$$

$$Y_n\phi_n = M^* \begin{bmatrix} \dot{Y}_1\hat{\phi}_1 + Y_1\Gamma_1Y_1^T e_1 + k_1\dot{e}_1 \\ \dot{Y}_2\hat{\phi}_2 + Y_2\Gamma_2Y_2^T e_2 + k_2\dot{e}_2 \end{bmatrix} + V_m^* \begin{bmatrix} Y_1\hat{\phi}_1 + k_1 e_1 \\ Y_2\hat{\phi}_2 + k_2 e_2 \end{bmatrix}$$

$$+ G^* + F^* + \begin{bmatrix} \dfrac{1}{\det(A)} e_1 \\ \left(\dfrac{A_3}{A_4} - \dfrac{A_2}{A_4 \det(A)} \right) e_1 \end{bmatrix}.$$

$$(4.204)$$

In (4.204), $Y_n(x, \dot{x}, v, \dot{v}, e) \in \mathbb{R}^{2 \times p}$ denotes a known regression matrix, $\phi_n \in \mathbb{R}^p$ denotes a vector containing unknown constant parameters associated with the camera and robot, and the time derivative of $Y_1(v_2, t)$ and $Y_2(t)$ of (4.196) are given by the following expressions

$$\dot{Y}_1 = \begin{bmatrix} \ddot{y}_{d1} & \ddot{y}_{d2} & -\dot{v}_2 \end{bmatrix} \qquad \dot{Y}_2 = \begin{bmatrix} \ddot{y}_{d1} & \ddot{y}_{d2} \end{bmatrix}. \qquad (4.205)$$

After substituting (4.202) into (4.200) for $\Psi(x, \dot{x}, v, \dot{v}, e)$, the following open-loop error system can be obtained for $\eta(t)$

$$M^*\dot{\eta} = -V_m^*\eta + Y_n\phi_n + \Lambda - T^T e - \tau^*. \qquad (4.206)$$

Based on the structure of (4.206) and the subsequent stability analysis, the control input $\tau^*(t)$ is designed as follows

$$\tau^* = \Lambda + Y_n\hat{\phi}_n + K_n\eta \qquad (4.207)$$

where $K_n \in \mathbb{R}^{2 \times 2}$ is a diagonal positive-definite control gain matrix and $\hat{\phi}_n(t) \in \mathbb{R}^p$ is a dynamic parameter estimate that is defined by the following gradient update law

$$\dot{\hat{\phi}}_n = \Gamma_n Y_n^T \eta \qquad (4.208)$$

where $\Gamma_n \in \mathbb{R}^{p \times p}$ is a diagonal positive-definite gain matrix. After substituting (4.207) into (4.206) for $\tau^*(t)$, the following closed-loop dynamics for $\eta(t)$ are obtained

$$M^*\dot{\eta} = -V_m^*\eta + Y_n\tilde{\phi}_n - K_n\eta - T^T e \qquad (4.209)$$

where the parameter estimation error $\tilde{\phi}_n(t) \in \mathbb{R}^p$ is defined as follows

$$\tilde{\phi}_n = \phi_n - \hat{\phi}_n. \qquad (4.210)$$

The stability of the adaptive controller in (4.197), (4.198), (4.207), and (4.208) can now be examined by using the following theorem.

Theorem 4.6 *The adaptive control law given in (4.197), (4.198), (4.207), and (4.208) ensures global asymptotic end-effector position tracking in the sense that*

$$\lim_{t \to \infty} e(t) = 0, \qquad (4.211)$$

provided the condition given in (4.159) is satisfied.

Proof: To prove Theorem 4.6, we define a nonnegative function $V(t) \in \mathbb{R}$ as follows

$$V = \frac{1}{2}e^T Z e + \frac{1}{2}\eta^T M^* \eta + \frac{1}{2}\phi_3^{-1}\tilde{\phi}_1^T \Gamma_1^{-1}\tilde{\phi}_1$$

$$+ \frac{1}{2}\tilde{\phi}_2^T \Gamma_2^{-1}\tilde{\phi}_2 + \frac{1}{2}\tilde{\phi}_n^T \Gamma_n^{-1}\tilde{\phi}_n. \qquad (4.212)$$

After taking the time derivative of (4.212), substituting (4.199) and (4.209) into the resulting expression for $Z\dot{e}(t)$ and $M^*(x)\dot{\eta}(t)$, respectively, substituting the adaptive update laws given in (4.198) and (4.208) for $\dot{\hat{\phi}}_1(t)$, $\dot{\hat{\phi}}_2(t)$, and $\dot{\hat{\phi}}_n(t)$, and then utilizing the skew-symmetry property in (4.167), the following expression can be obtained

$$\dot{V} = -\phi_3^{-1}k_1 e_1^2 - k_2 e_2^2 - \eta^T K_n \eta$$

$$\leq -\min\left\{\phi_3^{-1}k_1, k_2\right\} \|e\|^2 - \lambda_{\min}\left\{K_n\right\} \|\eta\|^2. \qquad (4.213)$$

The result given in (4.211) and boundedness of all the signals under closed-loop operation can be shown by utilizing (4.212), (4.213), and similar arguments as in the proof of Theorem 4.5. □

4.4.4   Redundant Robot Extension

Redundant manipulators provide increased flexibility and dexterity for the execution of complex tasks. The extra degrees of freedom of the redundant manipulator not only increase the workspace area, but also enable the execution of sub-control tasks. In this section, the adaptive controller of Section 4.4.3 is modified for redundant revolute planar manipulators (i.e., the restriction that $n = 2$ is removed). Given that the robot motion is limited to two degrees of freedom and the manipulator has n-joints, the manipulator Jacobian defined in (4.58) will not be square (i.e., $J(q) \in \mathbb{R}^{2 \times n}$). Since the nonsquare manipulator Jacobian is not invertible, a pseudo-inverse of $J(q)$, denoted by $J^+(q) \in \mathbb{R}^{n \times m}$, can be defined as follows

$$J^+ = J^T \left(JJ^T\right)^{-1} \qquad (4.214)$$

where $J^+(q)$ satisfies the following equality

$$JJ^+ = I_6. \tag{4.215}$$

As shown in [64], the pseudo-inverse defined by (4.214) satisfies the Moore-Penrose Conditions given below

$$
\begin{aligned}
JJ^+J &= J & J^+JJ^+ &= J^+ \\
(J^+J)^T &= J^+J & (JJ^+)^T &= JJ^+
\end{aligned} \tag{4.216}
$$

and the matrix $I_n - J^+J$, which projects vectors onto the null space of $J(q)$, satisfies the following properties

$$
\begin{aligned}
(I_n - J^+J)(I_n - J^+J) &= I_n - J^+J & J(I_n - J^+J) &= 0 \\
(I_n - J^+J)^T &= (I_n - J^+J) & (I_n - J^+J)J^+ &= 0
\end{aligned} \tag{4.217}
$$

where the minimum singular value of $J(q)$ is assumed to be greater than a known small positive constant $\delta > 0$, such that $\sup_q \{\|J^+(q)\|\}$ is known a priori, and hence, all kinematic singularities are always avoided.

Following a similar control design strategy as delineated in Section 4.4.3, the closed-loop dynamics for $e(t)$ given in (4.199) can be developed as follows

$$
Z\dot{e} = \begin{bmatrix} \phi_3^{-1}\left(W_1\tilde{\phi}_1 - k_1e_1\right) \\ W_2\tilde{\phi}_2 - k_2e_2 \end{bmatrix} + T(v - J\dot{q}) \tag{4.218}
$$

where (4.57) and (4.193) have been utilized. After using (4.215) and (4.217), the following expression can be obtained

$$
Z\dot{e} = \begin{bmatrix} \phi_3^{-1}\left(W_1\tilde{\phi}_1 - k_1e_1\right) \\ W_2\tilde{\phi}_2 - k_2e_2 \end{bmatrix} + TJ\varrho \tag{4.219}
$$

where $\varrho(t) \in \mathbb{R}^n$ is defined as follows

$$
\varrho = J^+v + \left(I_n - J^+J\right)g - \dot{q} \tag{4.220}
$$

where $I_n \in \mathbb{R}^{n \times n}$ denotes the $n \times n$ identity matrix and $g(t) \in \mathbb{R}^n$ denotes a bounded design vector that is used to control the self-motion of the redundant manipulator to accomplish sub-tasks. The open-loop error dynamics for $\varrho(t)$ can be obtained by taking the time derivative of (4.220) and pre-multiplying the resultant equation by the inertia matrix $M(q)$ of (4.163) as follows

$$
M\dot{\varrho} = -V_m\varrho + Y_r\phi_r - J^TT^Te - \tau \tag{4.221}
$$

where the linear parameterization $Y_r(q, \dot{q}, v, \dot{v}, e)\phi_r \in \mathbb{R}^2$ is defined as follows

$$Y_r\phi_r = \begin{aligned}[t] &M\frac{d}{dt}\{J^+v + (I_n - J^+J)g\} + J^TT^Te \\ &+V_m\{J^+v + (I_n - J^+J)g\} + G + F \end{aligned} \tag{4.222}$$

where $Y_r(\cdot) \in \mathbb{R}^{2 \times p}$ is a known regression matrix, and $\phi_r \in \mathbb{R}^p$ is a vector containing unknown, constant parameters associated with the camera system and mechanical dynamics. Based on the structure of (4.221), the joint space control input $\tau(t)$ can be designed in a similar fashion as in Section 4.4.3. By following steps similar to that used in the proof of Theorem 4.6, global asymptotic position tracking and global asymptotic tracking of the auxiliary variable $\varrho(t)$ (i.e., $\lim_{t \to \infty} e(t)$, $\varrho(t) = 0$) can be proven. Moreover, given that $\lim_{t \to \infty} \varrho(t) = 0$, the expression in (4.216) can be used to show that $\dot{q}(t)$ converges to $g(t)$ in the null space of $J(q)$ as shown below

$$\lim_{t \to \infty} \varrho = 0 \Rightarrow \lim_{t \to \infty} (I_n - J^+J)(g - \dot{q}) = 0. \tag{4.223}$$

Thus, the vector $g(t)$ can be used to control the self-motion of the redundant manipulator to perform subtasks [30] (e.g., obstacle avoidance, joint-limit avoidance, to maximize manipulability of the robot, to minimize energy consumption).

Remark 4.6 *A stability proof for the redundant robot extension cannot be formulated to prove that $q(t) \in \mathcal{L}_\infty$. However, the remaining signals and the control law can be proven to be bounded (i.e., all of the occurrences of $q(t)$ are within sinusoidal terms for revolute robots). For revolute redundant robots, the lack of a proof for the boundedness of $q(t)$ is typically not a concern since the self-motion effects can cause the actuators to continually turn.*

4.4.5   Camera-in-Hand Extension

The control laws developed in Section 4.4.3 are based on a fixed camera configuration. However, for some applications, the camera may be mounted directly on the end-effector of the robot (i.e., the camera-in-hand configuration) as illustrated in Figure 4.10. To address these applications, the control objective in this section is to regulate the task space position of a robot end-effector (and hence, the camera) so that the task space projection of the camera's optical center is equal to a constant desired setpoint. Given that the camera is now assumed to be mounted on the robot end-effector, the relationship between the task space position of the end-effector and the

image space coordinates of an object given in (4.156) is modified as follows
[41]

$$y = BR\left(x - x_d\right) \tag{4.224}$$

where $x_d \in \mathbb{R}^2$ denotes the constant desired task space setpoint. For the
fixed camera configuration, the rotation matrix $R\left(\cdot\right)$ was defined as a function of the unknown constant orientation of the camera. For the camera-in-hand configuration, the camera orientation, denoted by $\theta_2(t) \in \mathbb{R}$, will
change as a function of the robot link positions as follows

$$\theta_2 = \theta + \sum_{i=1}^{2} q_i \tag{4.225}$$

where θ is the unknown constant orientation of the camera given in (4.158),
and $q_i(t) \in \mathbb{R}$ denotes the ith link position. To facilitate the subsequent
control design, the camera-in-hand rotation matrix can be divided into a
unknown constant rotation matrix multiplied by a measurable time-varying
rotation matrix as follows

$$R(\theta_2) = R\left(\theta + \sum_{i=1}^{2} q_i\right) = R(\theta)R\left(\sum_{i=1}^{2} q_i\right). \tag{4.226}$$

In the subsequent development, the rotation matrices given in (4.226) are
distinguished by the following notation

$$R = R(\theta) \qquad R_k = R\left(\sum_{i=1}^{2} q_i\right). \tag{4.227}$$

Remark 4.7 *As with the fixed camera configuration, the following assumptions are made: (i) x_d is always located in the robot workspace, (ii) the
camera's image plane is parallel to the object plane, (iii) the camera can
capture images throughout the entire robot workspace, and (iv) kinematic
singularities are always avoided (i.e., $J^{-1}(q) \in \mathbb{R}^{2\times 2}$ always exists).*

Control Design

To quantify the control objective for the camera-in-hand problem, the task
space position error, denoted by $e_x(t) \in \mathbb{R}^2$, is defined as follows

$$e_x = x_d - x. \tag{4.228}$$

After taking the time derivative of (4.228) and utilizing the forward kinematic relationship given in (4.57), the following expression can be obtained

$$\dot{e}_x = J\eta_x - Ju \tag{4.229}$$

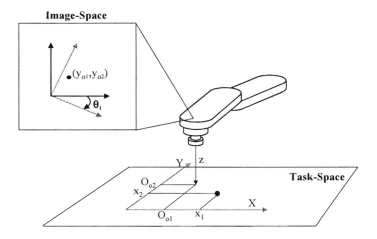

FIGURE 4.10. Camera-in-hand configuration.

where $\eta_x(t) \in \mathbb{R}^2$ is defined as follows

$$\eta_x = u - \dot{q}, \tag{4.230}$$

and $u(t) \in \mathbb{R}^2$ represents the kinematic control input. Based on the structure of (4.229) and the subsequent stability analysis, $u(t)$ is designed as follows

$$u = -J^{-1}k_c R_k^T(q)y \tag{4.231}$$

where $k_c \in \mathbb{R}$ is a positive control gain. After utilizing (4.224), (4.227), and (4.231), the following closed-loop error dynamics can be obtained

$$\dot{e}_x = -k_c R_k^T(q)BRR_k(q)e_x + J\eta_x. \tag{4.232}$$

To develop the open-loop dynamics for $\eta_x(t)$, we take the time derivative of (4.230), premultiply the resulting equation by $M(q)$, and then use (4.163) and (4.193) to obtain the following expression

$$M\dot{\eta}_x = -V_m\eta_x + Y_x\phi_x - \tau \tag{4.233}$$

where $Y_x(y, \dot{y}, q, \dot{q}) \in \mathbb{R}^{2 \times p}$ denotes a known regression matrix, $\phi_x \in \mathbb{R}^p$ denotes the constant unknown system parameters, and the linear parametrization $Y_x(y, \dot{y}, q, \dot{q})\phi_x$ is defined as follows

$$Y_x\phi_x = M\dot{u} + V_m u + G + F. \tag{4.234}$$

Based on the form of (4.233) and the subsequent stability analysis, the control torque input can be defined as follows

$$\tau = Y_x\hat{\phi}_x + k_s\eta_x + J^T e_x \tag{4.235}$$

where $k_s \in \mathbb{R}$ is a positive control gain, and $\hat{\phi}_x(t) \in \mathbb{R}^p$ is a dynamic parameter estimate for ϕ_x that is defined by the following gradient update law

$$\dot{\hat{\phi}}_x = \Gamma_x Y_x^T \eta_x \qquad (4.236)$$

where $\Gamma_x \in \mathbb{R}^{p \times p}$ is a diagonal positive-definite gain matrix. After substituting (4.235) into (4.233) for $\tau(t)$, the following closed-loop error dynamics can be obtained

$$M\dot{\eta}_x = -V_m \eta_x - k_s \eta_x + Y_x \tilde{\phi}_x - J^T e_x \qquad (4.237)$$

where the parameter estimation error vector $\tilde{\phi}_x(t) \in \mathbb{R}^p$ is defined as follows

$$\tilde{\phi}_x = \phi_x - \hat{\phi}_x. \qquad (4.238)$$

Stability Analysis

The stability of the adaptive camera-in-hand controller given in (4.231), (4.235), and (4.236) can now be examined by the following theorem.

Theorem 4.7 *Provided that the camera-space parameters satisfy the following inequality*

$$\cos(\theta) > \left| \frac{\beta_1 - \beta_2}{\beta_1 + \beta_2} \right| \qquad (4.239)$$

where θ is given in (4.225) and β_1, β_2 are defined in (4.157), the adaptive camera-in-hand controller given in (4.231), (4.235), and (4.236) guarantees global asymptotic regulation in the sense that

$$\lim_{t \to \infty} e_x(t) = 0. \qquad (4.240)$$

Proof: To prove Theorem 4.7, we define a nonnegative function $V(t) \in \mathbb{R}$ as follows

$$V = \frac{1}{2} e_x^T e_x + \frac{1}{2} \eta_x^T M \eta_x + \frac{1}{2} \tilde{\phi}_x^T \Gamma_x^{-1} \tilde{\phi}_x. \qquad (4.241)$$

After taking the time derivative of (4.241), substituting (4.232) and (4.237) into the resulting expression for $\dot{e}_x(t)$ and $M(q)\dot{\eta}_x(t)$, respectively, utilizing the skew-symmetry property of (4.167), and then cancelling common terms, the following expression can be obtained

$$\dot{V} = -k_c e_x^T R_k^T(q) \left(\frac{BR + (BR)^T}{2} \right) R_k(q) e_x - \eta^T k_s \eta_x \qquad (4.242)$$

where the following property has been utilized

$$\zeta^T BR \zeta = \zeta^T \left(\frac{BR + (BR)^T}{2} \right) \zeta \qquad \forall \zeta \in \mathbb{R}^2. \qquad (4.243)$$

Note that if (4.239) is satisfied, the symmetric matrix $BR + (BR)^T$ will be positive-definite (see Lemma B.18 of Appendix B), and hence, Lemma A.12 of Appendix A can be invoked to prove that

$$\dot{V} \leq -k_c \lambda_{\min} \left\{ \left(\frac{BR + (BR)^T}{2} \right) \right\} \|e_x\|^2 - k_s \|\eta_x\|^2 \qquad (4.244)$$

where $\lambda_{\min} \{\cdot\}$ denotes the minimum eigenvalue of a matrix. Given (4.241) and (4.244), $e_x(t)$, $\eta_x(t)$, $\tilde{\phi}_x(t) \in \mathcal{L}_\infty$ and $e_x(t)$, $\eta_x(t) \in \mathcal{L}_2$ can be proved. Standard signal chasing arguments can be used to show that all system and controller signals are bounded. Having proved that $e_x(t) \in \mathcal{L}_\infty \cap \mathcal{L}_2$ and $\dot{e}_x(t) \in \mathcal{L}_\infty$, Barbalat's Lemma (see Lemma A.16 of Appendix A) can be invoked to prove the result given in (4.240).   □

Remark 4.8 *With respect to (4.239), note that*

$$0 \leq \left| \frac{\beta_1 - \beta_2}{\beta_1 + \beta_2} \right| < 1,$$

and that the condition on θ can be written as $\theta \in \{ -|\theta_c|, |\theta_c| \}$ where θ_c is determined from a given set of values for β_1 and β_2. If the camera's scale factors have the same value (i.e., $\beta_1 = \beta_2$), then $\theta_c = 90°$ as in the fixed camera configuration examples. As the difference between the values of β_1 and β_2 increases, the ratio $\left| \frac{\beta_1 - \beta_2}{\beta_1 + \beta_2} \right| \to 1$, and hence, $\theta_c \to 0°$. Since in most camera systems β_1 and β_2 have similar values (i.e., $\left| \frac{\beta_1 - \beta_2}{\beta_1 + \beta_2} \right| << 1$, and hence, $\theta_c >> 0°$), the condition given in (4.239) is not difficult to satisfy in practice.

4.4.6   Experimental Setup and Results

To illustrate the performance of the controllers proposed in Section 4.4.3, an experimental testbed (see Figure 4.11) was constructed that consisted of the following components: (i) the 2-link IMI revolute direct-drive manipulator described in Section 4.2.3, (ii) a Dalsa (CA-D6-0256W) MotionVision area scan digital camera that captures 955 frames per second with 8-bit gray scale at a 256×256 resolution, (iii) a Road Runner Model 24 video capture board, and (iv) two Pentium II-based PCs operating under the real-time operating system QNX. The camera was mounted 1.2 [m] above the robot workspace, and an LED was placed at the tip of the robot's second link to provide an easily detectable feature to indicate the end-effector position. One of the PCs (600 [MHz]) was connected to the camera system and was used to capture the image, extract the desired feature, and transmit the

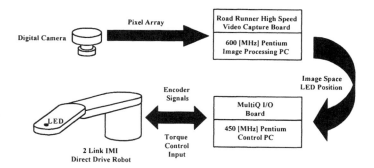

FIGURE 4.11. Schematic representation of the robot/camera experimental test-bed.

pixel coordinates to the shared memory of a second PC over a 100 [Mb/sec] dedicated Ethernet connection. A simple thresholding algorithm was used to detect the position of the LED in the camera frame.

The other PC (450 [MHz], operating under QNX) was connected to the robot manipulator and used to acquire the pixel coordinates provided by the first PC from a shared memory location and perform the real-time I/O and control computations. Link encoders were used to determine the link positions and an approximate link velocity was determined from the link position via a backward difference/filtering operation. The link position and velocity signals were utilized to calculate the task space position and velocity of the end-effector. The task space position and velocity signals are required for calculation of the feedforward terms in the control law associated with the robot dynamics (i.e., the inertia related terms, the friction terms, and the gravity terms). Data acquisition and control implementation were performed at 1 [kHz] via the Quanser MultiQ I/O board with custom interfacing circuitry. The control algorithms were executed using the real-time control environment Qmotor [16].

Since the output of the camera system is discrete, a low-pass filtering procedure was applied to the camera output signal to obtain a continuous differentiable signal for the control input.[10] Specifically, we used a second-order Butterworth filter with a 300 [Hz] cut-off frequency. By using

---

[10]The application of the filter on the pixel information means that "approximate" sub-pixel information is being used to close the control loop.

a bilinear transformation the following discretized filter was obtained

$$y(t) = \frac{Z_b - Z_a}{w^2 - 4w/t_s + 4/t_s^2} \tag{4.245}$$

where

$$Z_b = w^2 \, x(t - 2t_s) + 2w^2 \, x(t - t_s) + w^2 \, x(t)$$

$$Z_a = \left(2w^2 - 8/t_s\right) y(t - t_s) + \left(w^2 - 4w/t_s + 4w/t_s^2\right) y(t - 2t_s) \tag{4.246}$$

where $x(t), y(t) \in \mathbb{R}$ denote the filter input and output, respectively, $w = 1884.956$ [Hz] denotes the experimentally determined filter cut-off frequency (i.e., it was selected to remove the sharp edges of the camera input signal), and $t_s = 0.001$ [sec] denotes the sampling time. A standard backward difference/filtering process was utilized to enable calculation of the camera-space velocity via numerical differentiation of the position signal.[11]

From the dynamic model of (4.53), the unknown parameter vector given in (4.204) was constructed as follows

$$\phi_n = \begin{bmatrix} p_1 & p_2 & p_3 & F_{d1} & F_{d2} & \dfrac{1}{\det(A)} & \left(\dfrac{A_3}{A_4} - \dfrac{A_2}{A_4 \det(A)}\right) \end{bmatrix}^T \tag{4.247}$$

where the static friction effects are considered as unmodeled disturbances. To ensure that the manipulator was configured to eliminate mechanical singularities and to provide for the maximum camera view, the following desired trajectory signal was generated in the camera space for both experiments

$$y_d = \begin{bmatrix} 130 + a_x \cos\left(3.5t\right) - a_y \sin(3.5t) \\ 150 + a_y \cos\left(3.5t\right) - a_x \sin(3.5t) \end{bmatrix} \tag{4.248}$$

where $a_x = 72$ and $a_y = 150$ represent the distance between the initial position of the end-effector in the image plane and the center of the image plane. Figure 4.12 illustrates the desired task space and the resultant desired image space trajectory. To highlight the effects of the adaptation, the initial values of all the parameter estimates were set to zero.

---

[11] For a visual servoing control design that does not require camera space velocity measurements, see [66].

Experiment 1: The control and adaptation gains for the adaptive visual servo controller given in Section 4.4.3 were selected as follows

$$\alpha = \begin{bmatrix} 2.182 & 1.875 \end{bmatrix}^T \qquad K_r = \begin{bmatrix} 0.77 & 0.96 \end{bmatrix}^T$$

$$\Gamma_{c1} = \begin{bmatrix} 0.0001 & 0.0001 & 0.0001 \end{bmatrix}^T \qquad \Gamma_{c2} = \begin{bmatrix} 0.0001 & 0.0001 \end{bmatrix}^T .$$
$$(4.249)$$

The actual image space trajectory of the robot end-effector is illustrated in Figure 4.13. Figures 4.14, 4.15, and 4.16 illustrate the image space link position tracking error, the camera calibration parameter estimates, and the torque control inputs, respectively. As shown in Figure 4.14, the steady state image space link position tracking errors were between ±2 [pixels] for both links.

Experiment 2: The control and adaptation gains for the adaptive visual servo controller given in Section 4.4.3 were selected as follows

$$k_1 = 1.412 \qquad k_2 = 1.40 \qquad K_n = \begin{bmatrix} 1.39 & 1.40 \end{bmatrix}^T$$
$$\Gamma_1 = \begin{bmatrix} 0.009 & 0.009 & 0.009 \end{bmatrix}^T \qquad \Gamma_2 = \begin{bmatrix} 0.009 & 0.012 \end{bmatrix}^T$$
$$\Gamma_n = \begin{bmatrix} 0.0004 & 0.0001 & 0.0002 & 0.0002 & 0.0002 & 0.0002 & 0.0004 \end{bmatrix}^T .$$
$$(4.250)$$

The actual image space trajectory of the robot end-effector is illustrated in Figure 4.17, and the link position tracking error is illustrated in Figure 4.18. Figures 4.19 and 4.20 depict the parameter estimates for the camera parameters and the robot parameters, respectively, and Figure 4.21 illustrates the control torques.

Remark 4.9 *As can be observed from the tracking error plots (Figures 4.14 and 4.18), the adaptive camera calibration controller reaches its steady state in approximately 3 [sec] with an overshoot of 6%, while the adaptive robot controller reaches its steady state in approximately 10 [sec] with an overshoot of 23%. The main reason for this behavior is the adaptive robot controller has more parameter estimates. Hence, a higher overshoot (approximately 16 [pixels]) occurs during the transient period, and the system takes longer to reach the steady state. Note that the overshoot can be reduced if the parameter estimates are initialized to values closer to the actual parameter values.*

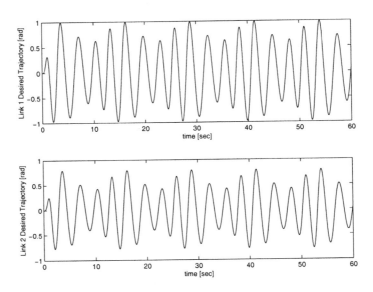

FIGURE 4.12. Desired image space trajectory for the robot end-effector.

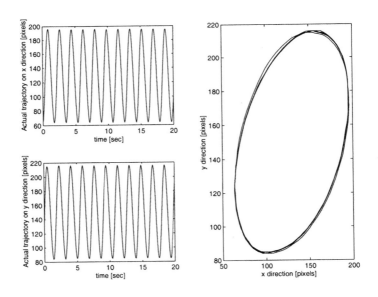

FIGURE 4.13. Actual image space trajectory of the robot end-effector.

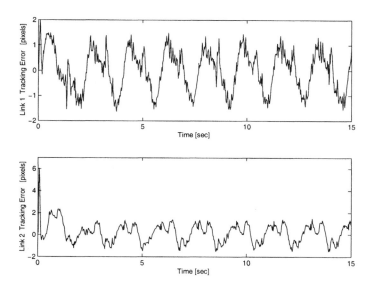

FIGURE 4.14. Link position tracking errors for the adaptive visual servo controller.

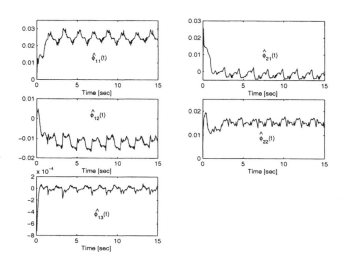

FIGURE 4.15. Parameter estimates for the adaptive visual servo controller.

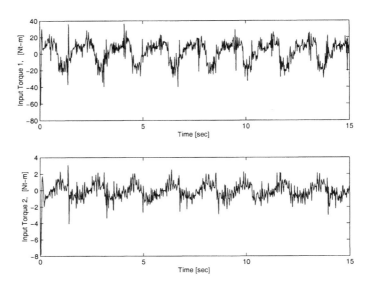

FIGURE 4.16. Joint torque control inputs for the adaptive visual servo controller.

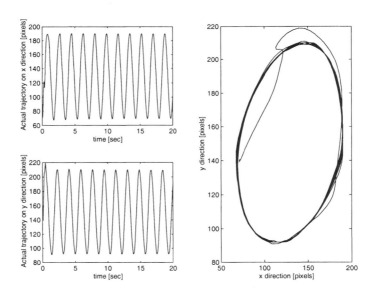

FIGURE 4.17. Actual image space trajectory of the robot end-effector.

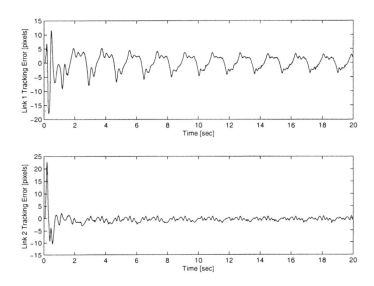

FIGURE 4.18. Link position tracking errors for the adaptive visual servo controller.

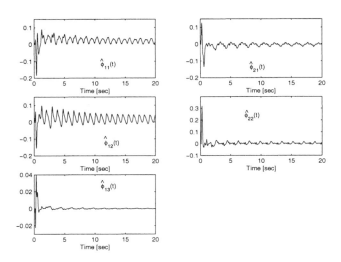

FIGURE 4.19. Camera parameter estimates for the adaptive robot controller.

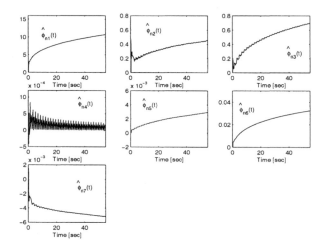

FIGURE 4.20. Mechanical system parameter estimates for the adaptive robot controller.

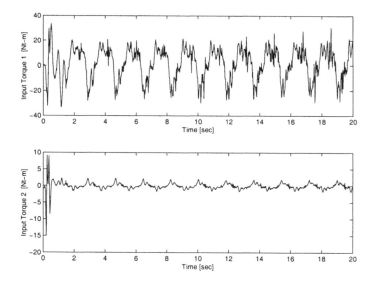

FIGURE 4.21. Control input torques for the adaptive robot controller.

4.5   Background and Further Reading

Based on characteristics such as a relatively simple structure and the ability to compensate for a broad class of periodic disturbances that are manifested in many robotic applications, learning-based control schemes have been the target of considerable effort. Some of the initial learning control research set as a goal the development of betterment learning controllers (see [2, 3, 4]). Unfortunately, one of the drawbacks of the betterment learning controllers is that the robot is required to return to the same initial configuration after each learning trial. In [28], Heinzinger et al. provided several examples that illustrated the lack of robustness of the betterment learning controllers to variations in the initial conditions of the robot. To address these robustness issues, Arimoto [1] incorporated a forgetting factor in the betterment learning algorithm given in [3, 4]. Motivated by the results from the betterment learning research, several researchers investigated the use of repetitive learning controllers. One of the advantages of the repetitive learning scheme is that the requirement that the robot return to the exact same initial condition after each learning trial is replaced by the less restrictive requirement that the desired trajectory of the robot be periodic. Some of the initial repetitive learning control research was performed in [27, 54, 55]. However, the asymptotic convergence of these basic repetitive control schemes could only be guaranteed under restrictive conditions on the plant dynamics that might not be satisfied. To enhance the robustness of these repetitive control schemes, researchers in [27, 54] modified the repetitive update rule to include the so-called Q-filter. Unfortunately, the use of the Q-filter did not lead to faster convergence of the tracking error to zero. In the search for new learning control algorithms, researchers in [29, 43] proposed an entirely new scheme that exploited the use of kernal and influence functions in the repetitive update rule. However, this class of controllers tends to be fairly complicated in comparison to the control schemes that utilize a standard repetitive update rule. In [20], the authors illustrate how the standard repetitive update rule could be partially saturated (ensuring the boundedness of the repetitive update rule) to yield global asymptotic tracking.

In [53, 57], iterative learning controllers (ILCs) were developed that do not require differentiation of the update rule, so that the algorithm can be applied to sampled data without introducing differentiation noise. In [9, 10, 11, 58], ILCs were developed to address the motion and force control problem for constrained robot manipulators. In [12], Cheah and Wang developed a model-reference learning control scheme for a class of nonlinear systems in which the performance of the learning system is specified by a reference model. In [61], Xu and Qu use a Lyapunov-based approach to illustrate how an ILC can be combined with a variable structure controller to handle a broad class of nonlinear systems. In [24], Ham, Qu, and Johnson utilized Lyapunov-based techniques to develop an ILC that is combined with a robust control design to achieve global uniformly ultimately bounded link position tracking for robot manipulators. The applicability of this design was extended to a broader class of nonlinear systems by Ham, Qu, and Kaloust in [25]. Recently, several researchers (see [13, 26, 34, 35]) have used a class of multiple-step "functional" iterative learning controllers to damp out steady state oscillations. As stated in [35], the fundamental difference between the previous learning controllers and the controllers proposed in [13, 26, 34, 35] are that the ILC is not updated continuously with time; rather, it is switched at iterations triggered by steady state oscillations. Han, Kim, and Ha used this iterative update procedure in [26] to damp out steady state oscillations in the velocity set-point problem for servo motors. The work in [26] was extended by [13] to compensate for friction effects and applied in [34] to video cassette recorder (VCR) servo motors (see [44, 45] for a comprehensive review and tutorial on ILC).

Motivated by the numerous industrial/manufacturing applications that require a robot manipulator to make contact with the environment, significant research efforts have targeted simultaneous control of the end-effector position and the interaction force. For example, a nonlinear reduced order transformation was introduced in [42] that allowed the constrained robot dynamics to be represented by a reduced-order model and hence, facilitated the separate design of position and force control strategies. In [21], Grabbe et al. compared and contrasted the transformation proposed in [42] with other position/force techniques. Most of the previous work on the simultaneous position/force tracking control problem requires feedback of link position, link velocity, and end-effector force. For example, results developed in [5, 63, 67] involved the design of other adaptive full-state feedback position/force tracking controllers that yielded asymptotic position/force tracking (see [8, 31, 37, 51, 52]). Motivated by the work in [7, 65], in which high-pass filters were designed to generate velocity-related signals for robust and adaptive position tracking controllers for unconstrained robot manipulators, the authors of [17] developed an adaptive output feedback controller that achieved semi-global position/force tracking. Colbaugh and Glass [14] also designed an adaptive position/force controller without velocity measurements that achieved semi-globally uniformly bounded position tracking/force regulation. Experimental results for position/force regulation or position tracking/force regulation are presented in [14, 56, 59, 60].

Inspired by the desire to provide robots with an improved sense of perception that would enable autonomous operation in unstructured environments, researchers have investigated vision-based sensing for over forty years. As stated in [15], most of this research has been aimed at image extraction and feature recognition, and the results of this research have been successfully commercialized. Within the last decade, researchers have investigated leveraging off of the previous image extraction and feature recognition technology to embed visual information in the feedback loop of the robot controller (i.e., visual servoing). From a review of literature (an overview of various robot visual servoing research is provided in [22, 46]), there seems to be a consensus that to extract high-level performance from vision-based robotic systems, the control system must incorporate information about the dynamics/kinematics of the robot and the calibration parameters of the camera system. With regard to the vision system, Bishop and Spong [6] emphasized the importance of adequate camera calibration with respect to the robot and the environment. As noted in [6], while a variety of techniques have been proposed for off-line camera calibration, only a few approaches were aimed at the more interesting problem of on-line calibration under closed-loop control.

Recently, some attention has been given to the design of vision-based control schemes (accompanied by stability analyses) that guarantee the convergence of the position error while accounting for uncalibrated camera effects and the mechanical dynamics of the robot. For example, Kelly and Marquez [33] designed a setpoint controller that compensated for unknown intrinsic camera parameters but required perfect knowledge of the camera orientation. Later, Kelly [32] redesigned the setpoint controller of [33] to also take into account uncertainties associated with the camera orientation, resulting in a local asymptotic stability result. In [6], Bishop and Spong developed an inverse dynamics, position tracking control scheme (i.e., exact model knowledge of the mechanical dynamics are required as in the first adaptive visual servo control example presented in this chapter) with an on-line adaptive camera calibration control loop that guaranteed global asymptotic position tracking. However, convergence of the position tracking error required that the desired position trajectory be persistently exciting.

References

[1] S. Arimoto, "Robustness of Learning Control for Robot Manipulators," *Proceedings of the IEEE International Conference on Robotics and Automation*, 1990.

[2] S. Arimoto, *Control Theory of Non-Linear Mechanical Systems: A Passivity-Based and Circuit-Theoretic Approach*, Oxford: Claredon Press, 1996.

[3] S. Arimoto, S. Kawamura, and F. Miyazaki, "Bettering Operation of Robots by Learning," *Journal of Robotic Systems*, Vol. 1, No. 2, 1984, pp. 123–140.

[4] S. Arimoto, S. Kawamura, and F. Miyazaki, "Realization of Robot Motion Based on a Learning Method," *IEEE Transactions on Systems, Man, and Cybernetics*, Vol. 18, No. 1, 1988, pp. 126–134.

[5] S. Arimoto, T. Naniwa, and Y. H. Liu, "Model-Based Adaptive Hybrid Control for Manipulators with Geometric Endpoint Constraint," *Advanced Robotics*, Vol. 9, No. 1, 1995, pp. 67–80.

[6] B. E. Bishop and M. W. Spong, "Adaptive Calibration and Control of 2D Monocular Visual Servo System," *IFAC Symposium on Robot Control*, Nantes, France, 1997, pp. 525–530.

[7] T. Burg, J. Hu, and D. Dawson, "A Redesigned DCAL Controller without Velocity Measurements: Theory and Experimentation," *Robotica*, Vol. 15, No. 3, May 1997, pp. 337–346.

[8] R. Carelli and R. Kelly, "Adaptive Control of Constrained Robots Modeled by Singular System," *Proceedings of the IEEE Conference on Decision and Control*, Tampa, FL, Dec. 1989, pp. 2635–2640.

[9] C. C. Cheah, D. Wang, and Y. C. Soh, "Convergence and Robustness of a Discrete-Time Learning Control Scheme for Constrained Manipulators," *Journal of Robotic Systems*, Vol. 11, No. 3, 1994, pp. 223–238.

[10] C. C. Cheah, D. Wang, and Y. C. Soh, "Learning Control of Motion and Force for Constrained Robotic Manipulators," *International Journal of Robotics and Automation*, Vol. 10, No. 3, 1995, pp. 79–88.

[11] C. C. Cheah, D. Wang, and Y. C. Soh, "Learning Control for a Class of Nonlinear Differential-Algebraic Systems with Application to Constrained Robots," *Journal of Robotic Systems*, Vol. 13, No. 3, 1996, pp. 141–151.

[12] C. C. Cheah and D. Wang, "A Model-Reference Learning Control Scheme for a Class of Nonlinear Systems," *International Journal of Control*, Vol. 66, No. 2, 1997, pp. 271–287.

[13] S.-I. Cho and I.-J. Ha, "A Learning Approach to Tracking in Mechanical Systems with Friction," *IEEE Transactions on Automatic Control*, Vol. 45, No. 1, 2000, pp. 111–116.

[14] R. Colbaugh and K. Glass, "Adaptive Compliant Motion Control of Manipulators without Velocity Measurements," *Proceedings of the IEEE Conference on Robotics and Automation*, Minneapolis, MN, Apr. 1996, pp. 2628–2635.

[15] P. Corke and M. Good, "Dynamic Effects in Visual Closed-Loop Systems," *IEEE Transactions on Robotics and Automation*, Vol. 12, No. 5, 1996, pp. 671–683.

[16] N. Costescu and D. Dawson, "QMotor 2.0 - A PC Based Real-Time Multitasking Graphical Control Environment," *Proceedings of the American Control Conference*, Philadelphia, PA, June 1998, pp. 1266–1270.

[17] M. S. de Queiroz, J. Hu, D. Dawson, T. Burg, and S. Donepudi, "Adaptive Position/Force Control of Robot Manipulators without Velocity Measurements: Theory and Experimentation," *IEEE Transactions on Systems, Man, and Cybernetics*, Vol. 27, No. 5 , Oct. 1997, pp. 796–809.

[18] M. S. de Queiroz, D. M. Dawson, S. P. Nagarkatti, and F. Zhang, *Lyapunov-Based Control of Mechanical Systems*, Boston, MA: Birkhäuser, 2000.

[19] *Direct Drive Manipulator Research and Development Package Operations Manual*, Berkeley, CA: Integrated Motion Inc., 1992.

[20] W. E. Dixon, E. Zergeroglu, D. M. Dawson, and B. T. Costic, "Repetitive Learning Control: A Lyapunov-Based Approach," *IEEE Transactions on Systems, Man, and Cybernetics-Part B: Cybernetics*, Aug. 2002, Vol. 32, No. 4, pp. 538–545.

[21] M. T. Grabbe, J. J. Carroll, D. M. Dawson, and Z. Qu, "Review and Unification of Reduced Order Force Control Methods," *Journal of Robotic Systems*, Vol. 10, No. 4, June 1993, pp. 481–504.

[22] G. D. Hager and S. Hutchinson (guest editors), Special Section on Vision-Based Control of Robot Manipulators, *IEEE Transactions on Robotics and Automation*, Vol. 12, No. 5, Oct. 1996.

[23] G. D. Hager, W. C. Chang, and A. S. Morse, "Robot Hand-Eye Coordination Based on Stereo Vision," *IEEE Control Systems Magazine*, Vol. 15, No. 1, Feb. 1995, pp. 30–39.

[24] C. Ham, Z. Qu, and R. Johnson, "A Nonlinear Learning Control for Robot Manipulators in the Presence of Actuator Dynamics," *International Journal of Robotics and Automation*, Vol. 15, No. 3, June 2000, pp. 119–130.

[25] C. Ham, Z. Qu, and J. Kaloust, "Nonlinear Learning Control for a Class of Nonlinear Systems," *Automatica*, Vol. 37, No. 3, 2001, pp. 419–428.

[26] S. H. Han, Y. H. Kim, and I.-J. Ha, "Iterative Identification of State-Dependent Disturbance Torque for High Precision Velocity Control of Servo Motors," *IEEE Transactions on Automatic Control*, Vol. 43, No. 5, 1998, pp. 724–729.

[27] S. Hara, Y. Yamamoto, T. Omata, and M. Nakano, "Control Systems: A New Type Servo System for Periodic Exogenous Signals," *IEEE Transactions on Automatic Control*, Vol. 33, No. 7, 1988, pp. 659–668.

[28] D. Heinzinger, B. Fenwick, B. Paden, and F. Miyazaki, "Robot Learning Control," *Proceedings of the IEEE Conference on Decision and Control*, Tampa, FL, Dec. 1989, pp. 436–440.

[29] R. Horowitz, "Learning Control of Robot Manipulators," *ASME Journal of Dynamic Systems, Measurement, and Control*, Vol. 115, 1993, pp. 402–411.

[30] P. Hsu, J. Hauser, and S. Sastry, "Dynamic Control of Redundant Manipulators," *Journal of Robotic Systems*, Vol. 6, 1989, pp. 133–148.

[31] J. H. Jean and L. C. Fu, "Efficient Adaptive Hybrid Control Strategies for Robots in Constrained Manipulation," *Proceedings of the IEEE Conference on Robotics and Automation*, Sacramento, CA, Apr. 1991, pp. 1681–1686.

[32] R. Kelly, "Robust Asymptotically Stable Visual Servoing of Planar Robots," *IEEE Transactions on Robotics and Automation*, Vol. 12, No. 5, Oct. 1996, pp. 759–766.

[33] R. Kelly and A. Marquez, "Fixed-Eye Direct Visual Feedback Control of Planar Robots," *Journal of Systems Engineering*, Vol. 4, No. 5, Nov. 1995, pp. 239–248.

[34] Y.-H. Kim and I.-J. Ha, "A Learning Approach to Precision Speed Control of Servomotors and Its Application to a VCR," *IEEE Transactions on Automatic Control*, Vol. 43, No. 5, 1998, pp. 724–729.

[35] Y.-H. Kim and I.-J. Ha, "Asymptotic State Tracking in a Class of Nonlinear Systems via Learning-Based Inversion," *IEEE Transactions on Automatic Control*, Vol. 45, No. 11, 2000, pp. 2011–2027.

[36] P. Kokotovic, "The Joy of Feedback: Nonlinear and Adaptive," *IEEE Control Systems Magazine*, Vol. 12, No. 3, June 1992, pp. 7–17.

[37] H. Krishnan and N. H. McClamroch, "Tracking in Nonlinear Differential-Algebraic Control Systems with Applications to Constrained Robot Systems," *Automatica*, Vol. 30, No. 12, 1994, pp. 1885–1897.

[38] R. K. Lenz and R. Y. Tsai, "Techniques for Calibration of the Scale Factor and Image Center for High Accuracy 3-D Machine Vision Metrology," *IEEE Transactions on Pattern Analysis and Machine Intelligence*, Vol. 10, No. 5, Sept. 1988, pp. 713–720.

[39] F. Lewis, C. Abdallah, and D. Dawson, *Control of Robot Manipulators*, New York, NY: MacMillan Publishing Co., 1993.

[40] R. Lozano and B. Brogliato, "Adaptive Control of Robot Manipulators with Flexible Joints," *IEEE Transactions on Automatic Control*, Vol. 37, No. 2, Feb. 1992, pp. 174–181.

[41] A. Maruyama and M. Fujita, "Robust Visual Servo Control for Planar Manipulators with Eye-In-Hand Configurations," *Proceedings of the Conference on Decision and Control*, San Diego, CA, Dec. 1997, pp. 2551–2552.

[42] N. McClamroch and D. Wang, "Feedback Stabilization and Tracking of Constrained Robots," *IEEE Transactions on Automatic Control*, Vol. 33, No. 5, May 1988, pp. 419–426.

[43] W. Messner, R. Horowitz, W.-W. Kao, and M. Boals, "A New Adaptive Learning Rule," *IEEE Transactions on Automatic Control*, Vol. 36, No. 2, 1991, pp. 188–197.

[44] K. L. Moore, "Iterative Learning Control – An Expository Overview," *Applied and Computational Controls, Signal Processing, and Circuits*, Vol. 14, 1999, pp. 151–171.

[45] K. L. Moore, *Iterative Learning Control for Deterministic Systems*, London: Springer-Verlag, 1993.

[46] B. Nelson and N. Papanikolopoulos (guest editors), Special Issue on Visual Servoing, *IEEE Robotics and Automation Magazine*, Vol. 5, No. 4, Dec. 1998.

[47] N. Sadegh and R. Horowitz, "Stability and Robustness Analysis of a Class of Adaptive Controllers for Robotic Manipulators," *International Journal of Robotic Research*, Vol. 9, No. 9, June 1990, pp. 74–92.

[48] N. Sadegh, R. Horowitz, W. W. Kao, and M. Tomizuka, "A Unified Approach to the Design of Adaptive Controllers for Robot Manipulators," *ASME Journal of Dynamic Systems, Measurement, and Control*, Vol. 112, 1990, pp. 618–629.

[49] M. W. Spong and R. Ortega, "On Adaptive Inverse Dynamics Control of Rigid Robots," *IEEE Transactions on Automatic Control*, Vol. 35, No. 1, Jan. 1990, pp. 92–95.

[50] M. Spong and M. Vidyasagar, *Robot Dynamics and Control*, New York, NY: John Wiley, 1989.

[51] C. Y. Su, T. P. Leung, and Q. J. Zhou, "Adaptive Control of Robot Manipulators under Constrained Motion," *Proceedings of the IEEE Conference on Decision and Control*, Honolulu, HI, Dec. 1990, pp. 2650–2655.

[52] C. Y. Su and Y. Stepanenko, "Adaptive Control of Constrained Robots without Using Regressor," *Proceedings of the IEEE Conference on Robotics and Automation*, Minneapolis, MN, Apr. 1996, pp. 264–269.

[53] M. Sun and D. Wang, "Sampled-Data Iterative Learning Control for Nonlinear Systems with Arbitrary Relative Degree," *Automatica*, Vol. 37, 2001, pp. 283–289.

[54] M. Tomizuka, T. Tsao, and K. Chew, "Discrete-Time Domain Analysis and Synthesis of Controllers," *ASME Journal of Dynamic Systems, Measurement, and Control*, Vol. 111, 1989, pp. 353–358.

[55] M. Tsai, G. Anwar, and M. Tomizuka, "Discrete-Time Control for Robot Manipulators," *Proceedings of the IEEE International Conference on Robotics and Automation*, 1988, pp. 1341–1347.

[56] R. Volpe and P. Khosla, "An Experimental Evaluation and Comparison of Explicit Force Control Strategies for Robotic Manipulators," *Proceedings of the IEEE Conference on Robotics and Automation*, Nice, France, May 1992, pp. 1387–1393.

[57] D. Wang, "On D-type and P-type ILC Designs and Anticipatory Approach," *International Journal of Control*, Vol. 23, No. 10, 2000, pp. 890–901.

[58] D. Wang, Y. C. Soh, and C. C. Cheah, "Robust Motion and Force Control of Constrained Manipulators by Learning," *Automatica*, Vol. 31, No. 2, 1995, pp. 257–262.

[59] L. Whitcomb, S. Arimoto, and F. Ozaki, "Experiments in Adaptive Model-Based Force Control," *Proceedings of the IEEE Conference on Robotics and Automation*, Nagoya, Japan, May 1995, pp. 1846–1853.

[60] L. S. Wilfinger, J. T. Wen, and S. H. Murphy, "Integral Force Control with Robustness Enhancement," *IEEE Control Systems Magazine*, Vol. 14, No. 1, Feb. 1994, pp. 31–40.

[61] J.-X. Xu and Z. Qu, "Robust Iterative Learning Control for a Class of Nonlinear Systems," *Automatica*, Vol. 34, No. 8, 1998, pp. 983–988.

[62] Z. Yao, N. P. Costescu, S. P. Nagarkatti, and D. M. Dawson, "Real-Time Linux Target: A MATLAB-Based Graphical Control Environment," *Proceedings of the IEEE Conference on Control Applications*, Anchorage, AK, Sept. 2000, pp. 173–178.

[63] B. Yao and M. Tomizuka, "Adaptive Control of Robot Manipulators in Constrained Motion-Controller Design," *ASME Journal of Dynamic Systems, Measurement, and Control*, Vol. 117, Sept. 1995, pp. 320–328.

[64] T. Yoshikawa, "Analysis and Control of Robot Manipulators with Redundancy," in *Robotics Research-The First International Symposium*, Cambridge, MA: MIT Press, 1984, pp. 735–747.

[65] J. Yuan and Y. Stepanenko, "Robust Control of Robotic Manipulators without Velocity Measurements," *International Journal of Robust and Nonlinear Control*, Vol. 1, 1991, pp. 203–213.

[66] E. Zergeroglu, D. M. Dawson, M. de Queiroz, and A. Behal, "Vision-Based Nonlinear Tracking Controllers in the Presence of Parametric Uncertainty," *IEEE/ASME Transactions on Mechatronics*, Vol. 6, No. 3, Sept. 2001, pp. 322–337.

[67] R. R. Y. Zhen and A. A. Goldenberg, "An Adaptive Approach to Constrained Robot Motion Control," *Proceedings of the IEEE Conference on Robotics and Automation*, Nagoya, Japan, May 1995, pp. 1833–1838.

5
Aerospace Systems

5.1 Introduction

In many aerospace applications, large amplitude maneuvers are performed that require a high degree of accuracy. Aerospace applications also typically require a system to track a time-varying reference trajectory rather than a simple setpoint regulation. These objectives motivate the need to incorporate the nonlinear dynamic effects of the system in the control system synthesis. However, the problem is further complicated because the mass and inertia are not exactly known due to fuel consumption, payload variation, appendage deployment, etc. Many existing control strategies for aerospace systems use singular (i.e., the Jacobian matrix in the kinematic equation is singular for some orientations) three-parameter attitude representations such as Euler angles, which are only locally valid. As described in Chapter 2, the unit quaternion is a four-parameter representation that can be used to globally represent the attitude of an object without singularities. However, an additional constraint equation is introduced. Along this line of reasoning, a full-state feedback quaternion-based attitude tracking controller is first developed for the nonlinear dynamics of a rigid spacecraft[1] with parametric uncertainty in the inertia matrix. Motivated by the desire to eliminate additional sensor payload, a second controller is devel-

---

[1] The words spacecraft and satellite are used interchangeably in this chapter.

oped under the additional constraint that angular velocity measurements are not available. The developed output feedback control strategy is proven to be globally valid with the exception of a mild restriction on the initial conditions of the system that can be easily satisfied.

Spacecrafts typically utilize separate devices to provide energy storage and attitude control. Conventionally, the energy collected from solar arrays is stored in chemical batteries, and the stored energy is then used when the spacecraft is in the earth's shadow. The typical configuration for spacecraft attitude control includes an array of reaction wheels or control moment gyros. An alternative configuration may include the use of flywheels. Flywheels store kinetic energy within a rapidly spinning wheel-like rotor or disk. Modern flywheels employ a high-strength composite rotor, which rotates in a vacuum chamber. A motor powered off the solar arrays can be mounted on the rotor's shaft to spin the rotor up to speed (also referred to as charging). The rotor's kinetic energy provides the torque required for the desired spacecraft maneuvers and sustains vital processing and communication procedures on board. A high-strength containment structure houses the rotating elements and low-energy-loss bearings can be used to stabilize the shaft (e.g., the magnetic bearings described in Chapter 2 could be used in tandem with the active autobalancing scheme described in Chapter 3). Flywheels have several advantages over traditional battery systems including (i) flywheels contain no acids, (ii) flywheels are not affected by temperature extremes, as most batteries are, and (iii) flywheels facilitate integrated power and attitude tracking applications. This last advantage allows the integration of the energy storage and attitude control functions via the use of four or more noncoplanar flywheels, and thereby, reduces the spacecraft bus mass, volume, cost, and maintenance requirements while maintaining or improving the spacecraft performance. This system is commonly referred to as an integrated power and attitude control system (IPACS) [3]. The integration of these two functions is achieved by decomposing the space of internal torques of the flywheel array into two orthogonal subspaces, the range space and null space of the torques, with decoupled control objectives (i.e., attitude tracking and energy/power tracking, respectively) [14, 17, 38]. As a result, the energy management function can be accomplished without affecting the attitude control function. Following this design concept, an adaptive IPACS is developed for a nonlinear spacecraft model where the body torque is produced by flywheels that are operated in a reaction wheel mode. The adaptive quaternion-based controller forces a spacecraft to track a desired attitude trajectory while simultaneously providing exponential energy/power tracking.

Another key technological concept in aerospace systems is the idea of distributed functionality of a large spacecraft among smaller, less expensive, cooperative spacecraft [4, 30]. Flying two or more spacecraft in a precise formation is typically referred to as multiple satellite formation flying (MSFF) and presents a number of complex challenges. The spacecraft must have a sensory and control system that enables it to attain and maintain a precise position relative to the other spacecraft. The spacecraft must also have a communication, sensory, and control system that allows it to attain a specified attitude, with all spacecraft targeting the desired object/trajectory. The United States Air Force and National Aeronautics and Space Administration (NASA) have identified MSFF as an enabling technology for future missions and have shown a keen interest in the development of a reliable autonomous formation keeping strategy to deploy multiple spacecraft for space missions (e.g., the Earth Orbiter-I (EO-I) and the New Millennium Interferometer, also known as Deep Space-3). Specifically, the EO-1 formation flying effort demonstrates the capability of satellites to react to each other and maintain a close proximity without human intervention. This advancement allows satellites to react autonomously to each other's orbit changes quickly and more efficiently. It permits scientists to obtain unique measurements by combining data from several satellites rather than flying all the instruments on one costly satellite. It also enables the collection of different types of scientific data unavailable from a single satellite, such as stereo views or simultaneously collecting data of the same ground scene at different angles. Following this paradigm, the relative position control problem for MSFF is considered. Specifically, the full nonlinear dynamics describing the relative positioning of MSFF is used to develop a Lyapunov-based nonlinear adaptive control law that guarantees global asymptotic convergence of the position tracking error in the presence of unknown constant, or slow-varying spacecraft masses, disturbance forces, and gravity forces.

5.2 Attitude Tracking

This section presents a full-state feedback adaptive control solution to the quaternion-based attitude tracking control problem that compensates for parametric uncertainty in the spacecraft inertia matrix. Based on the structure of the full-state feedback control design and the stability analysis, a second controller is then developed under the additional constraint that angular velocity measurements are not available. Specifically, a novel transformation is first applied to the open-loop quaternion tracking error dynamics. The transformed tracking error dynamics are then used to design a new adaptive full-state feedback controller that compensates for uncertainties in the inertia matrix. A Lyapunov-like function, which exploits the quaternion constraint equation, is used to prove asymptotic attitude tracking. To eliminate the requirement for velocity measurements, a nonlinear filtering scheme is employed to generate a velocity-related signal from attitude measurements. The output feedback controller is also proven to yield asymptotic attitude tracking.

5.2.1 System Model

The attitude motion of a rigid body is represented by a kinematic equation that relates the time derivatives of the orientation angles to an angular velocity vector and by Euler's dynamic equation that describes the time evolution of the angular velocity vector. As described in Chapter 2, several singular three-parameter representations (e.g., the Euler angles, Gibbs vector, Cayley-Rodrigues parameters, and modified Rodrigues parameters) exist to represent the attitude of a rigid body. However, the subsequent model development will be based on the four-parameter unit quaternion representation since it provides a nonsingular representation.

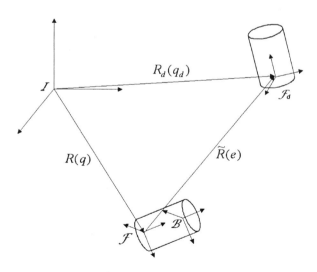

FIGURE 5.1. Relationship between coordinate frames.

Kinematic Model

Let $\mathcal{F}$ represent a body-fixed orthogonal coordinate frame attached to the center of mass of a rigid spacecraft. Let $\mathcal{B}$ denote a body-fixed coordinate frame that is offset from $\mathcal{F}$ by a translation as depicted in Figure 5.1. The spacecraft is equipped with actuators that provide body-fixed torques to each axis of $\mathcal{B}$ by a pair of equal and opposite forces that act in a direction perpendicular to the line joining the actuators [1]. Let $\mathcal{I}$ denote an inertial reference frame (e.g., attached to the earth center). As stated in Chapter 2, the position and orientation of these coordinate axes are commonly represented by homogeneous transformation matrices, from which several different representations can be utilized to develop the kinematic model. Recall from Chapter 2 that three-parameter representations are singular. Hence, in this section the unit quaternion vector $q(t) = [q_0(t), q_v^T(t)]^T \in \mathbb{R}^4$, with $q_0(t) \in \mathbb{R}$ and $q_v(t) \in \mathbb{R}^3$, is used to relate the attitude of $\mathcal{B}$ and $\mathcal{F}$ to $\mathcal{I}$ ($\mathcal{B}$ and $\mathcal{F}$ have the same orientation). As also stated in Chapter 2, the unit quaternion is subject to the following constraint

$$q_v^T q_v + q_0^2 = 1. \tag{5.1}$$

Given the unit quaternion parameterization, the rotation matrix that brings $\mathcal{I}$ onto $\mathcal{F}$, denoted by $R(q) \in SO(3)$, is defined as follows

$$R = \left(q_0^2 - q_v^T q_v\right) I_3 + 2 q_v q_v^T - 2 q_0 q_v^\times \tag{5.2}$$

where I_3 is the 3×3 identity matrix, and the notation $\zeta^\times$ denotes the following skew-symmetric matrix

$$\zeta^\times = \begin{bmatrix} 0 & -\zeta_3 & \zeta_2 \\ \zeta_3 & 0 & -\zeta_1 \\ -\zeta_2 & \zeta_1 & 0 \end{bmatrix} \qquad \forall \zeta = \begin{bmatrix} \zeta_1 & \zeta_2 & \zeta_3 \end{bmatrix}^T. \tag{5.3}$$

To facilitate the subsequent control development and stability analysis, we also exploit the fact that $q(t)$ is related to the angular velocity of $\mathcal{F}$ with respect to $\mathcal{I}$ expressed in $\mathcal{F}$, denoted by $\omega(t) \in \mathbb{R}^3$, via the following differential equation [3, 27]

$$\dot{q}_v = \frac{1}{2} \left(q_v^\times \omega + q_0 \omega\right) \tag{5.4}$$

$$\dot{q}_0 = -\frac{1}{2} q_v^T \omega. \tag{5.5}$$

Based on (5.4) and (5.5), the angular velocity can also be written as follows

$$\omega = 2 \left(q_0 \dot{q}_v - q_v \dot{q}_0\right) - 2 q_v^\times \dot{q}_v. \tag{5.6}$$

The desired attitude of the spacecraft is assumed to be described by a desired body-fixed reference frame $\mathcal{F}_d$ whose orientation with respect to the inertial frame $\mathcal{I}$ is specified by the desired unit quaternion $q_d(t) = [q_{0d}(t), q_{vd}^T(t)]^T \in \mathbb{R}^4$ that is constructed to satisfy the following constraint

$$q_{vd}^T q_{vd} + q_{0d}^2 = 1. \tag{5.7}$$

The objective in this section is to design a control input to ensure that the attitude of $\mathcal{F}$ tracks the attitude of $\mathcal{F}_d$ where the rotation matrix that brings $\mathcal{I}$ onto $\mathcal{F}_d$ is denoted by $R_d(q_d) \in SO(3)$, where $R_d(q_d)$ can be calculated from $q_d(t)$ as follows

$$R_d(q_d) = \left(q_{0d}^2 - q_{vd}^T q_{vd}\right) I_3 + 2 q_{vd} q_{vd}^T - 2 q_{0d} q_{vd}^\times. \tag{5.8}$$

As in (5.4) and (5.5), the desired angular velocity of $\mathcal{F}_d$ with respect to $\mathcal{I}$ expressed in $\mathcal{F}_d$, denoted by $\omega_d(t) \in \mathbb{R}^3$, can be obtained from the following dynamic equations

$$\dot{q}_{vd} = \frac{1}{2}\left(q_{vd}^\times \omega_d + q_{0d}\omega_d\right) \tag{5.9}$$

$$\dot{q}_{0d} = -\frac{1}{2} q_{vd}^T \omega_d. \tag{5.10}$$

That is, (5.9) and (5.10) can be used to compute $\omega_d(t)$ as shown below

$$\omega_d = 2\left(q_{0d}\dot{q}_{vd} - q_{vd}\dot{q}_{0d}\right) - 2 q_{vd}^\times \dot{q}_{vd} \tag{5.11}$$

where the standard assumption is made that $q_{0d}(t)$, $q_{vd}(t)$, and their first three time derivatives are bounded for all time. This assumption ensures that $\omega_d(t)$ of (5.11) and its first two time derivatives are bounded for all time.

To quantify the mismatch between the actual and desired spacecraft attitudes, the rotation matrix $\tilde{R}(e) \in SO(3)$ that brings $\mathcal{F}_d$ onto $\mathcal{F}$ can be defined as follows

$$\tilde{R} = R R_d^T = \left(e_0^2 - e_v^T e_v\right) I_3 + 2 e_v e_v^T - 2 e_0 e_v^\times \tag{5.12}$$

where $R(q)$ and $R_d(q_d)$ were defined in (5.2) and (5.8), respectively, and the quaternion tracking error $e(t) = [e_0(t), e_v^T(t)] \in \mathbb{R}^4$ is defined as follows

$$e_0 = q_0 q_{0d} + q_v^T q_{vd} \tag{5.13}$$

$$e_v = q_{0d} q_v - q_0 q_{vd} + q_v^\times q_{vd}. \tag{5.14}$$

Based on (5.9), (5.10), (5.13), and (5.14), the following open-loop quaternion tracking error kinematics can be developed

$$\dot{e}_v = \frac{1}{2}\left(e_v^\times + e_0 I_3\right)\tilde{\omega} \tag{5.15}$$

$$\dot{e}_0 = -\frac{1}{2} e_v^T \tilde{\omega} \tag{5.16}$$

where $\tilde{\omega}(t) \in \mathbb{R}^3$ denotes the angular velocity of $\mathcal{F}$ with respect to $\mathcal{F}_d$ expressed in $\mathcal{F}$, and is defined as follows

$$\tilde{\omega} = \omega - \tilde{R}\omega_d. \tag{5.17}$$

Remark 5.1 *The relations given in (5.13) and (5.14) can be explicitly calculated via quaternion algebra by noticing that the quaternion equivalent of (5.12) is the quaternion product (see [44] and Theorem 5.3 of [23])*

$$e = q_d^* q \tag{5.18}$$

where $q_d^*(t) = [q_{0d}(t),\ -q_{vd}(t)] \in \mathbb{R}^4$ is the unit quaternion representing the rotation matrix $R_d^T(q_d)$.

Dynamic Model

The dynamic model for the rigid spacecraft can be expressed as follows [18, 19]

$$J\dot{\omega} = -\omega^\times J\omega + u \tag{5.19}$$

where $J \in \mathbb{R}^{3\times 3}$ represents the constant positive-definite symmetric inertia matrix, $\omega(t)$ is introduced in (5.4–5.6), and $u(t) \in \mathbb{R}^3$ is a vector of control torques. In (5.19), the spacecraft inertia matrix is uncertain due to phenomena such as fuel consumption, payload variation, and appendage deployment. Based on (5.4), (5.5), (5.9), (5.10), (5.13), (5.14), and (5.17), the dynamic model given in (5.19) can be rewritten as follows

$$J\dot{\tilde{\omega}} = -\left(\tilde{\omega} + \tilde{R}\omega_d\right)^\times J\left(\tilde{\omega} + \tilde{R}\omega_d\right) + J\left(\tilde{\omega}^\times \tilde{R}\omega_d - \tilde{R}\dot{\omega}_d\right) + u \tag{5.20}$$

where the following fact has been used[2] [36]

$$\dot{\tilde{R}} = -\tilde{\omega}^\times \tilde{R}. \tag{5.21}$$

To facilitate the subsequent stability analysis, (5.15) can be rewritten as follows

$$\dot{e}_v = \frac{1}{2}T\tilde{\omega} \tag{5.22}$$

where the Jacobian-type matrix $T(e) \in \mathbb{R}^{3\times 3}$ is defined as follows

$$T = e_v^\times + e_0 I_3. \tag{5.23}$$

---

[2] The negative sign in equation (5.21) is because the coordinates of (5.17) are expressed in terms of the body-fixed coordinate frame rather than the inertial reference frame.

After taking the time derivative of (5.22) and premultiplying both sides of the resulting expression by $J^*(e)$, the following expression can be obtained

$$J^* \ddot{e}_v = \frac{1}{2} J^* \dot{T} \tilde{\omega} + \frac{1}{2} T^{-T} J \, \dot{\tilde{\omega}} \tag{5.24}$$

where $J^*(e) \in \mathbb{R}^{3 \times 3}$ is defined as follows

$$J^* = T^{-T} J T^{-1}. \tag{5.25}$$

By substituting (5.20) and (5.22) into (5.24), the following expression can be obtained

$$J^* \ddot{e}_v + C^* \dot{e}_v + N^* = u^* \tag{5.26}$$

where (5.3) has been used. In (5.26), the transformed control input $u^*(t) \in \mathbb{R}^3$ is defined as follows

$$u^* = \frac{1}{2} T^{-T} u \tag{5.27}$$

where the transformed dynamic terms, denoted by $C^*(e, \dot{e}_v) \in \mathbb{R}^{3 \times 3}$ and $N^*(e, \dot{e}_v, \omega_d, \dot{\omega}_d) \in \mathbb{R}^3$, are defined as

$$C^* = -J^* \dot{T} T^{-1} - 2 T^{-T} \left(J T^{-1} \dot{e}_v \right)^\times T^{-1} \tag{5.28}$$

$$N^* = T^{-T} \left((T^{-1} \dot{e}_v)^\times J \tilde{R} \omega_d \right) + T^{-T} \left(\left(\tilde{R} \omega_d \right)^\times J T^{-1} \dot{e}_v \right)$$

$$+ \frac{1}{2} T^{-T} \left(\left(\tilde{R} \omega_d \right)^\times J \tilde{R} \omega_d \right) - \frac{1}{2} T^{-T} J \left((2 T^{-1} \dot{e}_v)^\times \tilde{R} \omega_d - \tilde{R} \dot{\omega}_d \right). \tag{5.29}$$

The transformed dynamic model given in (5.26) satisfies the following properties that will be utilized in the subsequent control development and analysis.

Property 5.1: Skew-Symmetry

The transformed inertia and centripetal-Coriolis matrices satisfy the following skew-symmetric relationship (see Lemma B.19 of Appendix B)

$$\xi^T \left(\frac{1}{2} \dot{J}^* - C^* \right) \xi = 0 \quad \forall \xi \in \mathbb{R}^3. \tag{5.30}$$

Property 5.2: Bounded Inertia Matrix

The inertia matrix can be lower and upper bounded as follows

$$j_1 \|\xi\|^2 \le \xi^T J \xi \le j_2 \|\xi\|^2 \quad \forall \xi \in \mathbb{R}^n \tag{5.31}$$

where $j_1, j_2 \in \mathbb{R}$ denote positive bounding constants.

Remark 5.2 *To examine the conditions that ensure that $T(e)$ of (5.23) can be inverted, the determinant $T(e)$ can be computed as follows*

$$\det(T) = e_0(t). \tag{5.32}$$

From (5.32), it can be determined that $T(e)$ is invertible provided $e_0(t) \neq 0$ for any time. To ensure that $e_0(t) \neq 0$ for all time, the desired trajectory must be initialized to guarantee that $e_0(0) \neq 0$, and the subsequent control design must ensure that $e_0(t) \neq 0$ after the initial time.

5.2.2   Control Objective

The attitude tracking control objective is to force the actual attitude of a spacecraft to track a desired time-varying attitude trajectory. Based on (5.12), the attitude tracking control objective can be mathematically formulated as follows

$$\lim_{t \to \infty} \tilde{R}(e(t)) = I_3. \tag{5.33}$$

Based on (5.1), (5.7), (5.13), and (5.14), it is clear that the unit quaternion tracking error satisfies the following constraint

$$e_v^T e_v + e_0^2 = 1 \tag{5.34}$$

where

$$0 \leq \|e_v(t)\| \leq 1 \qquad 0 \leq |e_0(t)| \leq 1. \tag{5.35}$$

Given (5.34) and (5.35), the orientation tracking objective given in (5.33) can also be formulated in the terms of the unit quaternion error introduced in (5.13) and (5.14). Specifically, it is obvious from (5.34) that

$$\text{if} \quad \lim_{t \to \infty} e_v(t) = 0, \quad \text{then} \quad \lim_{t \to \infty} |e_0(t)| = 1. \tag{5.36}$$

Hence, (5.12) and (5.36) can be utilized to prove that

$$\text{if} \quad \lim_{t \to \infty} e_v(t) = 0, \quad \text{then} \quad \lim_{t \to \infty} \tilde{R}(e) = I_3 \tag{5.37}$$

(i.e., the attitude tracking objective given in (5.33) is achieved).

5.2.3   Adaptive Full-State Feedback Control

In this section, the control objective is to design an adaptive attitude controller for the open-loop tracking error dynamics given by (5.26) under the constraint that the spacecraft inertia matrix is unknown and under the assumption that $q(t)$ and $\dot{q}(t)$ (and hence, $e_v(t)$ and $\dot{e}_v(t)$) are measurable.

A filtered tracking error, denoted by $r(t) \in \mathbb{R}^3$, is defined to facilitate the controller design as follows

$$r = \dot{e}_v + \alpha e_v \tag{5.38}$$

where $e_v(t)$ and $\dot{e}_v(t)$ are defined in (5.14) and (5.15), respectively, and $\alpha \in \mathbb{R}^{3\times3}$ is a constant positive-definite diagonal control gain matrix.

Closed-Loop Error System

To develop the closed-loop tracking error system, we take the time derivative of (5.38) and then premultiply both sides of the resulting equation by J^* to obtain the following expression

$$J^* \dot{r} = J^* \ddot{e}_v + J^* \alpha \dot{e}_v. \tag{5.39}$$

After substituting (5.26) into (5.39), the following expression can be obtained

$$J^* \dot{r} = -C^* r + Y\theta + u^* \tag{5.40}$$

where (5.38) has been used and the linear parameterization $Y(\cdot)\theta$ is defined as follows

$$Y\theta = J^* \alpha \dot{e}_v + C^* \alpha e_v - N^*. \tag{5.41}$$

In (5.41), $Y(e, \dot{e}_v, \omega_d, \dot{\omega}_d) \in \mathbb{R}^{3\times6}$ denotes a known regression matrix and $\theta \in \mathbb{R}^6$ denotes a constant unknown vector of inertia parameters defined as follows

$$\theta = \begin{bmatrix} J_{11} & J_{12} & J_{13} & J_{22} & J_{23} & J_{33} \end{bmatrix}^T \tag{5.42}$$

where $J_{ij} \, \forall i, j = 1, 2, 3$ denote inertia matrix elements. Based on the open-loop tracking error system given by (5.40) and the subsequent stability analysis, the control input $u^*(t)$ is designed as follows

$$u^* = -Y\hat{\theta} - Kr - \frac{e_v}{(1 - e_v^T e_v)^2} \tag{5.43}$$

where $K \in \mathbb{R}^{3\times3}$ is a constant positive-definite diagonal control gain matrix, and $\hat{\theta}(t) \in \mathbb{R}^6$ denotes a parameter estimate vector for (5.42) that is generated by the following gradient update law

$$\dot{\hat{\theta}} = \Gamma Y^T (\cdot) r \tag{5.44}$$

where $\Gamma \in \mathbb{R}^{6\times6}$ is a constant positive-definite diagonal adaptation gain matrix. After substituting (5.43) into (5.40) for $u^*(t)$, the following closed-loop tracking error dynamics can be obtained

$$J^* \dot{r} = -C^* r + Y\tilde{\theta} - Kr - \frac{e_v}{(1 - e_v^T e_v)^2} \tag{5.45}$$

where $\tilde{\theta}(t) \in \mathbb{R}^6$ denotes the following parameter estimate error signal

$$\tilde{\theta}(t) = \theta - \hat{\theta}(t). \tag{5.46}$$

Stability Analysis

The adaptive full-state feedback controller of (5.43) and (5.44) provides asymptotic attitude tracking as stated by the following theorem.

Theorem 5.1 *The adaptive attitude controller given in (5.43) and (5.44) ensures asymptotic attitude tracking in the sense that*

$$\lim_{t \to \infty} e_v(t) = 0 \qquad and \qquad \lim_{t \to \infty} \tilde{\omega}(t) = 0 \tag{5.47}$$

provided that the initial conditions are selected such that

$$e_0(0) \neq 0. \tag{5.48}$$

Proof: To prove Theorem 5.1, we define a nonnegative function $V(t) \in \mathbb{R}$ as follows

$$V = \frac{1}{2} \left(\frac{e_v^T e_v}{1 - e_v^T e_v} \right) + \frac{1}{2} y^T J y + \frac{1}{2} \tilde{\theta}^T \Gamma^{-1} \tilde{\theta} \tag{5.49}$$

where $y(t) \in \mathbb{R}^3$ is defined as follows

$$y = T^{-1} r \tag{5.50}$$

where $T(e)$ and $r(t)$ are defined in (5.23) and (5.38), respectively. After taking the time derivative of (5.49), utilizing (5.25), (5.44), (5.45), and (5.50), and cancelling common terms, the following expression can be obtained

$$\dot{V}(t) = \frac{e_v^T \dot{e}_v}{(1 - e_v^T e_v)^2} - r^T C^* r - r^T K r - \frac{r^T e_v}{(1 - e_v^T e_v)^2} + \frac{1}{2} r^T \dot{J}^* r. \tag{5.51}$$

After substituting (5.38) into (5.51) for $\dot{e}_v(t)$, the following expression can be obtained

$$\dot{V}(t) = \frac{-e_v^T \alpha e_v}{(1 - e_v^T e_v)^2} - r^T K r \tag{5.52}$$

where the skew-symmetric property given in (5.30) has been used.

The development given in (5.49) and (5.52) can be used to prove that

$$0 \leq V(t) \leq V(0) < \infty, \tag{5.53}$$

and hence, $y(t)$, $\tilde{\theta}(t) \in \mathcal{L}_\infty$. Based on (5.53), the development given in (5.34), (5.35), (5.48), and (5.53) can be used to prove that

$$\|e_v(t)\| < 1. \tag{5.54}$$

The constraint introduced in (5.34) and the initial condition constraint given in (5.48) can be used in conjunction with the inequality given in (5.54) to prove that $e_0(t)$ is nonzero for all time. Hence, (5.32) can be used to prove that $T(e)$ is invertible for all time. Since $y(t) \in \mathcal{L}_\infty$ and $T(e)$ is invertible for all time, (5.50) can be utilized to prove that $r(t)$, $\tilde{\theta}(t) \in \mathcal{L}_\infty$. Given that $r(t) \in \mathcal{L}_\infty$, Lemma A.13 of Appendix A can be invoked to prove that $e_v(t)$, $\dot{e}_v(t) \in \mathcal{L}_\infty$. Hence, (5.25), (5.28), (5.29), (5.41), (5.45), and (5.54) can be used to prove that $Y(\cdot)$, $\dot{r}(t) \in \mathcal{L}_\infty$. The differential equation given in (5.44) can be used to prove that $\dot{\hat{\theta}}(t) \in \mathcal{L}_\infty$, and the expressions given in (5.27) and (5.43) can be used to prove that $u(t)$, $u^*(t) \in \mathcal{L}_\infty$. Based on (5.52) and (5.53), Lemma A.11 of Appendix A can be invoked to prove that $r(t) \in \mathcal{L}_2$. Based on the fact that $r(t), \dot{r}(t) \in \mathcal{L}_\infty$ and that $r(t) \in \mathcal{L}_2$, Barbalat's Lemma (see Lemma A.16 of Appendix A) can be invoked to prove that

$$\lim_{t \to \infty} r(t) = 0. \tag{5.55}$$

Based on (5.38) and the result obtained in (5.55), Lemma A.15 of Appendix A can be invoked to prove that

$$\lim_{t \to \infty} e_v(t), \dot{e}_v(t) = 0. \tag{5.56}$$

Based on (5.22) and the fact that $T(e)$ is invertible, (5.56) can be used to prove that

$$\lim_{t \to \infty} \tilde{\omega}(t) = 0. \quad \square \tag{5.57}$$

5.2.4 Adaptive Output Feedback Controller

The adaptive control development in the previous section is based on the assumption that full-state feedback is available. However, the cost of having accurate angular velocity sensors for each axis on a spacecraft may be prohibitive. The requirement for additional sensors (that are not in a redundant configuration) decreases the system reliability. Moreover, most angular velocity sensors yield noisy measurements that require costly processing before they can be usefully utilized in the control algorithm. The backwards difference algorithm is commonly employed to produce a surrogate for velocity measurements without requiring additional sensors. However, this approach is theoretically unsatisfying because the effects of the backwards difference algorithm are not examined in the stability analysis. Motivated by the desire to design a control torque input that eliminates the need for velocity measurements, a filtered tracking error signal, denoted by $e_{vf}(t) \in \mathbb{R}^3$, is defined as follows

$$e_{vf} = -ke_v + p \tag{5.58}$$

where $k \in \mathbb{R}$ is a positive, constant control gain, and $p(t) \in \mathbb{R}^3$ is generated by the following differential expression and initial condition

$$\dot{p} = -(k+1)\,p + k^2 e_v + \frac{e_v}{\left(1 - e_v^T e_v\right)^2} \qquad p(0) = k e_v(0). \qquad (5.59)$$

To facilitate the subsequent stability analysis, an error signal, denoted by $\eta(t) \in \mathbb{R}^3$, is defined as follows

$$\eta = e_v + e_{vf} + \dot{e}_v \qquad (5.60)$$

where $\eta(t)$ is unmeasurable because it is dependent on velocity measurements.

Closed-Loop Error System

To determine the closed-loop dynamics for $e_{vf}(t)$, we take the time derivative of (5.58) and then substitute (5.59) into the resulting expression as follows

$$\dot{e}_{vf} = -k\dot{e}_v - (k+1)\,p + k^2 e_v + \frac{e_v}{\left(1 - e_v^T e_v\right)^2}. \qquad (5.61)$$

After utilizing (5.58) and (5.60), the closed-loop dynamics for $e_{vf}(t)$ can be rewritten as follows

$$\dot{e}_{vf} = -k\eta - e_{vf} + \frac{e_v}{\left(1 - e_v^T e_v\right)^2}. \qquad (5.62)$$

To develop the closed-loop expression for $\eta(t)$, we take the time derivative of (5.60), premultiply the resulting expression by $J^*(\cdot)$ of (5.25), and then substitute (5.26) for $J^*(\cdot)\,\ddot{e}_v(t)$ to obtain the following expression

$$J^*\dot{\eta} = u^* - C^*\dot{e}_v - N^* + J^*\dot{e}_v + J^*\dot{e}_{vf}. \qquad (5.63)$$

To facilitate further development, the following linear parameterization is defined

$$W_d\theta = -J\dot{\omega}_d - \frac{1}{2}\omega_d^\times J\omega_d, \qquad (5.64)$$

where $W_d(\omega_d, \dot{\omega}_d) \in \mathbb{R}^{3\times 6}$ is a known regression matrix, and θ is the vector of unknown constant inertia parameters introduced in (5.42). After adding and subtracting $W_d\theta$ to the right side of (5.63) and utilizing (5.60) and (5.62), (5.63) can be rewritten as follows

$$J^*\dot{\eta} = u^* - kJ^*\eta - C^*\eta + \chi + W_d\theta \qquad (5.65)$$

where $\chi(e, e_{vf}, \eta, t) \in \mathbb{R}^3$ is defined as follows

$$\chi = C^*\left(e_{vf} + e_v\right) + J^*\left(\eta - e_v + \frac{e_v}{\left(1 - e_v^T e_v\right)^2}\right) - 2J^* e_{vf} - W_d\theta - N^*. \qquad (5.66)$$

The auxiliary signal $\overline{\chi}(e, e_{vf}, \eta, t) \in \mathbb{R}^3$ defined as follows facilitates the subsequent stability analysis

$$\overline{\chi} = T^T \chi \tag{5.67}$$

where $\chi(\cdot)$ is introduced in (5.66). Based on the boundedness assumptions for the desired attitude trajectory and the structure of (5.67), $\overline{\chi}(\cdot)$ can be upper bounded as follows (see Lemma B.20 of Appendix B)

$$\|\overline{\chi}\| \leq \rho(\|z\|) \|z\| \tag{5.68}$$

where $\rho(\cdot) \in \mathbb{R}$ is a positive, nondecreasing function, $z(t) \in \mathbb{R}^9$ is defined as

$$z = \left[\begin{array}{ccc} \dfrac{e_v^T}{\sqrt{1 - e_v^T e_v}} & e_{vf}^T & y^T \end{array} \right]^T, \tag{5.69}$$

and $y(t) \in \mathbb{R}^3$ is redefined for the output feedback case as follows

$$y = T^{-1}\eta. \tag{5.70}$$

Based on the structure of (5.26) and the subsequent stability analysis, the transformed control input is designed as follows

$$u^* = -W_d\hat{\theta} + ke_{vf} - \frac{e_v}{\left(1 - e_v^T e_v\right)^2} \tag{5.71}$$

where $W_d(\cdot)$ was introduced in (5.64), k is the same control gain introduced in (5.58), and $\hat{\theta}(t) \in \mathbb{R}^6$ is a parameter update law generated by the following differential expression

$$\dot{\hat{\theta}} = \Gamma W_d^T \eta, \tag{5.72}$$

where $\Gamma \in \mathbb{R}^{6 \times 6}$ is a positive-definite diagonal constant control gain matrix. To illustrate that $\hat{\theta}(t)$ can be calculated using only measurable signals, (5.72) can be rewritten as the following integral expression

$$\hat{\theta}(t) = \Gamma \int_0^t W_d^T\left(\omega_d(\sigma), \dot{\omega}_d(\sigma)\right)\left(\dot{e}_v(\sigma) + e_v(\sigma) + e_{vf}(\sigma)\right) d\sigma + \hat{\theta}(0). \tag{5.73}$$

After performing integration by parts, (5.73) can be written in the following velocity-independent output feedback form

$$\hat{\theta}(t) = \Gamma \int_0^t \left[W_d^T(\cdot)\left(e_v(\sigma) + e_{vf}(\sigma)\right) - \dot{W}_d^T(\cdot) e_v(\sigma) \right] d\sigma + \Gamma W_d^T e_v + \hat{\theta}(0). \tag{5.74}$$

Based on the subsequent stability analysis, the control gain given in (5.58) and (5.71) is selected as follows

$$k = \frac{1}{j_1} (k_n + 1) \tag{5.75}$$

where j_1 was defined in (5.31), and $k_n \in \mathbb{R}$ denotes a positive, constant control gain. After substituting (5.71) into (5.65) for $u^*(t)$, the following closed-loop error system can be obtained

$$J^* \dot{\eta} = \chi + W_d \tilde{\theta} + \frac{1}{j_1} (k_n + 1) (e_{vf} - J^* \eta) - \frac{e_v}{(1 - e_v^T e_v)^2} - C^* \eta \tag{5.76}$$

where $\tilde{\theta}(t)$ was defined in (5.46) and (5.75) was used.

Stability Analysis

The adaptive output feedback control strategy of (5.58), (5.71) and (5.74) yields asymptotic attitude tracking as stated by the following theorem.

Theorem 5.2 *Given the closed-loop error systems of (5.60), (5.62), and (5.76), the adaptive controller of (5.58–5.71) and (5.74) ensures asymptotic attitude tracking in the sense that*

$$\lim_{t \to \infty} e_v(t) = 0 \qquad and \qquad \lim_{t \to \infty} \tilde{\omega}(t) = 0 \tag{5.77}$$

provided that the desired trajectory is selected to ensure that $e_0(0) \neq 0$, and the control gain k_n introduced in (5.75) is selected according to the following inequality

$$k_n > \frac{1}{4} \rho^2 \left(\sqrt{\frac{\lambda_2}{\lambda_1}} \, \|x(0)\| \right) \tag{5.78}$$

where $\rho(\cdot)$ was defined in (5.68), $x(t) \in \mathbb{R}^{15}$ is defined as follows

$$x = \begin{bmatrix} z^T & \tilde{\theta}^T \end{bmatrix}^T, \tag{5.79}$$

and λ_1, λ_2 are positive constants defined as

$$\lambda_1 = \frac{1}{2} \min \left\{ 1, j_1, \lambda_{\min} \left\{ \Gamma^{-1} \right\} \right\} \qquad \lambda_2 = \frac{1}{2} \max \left\{ 1, j_2, \lambda_{\max} \left\{ \Gamma^{-1} \right\} \right\}, \tag{5.80}$$

where $\lambda_{\min} \{\cdot\}$ and $\lambda_{\max} \{\cdot\}$ represent the minimum and maximum eigenvalue of a matrix, respectively.

Proof: To prove Theorem 5.2, we define a nonnegative function $V(t) \in \mathbb{R}$ as follows

$$V = \frac{1}{2} \left(\frac{e_v^T e_v}{1 - e_v^T e_v} \right) + \frac{1}{2} e_{vf}^T e_{vf} + \frac{1}{2} y^T J y + \frac{1}{2} \tilde{\theta}^T \Gamma^{-1} \tilde{\theta}. \tag{5.81}$$

Based on the structure of (5.81) and the boundedness property given in (5.31) for the inertia matrix, $V(t)$ can be lower and upper bounded as follows

$$\lambda_1 \|x\|^2 \leq V \leq \lambda_2 \|x\|^2 \tag{5.82}$$

where $x(t)$ was defined in (5.79), and λ_1, λ_2 were defined in (5.80). After taking the time derivative of (5.81), substituting (5.70) for $y(t)$, using (5.25) and (5.76), and then cancelling common terms, the following expression can be obtained

$$\dot{V} = \frac{e_v^T \dot{e}_v}{(1 - e_v^T e_v)^2} + e_{vf}^T \dot{e}_{vf} + \frac{1}{2} \eta^T \dot{J}^* \eta + \tilde{\theta}^T \Gamma^{-1} \dot{\tilde{\theta}}$$

$$+ \eta^T \left(\chi + W_d \tilde{\theta} + \frac{1}{j_1} (k_n + 1)(e_{vf} - J^* \eta) - \frac{e_v}{(1 - e_v^T e_v)^2} - C^* \eta \right). \tag{5.83}$$

After utilizing (5.60), (5.62), (5.67), (5.70), and the skew-symmetric property of (5.30), (5.83) can be rewritten as follows

$$\dot{V} = -\frac{e_v^T e_v}{(1 - e_v^T e_v)^2} - e_{vf}^T e_{vf} + y^T \overline{\chi}$$

$$- \frac{1}{j_1} (k_n + 1) \eta^T J^* \eta + \tilde{\theta}^T \left(W_d^T \eta + \Gamma^{-1} \dot{\tilde{\theta}} \right). \tag{5.84}$$

After utilizing (5.25), (5.72), and (5.70), the following inequality can be developed

$$\dot{V} \leq -\frac{e_v^T e_v}{(1 - e_v^T e_v)^2} - e_{vf}^T e_{vf} - (k_n + 1) \|y\|^2 + \|y\| \|\overline{\chi}\|. \tag{5.85}$$

After substituting (5.68) into (5.85) for $\overline{\chi}(\cdot)$, the following expression can be obtained

$$\dot{V} \leq -\|z\|^2 + \left[\|y\| \|z\| \rho(\|z\|) - k_n \|y\|^2 \right] \tag{5.86}$$

where (5.69) was utilized. By completing the squares on the bracketed term in (5.86) (see also the nonlinear damping tool in Lemma A.17 of Appendix A) the following expression can be obtained

$$\dot{V} \leq -\left(1 - \frac{\rho^2(\|z\|)}{4k_n} \right) \|z\|^2. \tag{5.87}$$

By using the definition given in (5.79), the inequality given in (5.87) can be written as follows

$$\dot{V} \leq -\beta \|z\|^2 \quad \text{for} \quad k_n > \frac{1}{4} \rho^2(\|x\|) \tag{5.88}$$

where $\beta \in \mathbb{R}$ is some positive constant. Given (5.82), the following sufficient condition can be developed

$$\dot{V} \leq -\beta \|z\|^2 \quad \text{for} \quad k_n > \frac{1}{4}\rho^2 \left(\sqrt{\frac{V(t)}{\lambda_1}} \right). \tag{5.89}$$

From (5.82) and (5.89), the following inequalities can be obtained

$$0 \leq V(t) \leq V(0) < \infty, \tag{5.90}$$

and hence, $y(t)$, $\tilde{\theta}(t)$, $e_{vf}(t) \in \mathcal{L}_\infty$. The inequalities given in (5.82) and (5.90) can now be used to formulate the following sufficient condition for (5.89)

$$\dot{V} \leq -\beta \|z\|^2 \quad \text{for} \quad k_n > \frac{1}{4}\rho^2 \left(\sqrt{\frac{\lambda_2}{\lambda_1}} \|x(0)\| \right). \tag{5.91}$$

Based on (5.90), the development given in (5.34), (5.35), (5.48), and (5.53) can be used to prove that

$$\|e_v(t)\| < 1. \tag{5.92}$$

Given (5.34) and the assumption that the desired trajectory is selected to ensure that $e_0(0) \neq 0$, the inequality given in (5.92) can be used to prove that $e_0(t)$ is nonzero for all time. Hence, (5.32) can be used to prove that $T(e)$ is invertible for all time. Since $y(t) \in \mathcal{L}_\infty$ and $T(e)$ is invertible, (5.70) can be used to prove that $\eta(t) \in \mathcal{L}_\infty$. It is now possible to utilize the previous development and (5.60) to prove that $\dot{e}_v(t) \in \mathcal{L}_\infty$. Based on (5.92) and the facts that $e_v(t)$, $\tilde{\theta}(t)$, $e_{vf}(t) \in \mathcal{L}_\infty$, (5.69) and (5.79) can be used to prove that $z(t)$, $x(t) \in \mathcal{L}_\infty$. Since $\eta(t)$, $e_{vf}(t)$, $e_v(t) \in \mathcal{L}_\infty$ and $\|e_v(t)\| < 1$, (5.62) can be used to prove that $\dot{e}_{vf}(t) \in \mathcal{L}_\infty$. Given $z(t) \in \mathcal{L}_\infty$, (5.68) can be used to determine that $\overline{\chi}(\cdot) \in \mathcal{L}_\infty$. Based on the fact that $\overline{\chi}(\cdot), \eta(t), e_{vf}(t), e_v(t) \in \mathcal{L}_\infty$, (5.76) and the result given in (5.92) can be used to prove that $\dot{\eta}(t) \in \mathcal{L}_\infty$. Based on (5.92) and the facts $\dot{\eta}(t)$, $\dot{e}_v(t)$, $\dot{e}_{vf}(t) \in \mathcal{L}_\infty$, the definition given in (5.69) can be used to determine that $\dot{z}(t) \in \mathcal{L}_\infty$. Based on (5.88) and (5.90), Lemma A.11 of Appendix A can be invoked to prove that $z(t) \in \mathcal{L}_2$. Based on the fact that $z(t), \dot{z}(t) \in \mathcal{L}_\infty$ and that $z(t) \in \mathcal{L}_2$, Barbalat's Lemma (see Lemma A.16 of Appendix A) can be invoked to prove that

$$\lim_{t \to \infty} z(t) = 0 \quad \text{for} \quad k_n > \rho^2 \left(\sqrt{\frac{\lambda_2}{\lambda_1}} \|x(0)\| \right). \tag{5.93}$$

Since $T(e)$ is invertible, (5.22), (5.60), (5.69), (5.70), and (5.93) can be used to obtain the result given in (5.77). $\quad \square$

5.2.5   Simulation Results

Numerical simulation results were obtained to validate the adaptive output feedback controller developed in (5.58–5.71) and (5.74). The simulation is based on the following inertia matrix [1]

$$
J = \begin{bmatrix} 20 & 1.2 & 0.9 \\ 1.2 & 17 & 1.4 \\ 0.9 & 1.4 & 15 \end{bmatrix} \ [\mathrm{kg \cdot m^2}]. \tag{5.94}
$$

The initial spacecraft attitude was selected as

$$
q_0(0) = \sqrt{0.1} \qquad q_v(0) = \begin{bmatrix} 0 & -\sqrt{0.45} & -\sqrt{0.45} \end{bmatrix}^T \tag{5.95}
$$

and the parameter estimates were initialized to zero (typically, the parameter estimates would be initialized to the best-guess values, however, the simulation is based on the assumption that no knowledge of the parameter values is available). The desired attitude trajectory was selected as follows

$$
q_{0d} = \sqrt{0.1} \qquad q_{vd}(t) = \begin{bmatrix} \sqrt{0.9}\sin(t) & \sqrt{0.45}\cos(t) & \sqrt{0.45}\cos(t) \end{bmatrix}^T. \tag{5.96}
$$

The control gain introduced in (5.58) and the adaptation gain matrix introduced in (5.72) were selected as follows

$$
k = 50 \qquad \Gamma_{ii} = 5000 \quad \forall i = 1, 2, ...6. \tag{5.97}
$$

The attitude and angular velocity tracking errors are depicted in Figures 5.2 and 5.3. The dynamic estimates of the inertia matrix are shown in Figure 5.4. The resulting control torque input is shown in Figure 5.5.

5.3   Energy/Power and Attitude Tracking

In the previous section, adaptive controllers were developed to achieve attitude tracking. However, the energy/power required to achieve the desired attitude maneuvers was not considered. Motivated by the economics associated with spaceflight and the desire to achieve extended mission life cycles, the need for a power management system that operates in tandem with the controller (i.e., an IPACS) is strong. In this section, an adaptive IPACS is developed that forces a spacecraft to track a desired attitude trajectory while simultaneously providing exponential energy/power tracking.

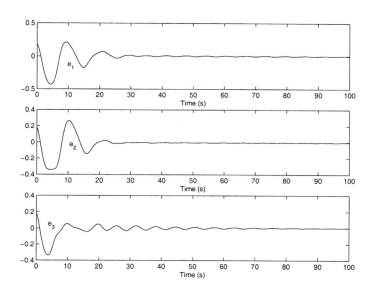

FIGURE 5.2. Position tracking error $e(t)$.

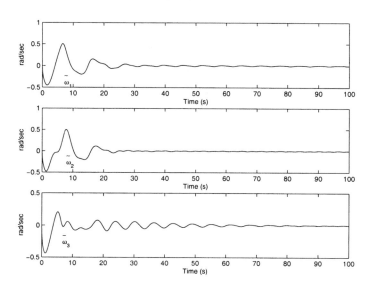

FIGURE 5.3. Angular velocity tracking error $\tilde{\omega}(t)$.

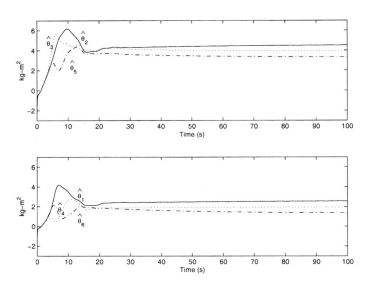

FIGURE 5.4. Parameter estimates $\hat{\theta}(t)$.

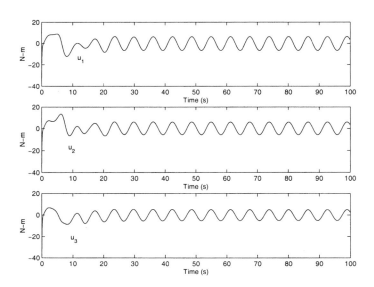

FIGURE 5.5. Control torques $u(t)$.

5.3.1 System Model

In this section, the same spacecraft system described in the previous section is used (see Figure 5.1). However, in this section, the body-fixed torques are assumed to be applied by an array of n (where $n > 3$) noncoplanar flywheels with a fixed axis of rotation with respect to $\mathcal{B}$ (i.e., reaction wheel-type mode of operation). A four noncoplanar flywheel cluster operating in reaction wheel mode is depicted in Figure 5.6 where W_1, W_2, W_3, W_4 denote flywheel clusters. The same differential equations introduced in (5.4) and (5.5) govern the attitude kinematics. As stated in [18, 38], the nonlinear dynamic model for the flywheel driven system can be written as follows

$$J\dot{\omega} = -\omega^\times J\omega - \omega^\times AJ_f\omega_f - A\tau_f \tag{5.98}$$

$$J_f\dot{\omega}_f = \tau_f. \tag{5.99}$$

In (5.98) and (5.99), $\omega(t) \in \mathbb{R}^3$ denotes the spacecraft angular velocity with respect to the inertial reference frame $\mathcal{I}$ expressed in terms of the body-fixed coordinate frame $\mathcal{B}$, $\omega_f(t) \in \mathbb{R}^n$ is the axial angular velocity of the flywheels[3] with respect to $\mathcal{B}$, $J \in \mathbb{R}^{3\times3}$ represents the unknown constant positive-definite symmetric inertia matrix inclusive of the flywheels, $J_f \in \mathbb{R}^{n\times n}$ is the unknown constant positive-definite diagonal matrix of the axial moments of inertia of the flywheels, $A \in \mathbb{R}^{3\times n}$ is a constant torque transmission matrix whose columns contain the axial unit vectors of the n flywheels, and $\tau_f(t) \in \mathbb{R}^n$ is the flywheel torque control input. Based on (5.99), the kinetic energy of the flywheel array can be determined as follows

$$E(t) = \frac{1}{2}\omega_f^T(t)J_f\omega_f(t). \tag{5.100}$$

5.3.2 Control Objective

The control objective in this section is to design the flywheel torque control input such that the flywheels track a desired energy/power profile, and the spacecraft tracks a desired attitude trajectory. The desired kinetic energy profile is denoted by $E_d(t) \in \mathbb{R}$ and the desired power requirement is denoted by $P_d(t) \in \mathbb{R}$. The energy/power profiles are assumed to be selected so that $E_d(t)$, $P_d(t) \in \mathcal{L}_\infty$, and the profiles are related through the

---

[3] In formulating (5.98) and (5.99), the flywheels are assumed to spin at a much higher speed than the spacecraft [38].

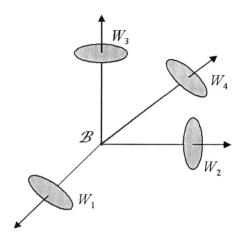

FIGURE 5.6. Schematic for flywheels in reaction wheel mode.

following integral expression

$$E_d(t) = \int_0^t P_d(\sigma)d\sigma. \tag{5.101}$$

To quantify the energy tracking objective, a kinetic energy tracking error $\eta_E(t) \in \mathbb{R}$ is defined as

$$\eta_E = E_d - E. \tag{5.102}$$

Based on (5.101) and (5.102), the power tracking error can be quantified by the following expression

$$\dot{\eta}_E = P_d - \dot{E}. \tag{5.103}$$

As stated in the previous section, the open-loop quaternion attitude tracking error system is given by (5.15) and (5.16), where it is assumed that the desired attitude trajectory is selected such that $\omega_d(t)$, $\dot{\omega}_d(t) \in \mathcal{L}_\infty$. To facilitate the subsequent IPACS development and stability analysis, an auxiliary attitude tracking error signal, denoted by $\eta(t) \in \mathbb{R}^3$, is defined as follows

$$\eta = \tilde{\omega} + K_1 e_v \tag{5.104}$$

where $K_1 \in \mathbb{R}^{3 \times 3}$ is a constant positive-definite diagonal control gain matrix. The subsequent control development will target the control objectives under the constraint that the spacecraft inertia matrix J is unknown and under the assumption that the spacecraft attitude $q(t)$, spacecraft angular velocity $\dot{q}(t)$, and flywheel axial angular velocities $\omega_f(t)$ are measurable.

5.3.3 Adaptive IPACS

To develop the closed-loop dynamics for the attitude tracking error, we take the time derivative of (5.104), premultiply by the inertia matrix, and then utilize (5.17) to obtain the following expression

$$J\dot{\eta} = J\dot{\omega} - J\,\dot{\tilde{R}}\,\omega_d - J\tilde{R}\dot{\omega}_d + JK_1\dot{e}_v. \tag{5.105}$$

After substituting (5.15) and (5.98) into (5.105) for $\dot{e}_v(t)$ and $\dot{\omega}(t)$, respectively, and then using the following fact[4] [36]

$$\dot{\tilde{R}} = -\tilde{\omega}^\times \tilde{R}, \tag{5.106}$$

the open-loop dynamics for $\eta(t)$ can be obtained as follows

$$J\dot{\eta} = Y\theta - \omega^\times AJ_f\omega_f - A\tau_f. \tag{5.107}$$

In (5.107), the linear parameterization $Y(\cdot)\theta$ is defined as follows

$$Y\theta = -\omega^\times J\omega + J\left(\tilde{\omega}^\times \tilde{R}\omega_d - \tilde{R}\dot{\omega}_d\right) + \frac{JK_1}{2}\left(e_v^\times \tilde{\omega} + e_o\tilde{\omega}\right) \tag{5.108}$$

where $Y(e,\omega,\omega_d,\dot{\omega}_d) \in \mathbb{R}^{3\times 6}$ is a measurable regression matrix, and θ is the vector of unknown constant inertia parameters introduced in (5.42). Since the torque transmission matrix A introduced in (5.98) and given in (5.107) is nonsquare, a pseudo-inverse of A must be used. Specifically, the pseudo-inverse of A, denoted by $A^+ \in \mathbb{R}^{n\times 3}$, is defined as follows

$$A^+ = A^T\left(AA^T\right)^{-1} \quad \text{such that} \quad AA^+ = I_3 \tag{5.109}$$

where I_n denotes the $n \times n$ identity matrix. As shown in [28], the pseudo-inverse given in (5.109) satisfies the Moore-Penrose Conditions given below

$$AA^+A = A \qquad A^+A\,A^+ = A^+$$
$$\left(A^+A\right)^T = A^+A \qquad \left(AA^+\right)^T = AA^+. \tag{5.110}$$

---

[4]The negative sign in (5.106) exists because the coordinates of (5.17) are expressed in terms of the body-fixed coordinate frame rather than the inertial reference frame.

In addition, the matrix $I_n - A^+A$, which projects vectors onto the null space of A, satisfies the following properties

$$
\begin{aligned}
(I_n - A^+A)(I_n - A^+A) &= I_n - A^+A & A(I_n - A^+A) &= 0 \\
(I_n - A^+A)^T &= (I_n - A^+A) & (I_n - A^+A)A^+ &= 0.
\end{aligned} \tag{5.111}
$$

Based on the structure of (5.107) and the subsequent stability analysis, the flywheel torque control input is designed as follows

$$
\tau_f = A^+ u_c + (I_n - A^+A)g \tag{5.112}
$$

where $u_c(t) \in \mathbb{R}^3$ is defined as follows

$$
u_c = Y\hat{\theta} + K_2\eta + e_v - \omega^\times A J_f \omega_f \tag{5.113}
$$

and $g(t) \in \mathbb{R}^n$ is an auxiliary function that will be designed to facilitate the energy/power tracking objective. In (5.113), $K_2 \in \mathbb{R}^{3\times3}$ denotes a constant positive-definite diagonal control gain matrix, and the parameter estimate $\hat{\theta}(t) \in \mathbb{R}^6$ is generated via the following gradient adaptive update law

$$
\dot{\hat{\theta}} = \Gamma Y^T \eta \tag{5.114}
$$

where $\Gamma \in \mathbb{R}^{6\times6}$ denotes a constant positive-definite diagonal adaptation gain matrix. The subspace of the control input vector $\tau_f(t)$ that lies in the range space of the torque transmission matrix A can be used for attitude control, while the remaining degrees of freedom of the flywheel cluster can be used to track a desired energy or power profile. That is, since the vector $(I_n - A^+A)g(t)$ lies in the null space of the torque transmission matrix, $g(t)$ will be subsequently designed independently of the attitude controller in order to track the desired energy/profile. After substituting (5.112) and (5.113) into (5.107), the following closed-loop dynamics for $\eta(t)$ can be obtained

$$
J\dot{\eta} = -K_2\eta + Y\tilde{\theta} - e_v \tag{5.115}
$$

where (5.109), (5.111), and (5.116) have been utilized, and $\tilde{\theta}(t) \in \mathbb{R}^6$ denotes the following parameter estimation error

$$
\tilde{\theta} = \theta - \hat{\theta}. \tag{5.116}
$$

To develop the closed-loop dynamics for the power tracking error, we substitute the time derivative of (5.100) into (5.103) for $\dot{E}(t)$ as follows

$$
\dot{\eta}_E = P_d - \omega_f^T \tau_f \tag{5.117}
$$

where (5.99) has been utilized. After substituting (5.112) into (5.117), the following expression can be obtained

$$\dot{\eta}_E = P_d - \omega_f^T A^+ u_c - \omega_f^T \left(I_n - A^+ A \right) g. \tag{5.118}$$

Based on the structure of (5.118), the signal $g(t)$ must be designed to satisfy the following equation

$$\omega_f^T \left(I_n - A^+ A \right) g = P_d - \omega_f^T A^+ u_c + k_E \eta_E \tag{5.119}$$

where $k_E \in \mathbb{R}$ is a constant positive control gain. Based on the Moore-Penrose pseudo-inverse properties introduced in (5.111), the minimum norm solution of (5.119) is given by the following expression

$$g = \left(I_n - A^+ A \right) \omega_f \left[\omega_f^T \left(I_n - A^+ A \right) \omega_f \right]^{-1} \left(P_d - \omega_f^T A^+ u_c + k_E \eta_E \right). \tag{5.120}$$

The solution given in (5.120) exists if $\left(I_n - A^+ A \right) \omega_f \neq 0$, which implies that $\omega_f(t) \neq 0$ and that $\omega_f(t)$ is not included in the null space of $I_n - A^+ A$. The control singularity in (5.120) is similar to the one encountered in [38]. As noted in [38], the practical implication of this singularity to the IPACS is that the controller will lose the capability to track the desired energy/power function. One method to avoid this problem is the singularity-avoidance scheme described in [38]. After substituting (5.120) into (5.118) for $g(t)$ the following closed-loop error system can be obtained

$$\dot{\eta}_E = -k_E \eta_E. \tag{5.121}$$

5.3.4 Stability Analysis

Given the adaptive tracking controller of (5.112), (5.114), and (5.120), asymptotic attitude tracking and exponential energy/power tracking can be obtained as described by the following theorem.

Theorem 5.3 *The flywheel torque control input of (5.112), (5.114), and (5.120) ensures asymptotic attitude tracking in the sense that*

$$\lim_{t \to \infty} e_v(t) = 0 \tag{5.122}$$

and exponential energy/power tracking in the sense that

$$\eta_E(t) = \eta_E(0) \exp\left(-k_E t\right). \tag{5.123}$$

Proof: Based on the differential expression given in (5.121), Lemma A.10 of Appendix A can be used to prove that $\eta_E(t) \in \mathcal{L}_\infty$ and that the exponential energy/power tracking result given in (5.123) is obtained with a rate of convergence that can be made arbitrarily fast by adjusting the control gain k_E. To prove the asymptotic attitude tracking result of Theorem 5.3, we define the nonnegative function $V(t) \in \mathbb{R}$ as follows

$$V = (e_o - 1)^2 + e_v^T e_v + \frac{1}{2}\eta^T J\eta + \frac{1}{2}\tilde{\theta}^T \Gamma^{-1}\tilde{\theta}. \tag{5.124}$$

After taking the time derivative of (5.124) and utilizing (5.15), (5.16), (5.114), and (5.115), the following negative semi-definite expression can be obtained after algebraic simplification

$$\dot{V} = -e_v^T K_1 e_v - \eta^T K_2 \eta. \tag{5.125}$$

Based on the structure of (5.124) and (5.125), $\tilde{\theta}(t) \in \mathcal{L}_\infty$ and $e_v(t)$, $\eta(t) \in \mathcal{L}_\infty \cap \mathcal{L}_2$ (See Lemma A.11 of Appendix A). Based on the fact that $e_v(t)$, $\eta(t)$, $\tilde{\theta}(t) \in \mathcal{L}_\infty$, (5.12), (5.34), (5.104), and (5.116) can be used to prove that $\tilde{R}(t)$, $\tilde{w}(t)$, $\hat{\theta}(t) \in \mathcal{L}_\infty$. The expressions given in (5.15–5.17) can be used to determine that $\dot{e}_o(t)$, $\dot{e}_v(t)$, $w(t) \in \mathcal{L}_\infty$. Based on the fact that $\eta_E(t) \in \mathcal{L}_\infty$, and due to the assumption that $E_d(t) \in \mathcal{L}_\infty$, (5.100) and (5.102) can be used to prove that $E(t)$, $w_f(t) \in \mathcal{L}_\infty$. From the previous boundedness statements, (5.108) can now be used to prove that $Y(t) \in \mathcal{L}_\infty$. Hence, from (5.113), $u_c(t) \in \mathcal{L}_\infty$. Due to the assumption that $P_d(t) \in \mathcal{L}_\infty$, (5.120) can now be used to determine that $g(t) \in \mathcal{L}_\infty$ if its minimum norm solution exists. Hence, from (5.112), $\tau_f(t) \in \mathcal{L}_\infty$. Based on the facts that $e_v(t)$, $\dot{e}_v(t) \in \mathcal{L}_\infty$ and that $e_v(t) \in \mathcal{L}_2$, Barbalat's Lemma (see Lemma A.16 of Appendix A) can be invoked to prove asymptotic attitude tracking in the sense of (5.122). $\square$

5.3.5   Simulation Results

Numerical simulation results were obtained to validate the adaptive IPACS developed in (5.112), (5.114), and (5.120). The inertia matrix for the spacecraft was selected as follows [38]

$$J = \operatorname{diag}\{200, 200, 175\} \; [\text{kg} \cdot \text{m}^2]. \tag{5.126}$$

The four-flywheel cluster described in [38] was used for the simulation, where the flywheel inertia and torque transmission matrix were selected as follows

$$J_f = 0.7I_4 \; [\text{kg} \cdot \text{m}^2] \qquad A = \begin{bmatrix} 1 & 0 & 0 & \frac{\sqrt{3}}{3} \\ 0 & 1 & 0 & \frac{\sqrt{3}}{3} \\ 0 & 0 & 1 & \frac{\sqrt{3}}{3} \end{bmatrix}. \tag{5.127}$$

The initial attitude of the spacecraft and the initial angular velocity of the flywheels were set as follows

$$q_o(0) = 1, \qquad q_v(0) = [0,0,0]^T,$$

$$\omega_f(0) = [100,100,100,100]^T \, [\text{rad} \cdot \text{s}^{-1}]. \tag{5.128}$$

The desired angular velocity of $\mathcal{F}_d$ with respect to $\mathcal{I}$ expressed in $\mathcal{F}_d$ was selected to be a smooth rotation as follows

$$\omega_d(t) = \begin{bmatrix} 3.33 \times 10^{-3} \exp\left(-1.6 \times 10^{-4}t\right) \\ 3.33 \times 10^{-3} \exp\left(-1.6 \times 10^{-4}t\right) \\ 3.33 \times 10^{-3} \exp\left(-1.6 \times 10^{-4}t\right) \end{bmatrix} \, [\text{rad} \cdot \text{s}^{-1}]. \tag{5.129}$$

The desired quaternion is related to $\omega_d(t)$ through the differential equations introduced in (5.9) and (5.10). The expressions given in (5.9) and (5.10) were integrated with $\omega_d(t)$ of (5.129) and the following initial conditions

$$q_{od}(0) = 0.9998 \quad \text{and} \quad q_{vd}(0) = [0.01, -0.01, 0.01]^T \tag{5.130}$$

to obtain the desired attitude trajectory in terms of the quaternion parameterization.

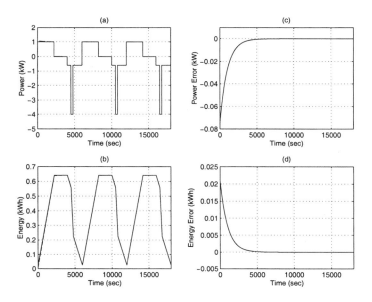

FIGURE 5.7. (a) Desired power profile P_d (solid line) and actual power P (dashed line), (b) desired energy profile E_d (solid line) and actual energy E (dashed line), (c) power tracking error $\dot{\eta}_E$, and (d) energy tracking error η_E.

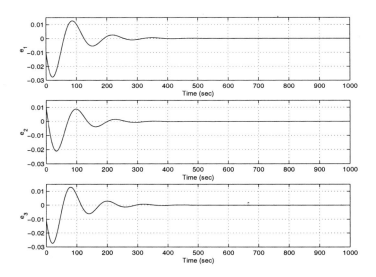

FIGURE 5.8. Attitude tracking error $e_v(t)$.

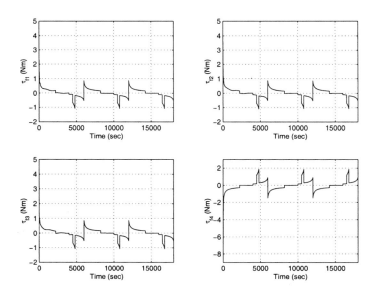

FIGURE 5.9. Control torque input $\tau_f(t)$.

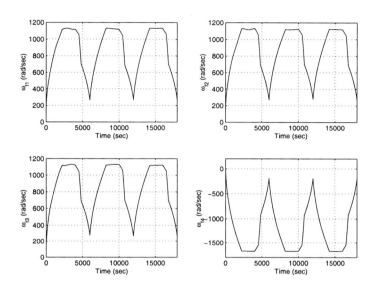

FIGURE 5.10. Flywheel angular velocity $\omega_f\,(t)$.

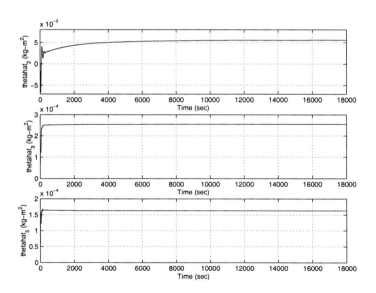

FIGURE 5.11. Elements 2, 3, and 5 of the parameter estimate vector $\hat{\theta}\,(t)$.

To simulate the energy management function of the IPACS, a desired power profile similar to the one considered in [38] was used. The desired energy/power profiles are shown in Figure 5.7. The desired energy/power profiles represent the requirements of a spacecraft undergoing a 100 [min] orbit with a sunlight duration of 66 [min] and an eclipse duration of 34 [min]. During the sunlight period the flywheels are charged by the solar panels (i.e., $P_d(t) \geq 0$) to supply the energy/power needs of the spacecraft during the eclipse period (i.e., $P_d(t) < 0$).

The control and adaptation gains in (5.112), (5.114), and (5.120) were selected as follows

$$K_1 = K_2 = 10^{-1}I_3, \quad k_e = 10^{-3}, \quad \text{and} \quad \Gamma = 10I_6. \tag{5.131}$$

The parameter estimate vector $\hat{\theta}(t)$ was initialized to 50% of the actual parameter values. The simulation results are depicted in Figures 5.7–5.11 over approximately three orbits of the spacecraft. Specifically, the actual energy/power are shown in Figure 5.7, along with their respective desired profiles. The energy/power tracking errors are also shown in Figure 5.7. The attitude tracking error $e_v(t)$ is depicted in Figure 5.8 while the control torque $\tau_f(t)$ is shown in Figure 5.9. The flywheel angular velocity $\omega_f(t)$ and some of the elements of the parameter estimate vector $\hat{\theta}(t)$ are depicted in Figures 5.10 and 5.11, respectively.

5.4   Formation Flying

MSFF refers to collaborative clusters of interdependent microsatellites that talk to each other and share data processing, payload, and mission functions. These satellites flying in formation are smaller, lighter, less expensive to launch, and offer more immediate information-gathering versatility. They also cost less than current satellites because they can be mass produced and placed in orbit using smaller launch vehicles. As mentioned earlier, formation flying permits scientists to obtain comprehensive data by combining measurements from several microsatellites.

For ease of presentation, the adaptive MSFF controller developed here is based on simplified system dynamics that model the spacecraft as point-masses (i.e., the spacecraft attitude dynamics are neglected). As a result, only a translational position controller is required. A more complete description of MSFF would not only use the attitude dynamics of the spacecraft but also the nonlinear couplings between the translational and attitude dynamics. The resulting six degree-of-freedom coupled nonlinear equations of motion would then dictate the development of a control system where the torque and force inputs must be designed in tandem to accurately control the translation and orientation of the follower spacecraft relative to the leader spacecraft.

This section focuses on the full nonlinear dynamics describing only the relative positioning of MSFF for control design purposes. Using Lyapunov-based control design and stability analysis techniques, we develop a nonlinear adaptive control law that guarantees global asymptotic convergence of the spacecraft relative position to any sufficiently smooth desired trajectory, despite the presence of unknown constant or slow varying spacecraft masses, disturbance forces, and gravity forces. In the case when the parameters are exactly known, the control strategy yields global exponential convergence of the tracking errors. As in [16, 40, 41], we will consider in this section the idealized scenario where the spacecraft actuators are capable of providing continuous-time control efforts, as opposed to being of pulse type [20].

5.4.1   System Model

In this section,[5] the MSFF fleet is assumed to be composed of two spacecraft: a leader spacecraft that provides the reference motion trajectory and a follower spacecraft that navigates in the neighborhood of the leader spacecraft according to a desired relative trajectory. The leader spacecraft is assumed to be in orbit around the earth with a constant angular velocity ω. A schematic representation of the MSFF system is depicted in Figure 5.12 where the following considerations are made: (i) the inertial coordinate system $\mathcal{I}$ is attached to the center of the earth, (ii) $\rho(t) \in \mathbb{R}^3$ denotes a position vector from the origin of the inertial coordinate system to the leader spacecraft, (iii) the coordinate frame $\mathcal{L}$ is attached to the leader spacecraft with the x_ℓ-axis pointing in the opposite direction as the tangential velocity, the y_ℓ-axis aligned in the direction of $\rho(t)$, and the z_ℓ-axis is mutually perpendicular to the x_ℓ and y_ℓ axes, and (iv) $q_{\mathcal{I}}(t) \in \mathbb{R}^3$ is expressed in $\mathcal{I}$ and denotes the relative position vector from $\mathcal{L}$ to the origin of the follower spacecraft coordinate system denoted by $\mathcal{F}$.

The nonlinear position dynamics of the leader and follower spacecraft with respect to $\mathcal{I}$ can be developed as follows [20, 39]

$$m_\ell \ddot{\rho} + m_\ell \left(M + m_\ell \right) G \frac{\rho}{\|\rho\|^3} + F_{d\ell} = u_\ell \qquad (5.132)$$

----

[5] The control development in this section can be easily extended to the case of an arbitrary (possibly non-Keplerian) motion of the leader spacecraft.

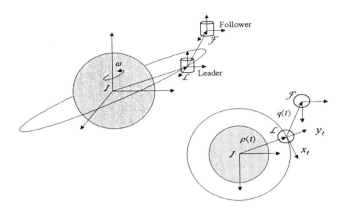

FIGURE 5.12. Schematic representation of the MSFF system.

$$m_f \left(\ddot{\rho} + \ddot{q}_{\mathcal{I}} \right) + m_f \left(M + m_f \right) G \frac{\rho + q_{\mathcal{I}}}{\|\rho + q_{\mathcal{I}}\|^3} + F_{df} = u_f. \tag{5.133}$$

In (5.132) and (5.133), m_ℓ, $m_f \in \mathbb{R}$ denote the masses, $F_{d\ell}$, $F_{df} \in \mathbb{R}^3$ denote disturbance force vectors, and $u_\ell(t) = [u_{\ell x} \quad u_{\ell y} \quad u_{\ell z}]$, $u_f(t) \in \mathbb{R}^3$ denote control input vectors of the leader and follower spacecraft, respectively, $M \in \mathbb{R}$ denotes the earth's mass, and $G \in \mathbb{R}$ denotes the universal gravity constant. In practice, the spacecraft masses vary slowly in time due to fuel consumption and payload variations. The disturbance forces also slowly vary in time because of solar radiation, aerodynamics, and magnetic fields [16]. Given the slow time-varying nature of these parameters, the subsequent development is based on the assumption that m_ℓ, m_f, $F_{d\ell}$, F_{df} remain constant. Since $M >> m_\ell, m_f$, (5.132) and (5.133) can be simplified as follows

$$m_\ell \ddot{\rho} + m_\ell M G \frac{\rho}{\|\rho\|^3} + F_{d\ell} = u_\ell \tag{5.134}$$

$$m_f \left(\ddot{\rho} + \ddot{q}_{\mathcal{I}} \right) + m_f M G \frac{\rho + q_{\mathcal{I}}}{\|\rho + q_{\mathcal{I}}\|^3} + F_{df} = u_f. \tag{5.135}$$

After solving (5.134) for $\ddot{\rho}(t)$ and then substituting the result into (5.135), the following dynamic equation can be determined, which describes the position of $\mathcal{F}$ relative to $\mathcal{L}$, expressed in $\mathcal{I}$,

$$m_f \ddot{q}_{\mathcal{I}} + m_f M G \left(\frac{\rho + q_{\mathcal{I}}}{\|\rho + q_{\mathcal{I}}\|^3} - \frac{\rho}{\|\rho\|^3} \right) + \frac{m_f}{m_\ell} u_\ell + F_{df} - \frac{m_f}{m_\ell} F_{d\ell} = u_f. \tag{5.136}$$

To rewrite (5.136) in terms of the moving coordinate system $\mathcal{L}$, the dynamics for $q_{\mathcal{I}}(t)$ must be written in terms of $\mathcal{L}$. Since $q_{\mathcal{I}}(t)$ is a time-varying vector in $\mathcal{I}$, and the coordinate frame $\mathcal{L}$ is moving relative to $\mathcal{I}$, the second time derivative of $q_{\mathcal{I}}(t)$ expressed in $\mathcal{L}$ (henceforth denoted by $q(t) = [x \quad y \quad z]^T$) is given by the following expression [36]

$$
\ddot{q} = \begin{bmatrix} 0 \\ 0 \\ \omega \end{bmatrix} \times \left(\begin{bmatrix} 0 \\ 0 \\ \omega \end{bmatrix} \times q_{\mathcal{I}} \right) + \begin{bmatrix} 0 \\ 0 \\ 2\omega \end{bmatrix} \times \dot{q}_{\mathcal{I}} + \ddot{q}_{\mathcal{I}} \tag{5.137}
$$

where the fact that ω is a constant angular acceleration was used. The first term in (5.137) represents the centripetal acceleration, the second term represents the Coriolis acceleration, and the last term represents the linear acceleration. After calculating the cross product terms, (5.137) can be simplified as follows

$$
\ddot{q} = \begin{bmatrix} \ddot{x} - 2\omega\dot{y} - \omega^2 x \\ \ddot{y} + 2\omega\dot{x} - \omega^2 y \\ \ddot{z} \end{bmatrix}. \tag{5.138}
$$

After substituting (5.138) into (5.136), the nonlinear position dynamics of the follower spacecraft relative to $\mathcal{L}$ can be written as follows

$$
m_f \ddot{q} + C\dot{q} + N + F_d = u_f. \tag{5.139}
$$

In (5.139), $C(\omega) \in \mathbb{R}^{3\times 3}$ denotes the following Coriolis-like matrix

$$
C = 2m_f \omega \begin{bmatrix} 0 & -1 & 0 \\ 1 & 0 & 0 \\ 0 & 0 & 0 \end{bmatrix}, \tag{5.140}
$$

$N(q, \omega, \rho, u_\ell) \in \mathbb{R}^3$ denotes the following nonlinear vector

$$
N = \begin{bmatrix} m_f M G \dfrac{x}{\|\rho + q\|^3} - m_f \omega^2 x + \dfrac{m_f}{m_\ell} u_{\ell x} \\[3mm] m_f M G \left(\dfrac{y + \|\rho\|}{\|\rho + q\|^3} - \dfrac{1}{\|\rho\|^2} \right) - m_f \omega^2 y + \dfrac{m_f}{m_\ell} u_{\ell y} \\[3mm] m_f M G \dfrac{z}{\|\rho + q\|^3} + \dfrac{m_f}{m_\ell} u_{\ell z} \end{bmatrix}, \tag{5.141}
$$

and $F_d \in \mathbb{R}^3$ is the total constant disturbance force vector defined as

$$
F_d = F_{df} - \frac{m_f}{m_\ell} F_{d\ell} \tag{5.142}
$$

where the fact that $\rho = [0 \quad \|\rho\| \quad 0]$ and is constant (see Figure 5.12) in the moving coordinate system $\mathcal{L}$ was utilized, and $u_\ell(t)$, $u_f(t)$ are expressed in $\mathcal{L}$. The dynamic model of (5.139–5.142) has the following property which will be exploited in the subsequent adaptive control design.

The dynamic equation given in (5.139) can be linearly parameterized as

$$m_f \xi + C(\omega)\dot{q} + N(q, \omega, \rho, u_\ell) + F_d = W(\xi, \dot{q}, q, \omega, \rho, u_\ell)\theta \quad \forall \xi \in \mathbb{R}^3 \tag{5.143}$$

where $W(\xi, \dot{q}, q, \omega, \rho, u_\ell) \in \mathbb{R}^{3\times6}$ denotes a regression matrix that is composed of known functions, and $\theta \in \mathbb{R}^6$ is a constant parameter vector. From the form of (5.139–5.142), the regression matrix and constant parameter vector can be determined as follows

$$W = \begin{bmatrix} \xi_x - 2\omega\dot{y} - \omega^2 x & \dfrac{x}{\|\rho + q\|^3} & u_{\ell x} & 1 & 0 & 0 \\[3ex] \xi_y + 2\omega\dot{x} - \omega^2 y & \dfrac{y + \|\rho\|}{\|\rho + q\|^3} - \dfrac{1}{\|\rho\|^2} & u_{\ell y} & 0 & 1 & 0 \\[3ex] \xi_z & \dfrac{z}{\|\rho + q\|^3} & u_{\ell z} & 0 & 0 & 1 \end{bmatrix} \tag{5.144}$$

and

$$\theta = \begin{bmatrix} m_f & m_f MG & \dfrac{m_f}{m_\ell} & F_{dx} & F_{dy} & F_{dz} \end{bmatrix}^T \tag{5.145}$$

where ξ_x, ξ_y, ξ_z and F_{dx}, F_{dy}, F_{dz} are the components of the vectors ξ and F_d, respectively.

5.4.2   Control Objective

Given a desired position trajectory $q_d(t) \in \mathbb{R}^3$ for the follower with respect to the leader (assuming that $q_d(t)$ and its first two time derivatives are bounded functions of time), the objective of MSFF is to design the control input $u_f(t)$ such that the actual position trajectory tracks the desired position trajectory. To quantify the MSFF position tracking objective, a position tracking error, denoted by $e(t) \in \mathbb{R}^3$, is defined as follows

$$e(t) = q_d(t) - q(t). \tag{5.146}$$

In addition, a filtered tracking error, denoted by $r(t) \in \mathbb{R}^3$, is defined as follows

$$r(t) = \dot{e}(t) + \alpha e(t) \tag{5.147}$$

to simplify the subsequent control design/analysis, where $\alpha \in \mathbb{R}^{3\times3}$ denotes a constant diagonal positive-definite control gain matrix. The subsequent control development is based on the assumption that $q(t)$ and $\dot{q}(t)$ are measurable and under the constraint that the spacecraft masses, disturbance forces, and gravitational force are unknown.

5.4.3   MSFF Control

To develop the open-loop error system for $r(t)$, we take the time derivative of (5.147) and multiply both sides of the resulting equation by m_f as follows

$$m_f \dot{r} = m_f (\ddot{q}_d + \alpha \dot{e}) + C(\omega)\dot{q} + N(q, \omega, R, u_\ell) + F_d - u_f \qquad (5.148)$$

where (5.139) and (5.146) were utilized. After using (5.143) and substituting the following expression

$$\xi = \ddot{q}_d + \alpha \dot{e} \qquad (5.149)$$

into the definition of (5.144) for $\xi(t)$, the expression given in (5.148) can be rewritten as follows

$$m_f \dot{r} = W\theta - u_f. \qquad (5.150)$$

Based on the form of the open-loop dynamics of (5.148), the control input $u_f(t)$ is designed as follows

$$u_f = W(\cdot)\hat{\theta} + Kr \qquad (5.151)$$

where $K \in \mathbb{R}^{3 \times 3}$ is a constant diagonal positive-definite control gain matrix, and the parameter estimate vector $\hat{\theta}(t)$ is generated by the following gradient update law

$$\dot{\hat{\theta}} = \Gamma W^T(\cdot)r \qquad (5.152)$$

where $\Gamma \in \mathbb{R}^{6 \times 6}$ is a constant diagonal positive-definite adaptation gain matrix. After substituting (5.151) into (5.148), the closed-loop dynamics for $r(t)$ can be determined as follows

$$m_f \dot{r} = -Kr + W(\cdot)\tilde{\theta} \qquad (5.153)$$

where $\tilde{\theta}(t) \in \mathbb{R}^6$ quantifies the difference between the actual and estimated parameters as follows

$$\tilde{\theta}(t) = \theta - \hat{\theta}(t). \qquad (5.154)$$

By differentiating (5.154) and utilizing (5.152), the closed-loop dynamics for $\tilde{\theta}(t)$ can be determined as follows

$$\dot{\tilde{\theta}} = -\Gamma W^T(\cdot)r. \qquad (5.155)$$

5.4.4   Stability Analysis

The MSFF controller given in (5.151) and (5.152) yields global asymptotic tracking as described by the following theorem.

Theorem 5.4 *The adaptive MSFF controller given in (5.151) and (5.152) ensures the follower spacecraft globally asymptotically tracks a desired trajectory with respect to a leader in the sense that*

$$\lim_{t \to \infty} e(t), \dot{e}(t) = 0. \tag{5.156}$$

Proof: To prove Theorem 5.4, we define a nonnegative function $V(t) \in \mathbb{R}$ as follows

$$V = \frac{1}{2} r^T m_f r + \frac{1}{2} \tilde{\theta}^T \Gamma^{-1} \tilde{\theta}. \tag{5.157}$$

After taking the time derivative of (5.157) and substituting the closed-loop dynamics developed in (5.153) and (5.155) into the resulting expression, the following expression can be obtained

$$\dot{V} = -r^T K r \le -\lambda_{\min}\{K\} \|r\|^2 \le 0 \tag{5.158}$$

where $\lambda_{\min}\{\cdot\}$ denotes the minimum eigenvalue of a matrix. The expressions given in (5.157) and (5.158) can now be used to prove that $r(t), \tilde{\theta}(t) \in \mathcal{L}_\infty$ and Lemma A.11 of Appendix A can be used to prove that $r(t) \in \mathcal{L}_2$. Since $r(t) \in \mathcal{L}_\infty$, Lemma A.13 of Appendix A can be utilized to prove that $e(t), \dot{e}(t) \in \mathcal{L}_\infty$. Hence, due to the boundedness of $q_d(t)$ and $\dot{q}_d(t)$, the definition of (5.146) can be utilized to conclude that $q(t), \dot{q}(t) \in \mathcal{L}_\infty$. Since $\tilde{\theta}(t) \in \mathcal{L}_\infty$ and θ is a constant vector, (5.154) can be used to determine that $\hat{\theta}(t) \in \mathcal{L}_\infty$. From the above boundedness statements and the fact that $\ddot{q}_d(t)$ is assumed to be bounded, the definitions of (5.144) and (5.149) can be used to state that $W(\cdot) \in \mathcal{L}_\infty$. Based on the preceding facts, (5.139) and (5.151) can be used to prove that $u_f(t), \ddot{q}(t), \dot{r}(t) \in \mathcal{L}_\infty$. Thus, all signals in the adaptive controller and system remain bounded during closed-loop operation. Based on the fact that $r(t), \dot{r}(t) \in \mathcal{L}_\infty$ and that $r(t) \in \mathcal{L}_2$, Barbalat's Lemma (see Lemma A.16 of Appendix A) can be invoked to prove that

$$\lim_{t \to \infty} r(t) = 0. \tag{5.159}$$

Based on (5.147) and (5.159), Lemma A.15 of Appendix A can be applied to obtain (5.156). $\square$

Remark 5.3 *In the case where the parameter vector θ of (5.145) is perfectly known, the control law introduced in (5.151) with $\hat{\theta}(t) = \theta$ (i.e., an exact model knowledge controller) could be developed to ensure global exponential convergence of the position and velocity tracking errors in the sense that*

$$\|r(t)\| \le \|r(0)\| \exp\left(-\frac{\lambda_{\min}\{K\}}{m_f} t\right). \tag{5.160}$$

This result can be proven by defining the following nonnegative function

$$V = \frac{1}{2}r^T m_f r \tag{5.161}$$

and utilizing similar arguments as in the proof of Theorem 5.4 along with Lemma A.14 of Appendix A.

Remark 5.4 *To minimize fuel consumption, many MSFF control systems utilize pulse-type actuators (e.g., Hall thrusters and pulse plasma thrusters [33]). Hence, the application of continuous-time control inputs in these cases for long periods of time is not practical as noted in [39, 40]. This constraint would require the implementation of a "pulse-based" version of the nonlinear controller developed in (5.151) and (5.152). The effect of this noncontinuous nonlinear control law on the closed-loop stability has not been considered and constitutes an interesting open problem in MSFF control. However, as mentioned in [39], continuous MSFF control laws can provide an idealized system response for comparison purposes with the actual responses obtained from the noncontinuous implementation of the controllers.*

5.4.5   Simulation Results

The adaptive MSFF controller developed in (5.151) and (5.152) was simulated for the two spacecraft MSFF problem depicted in Figure 5.12, where the following parameters were used to define the system [20, 39]

$$M = 5.974 \times 10^{24} \text{ [kg]}, \quad m_f = 410 \text{ [kg]}, \quad m_\ell = 1550 \text{ [kg]},$$

$$G = 6.673 \times 10^{-11} \text{ [kg} \cdot \text{m}^3 \cdot \text{s}^2],$$

$$\rho = \begin{bmatrix} 0, 4.224 \times 10^7, 0 \end{bmatrix}^T \text{ [m]}, \quad \omega = 7.272 \times 10^{-5} \text{ [rad} \cdot \text{s}^{-1}], \tag{5.162}$$

$$u_\ell = 0 \text{ [N]}, \quad F_d = [-1.025, 6.248, -2.415] \times 10^{-5} \text{ [N]}.$$

The parameter estimates were initialized to 50% of the actual parameter values defined in (5.145) and (5.162) (i.e., $\hat{\theta}(0) = 0.5\theta$). The desired relative trajectory was selected as follows

$$q_d(t) = \begin{bmatrix} 100\sin(4\omega t)\left(1 - \exp\left(-0.05t^3\right)\right) \\ 100\cos(4\omega t)\left(1 - \exp\left(-0.05t^3\right)\right) \\ 0 \end{bmatrix} \text{ [m]} \tag{5.163}$$

where the exponential term was included to ensure that $\dot{q}_d(0) = \ddot{q}_d(0) = 0$. The relative position and relative velocity were initialized to the following values

$$q(0) = [0, 0, -200]^T \text{ [m]} \qquad \dot{q}(0) = 0 \text{ [m} \cdot \text{s}^{-1}] \qquad (5.164)$$

while the desired trajectory of (5.163) and the initial conditions of (5.164) represent a follower spacecraft that is initially stationary with respect to the leader spacecraft, and is then commanded to move around the leader spacecraft in a circular orbit of radius 100 [m] in the plane formed by the x_ℓ, y_ℓ axes of $\mathcal{L}$ with a constant angular velocity of 4ω. The selection of the above desired trajectory does not account for fuel consumption considerations. However, it illustrates the capability of the controller to track demanding trajectories that may occur during formation reconfiguration maneuvers.

After selecting the control and adaptation gains as follows

$$\Lambda = \text{diag}\,(2, 2, 1)\,, \quad K = \text{diag}\,(100, 200, 120)\,,$$
$$\Gamma = \text{diag}\,(500, 50, 300, 600, 850, 480) \qquad (5.165)$$

the resulting position tracking error illustrated in Figure 5.13 was obtained. The phase portrait of the trajectory $q(t)$ of the follower spacecraft relative to the leader spacecraft is illustrated in Figure 5.14 where $*$ represents the leader spacecraft at the origin. The control input $u_f(t)$ is given in Figure 5.15, while four sample components of the parameter estimate vector $\hat{\theta}(t)$ are presented in Figure 5.16.

5.5   Background and Further Reading

Several solutions to the attitude control problem have been presented in the literature since the early 1970s [27]. See [42] for a comprehensive literature review of earlier work. The authors of [42] presented a general attitude control design framework which includes linear, model-based, and adaptive set-point controllers. Adaptive tracking control schemes based on three-parameter, kinematic representations were presented in [32, 35] to compensate for the unknown spacecraft inertia matrix. In [1], an adaptive attitude tracking controller based on the unit quaternion was proposed that identified the inertia matrix via periodic command signals. The work of [1] was later applied to the angular velocity tracking problem in [2]. An $\mathcal{H}_\infty$-suboptimal state feedback controller was developed for the quaternion representation in [11]. In [22], the authors designed an inverse optimal control law for attitude regulation using the backstepping method for a three-parameter representation. The authors of [5] recently presented a variable

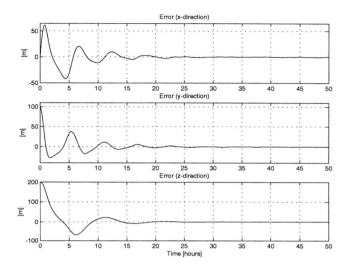

FIGURE 5.13. Position tracking error $e\,(t)$.

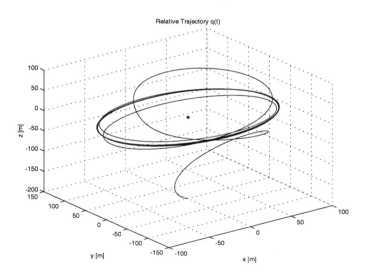

FIGURE 5.14. Trajectory of follower spacecraft relative to leader spacecraft ($*$ denotes the leader spacecraft).

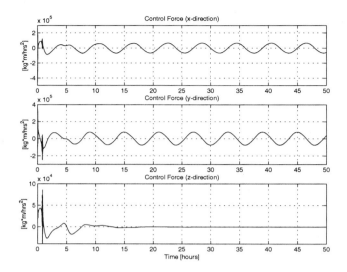

FIGURE 5.15. Control input $u_f(t)$.

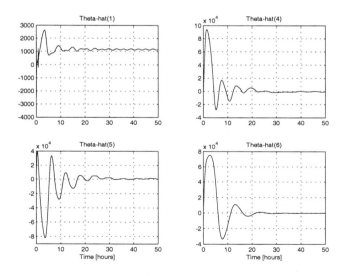

FIGURE 5.16. Sample of parameter estimates.

structure tracking controller using quaternions in the presence of spacecraft inertia uncertainties and external disturbances.

A typical feature in all the above-mentioned attitude control schemes is that angular velocity measurements are required. Unfortunately, this requirement is not always satisfied in reality. Thus, a common practice is to approximate the angular velocity signal through an ad hoc numerical differentiation of the attitude angles, and directly use this surrogate signal for control design with no guarantee of closed-loop stability. With this in mind, an angular velocity observer was developed in [31] for the quaternion representation. However, the observer was based on an unproven separation principle argument. In [26], a passivity approach was used to develop an asymptotically stabilizing setpoint controller that eliminated velocity measurements via the filtering of the unit quaternion. The passivity-based velocity-free setpoint controller of [26] was later applied to the simpler, three-parameter problem in [37]. Wong, de Queiroz, and Kapila [43] recently proposed an adaptive attitude tracking controller without angular velocity measurements using the modified Rodrigues parameters. In [10], Costic et al. proposed an output feedback quaternion-based attitude tracking controller.

A comprehensive literature review of the IPACS concept is presented in [17]. As noted in this work, the IPACS concept has been investigated since the 1970s [3]. However, the enabling technologies have only recently reached a level of maturity that facilitates on-board evaluation. Also noted in [17] is that most control designs for the IPACS problem are based on linearization of the spacecraft dynamic equation. With this fact in mind, Tsiotras, Shen, and Hall [38] used the nonlinear dynamic equation presented in [17] along with a modified Rodrigues parameters-based kinematic representation to design an attitude and power tracking control scheme using an array of reaction wheels. Recently, Fausz and Richie [14] extended the approach of [38] for flywheels operating in a variable-speed control moment gyro mode. To compensate for parametric uncertainty in the spacecraft inertia (the IPACS results of [14, 38] require exact knowledge of the spacecraft inertia), adaptive attitude controllers were developed in [1, 5, 10, 32, 43]. However, these results do not deal with the problem of simultaneous attitude and energy/power tracking, and hence, do not account for the flywheel dynamics. A recent result that does address simultaneous attitude and energy/power tracking while accounting for the flywheel dynamics is given in [9].

The interest demonstrated by NASA and the U. S. Air Force has led to several studies on autonomous MSFF reported in the literature. For example, in early research, [39] presented the concept of formation keeping of spacecraft for a ground-based terrestrial laser communication system, whereas [29] considered station keeping for the space shuttle orbiter. Recently, [24] considered MSFF control for NASA's NMI mission using separated spacecraft interferometry. Similarly, [15] considered MSFF control for NASA's EO-I mission which was scheduled to demonstrate a stereo imaging concept in 1999.

The practical implementation of the MSFF concept relies on the accurate control of the relative positions and orientations between the participating spacecraft for formation configuration and collision avoidance. Most MSFF control designs utilize simplifying modeling assumptions to aid the control synthesis due to inherent difficulties associated with the structure of the full nonlinear dynamic model of MSFF. These simplifications result in the well-known Clohessy-Wiltshire linear dynamic equations [7, 8, 20] for the relative positioning of MSFF. The Clohessy-Wiltshire model has formed the basis for the application of various linear control techniques to the MSFF position control problem [20, 25, 29, 39]. Reviewing the current state of MSFF control, it seems that in contrast to linear control, nonlinear control theory has not been exploited to its full potential. A few results investigating nonlinear control of MSFF can be found in [16, 40, 41] using a Lyapunov-based approach. In particular, [40] designed a class of control laws based on exact knowledge of the MSFF model that yielded local asymptotic position tracking and global exponential attitude tracking. The application of the controllers proposed in [40] to formation rotation of MSFF about a given axis and synchronization of individual spacecraft rotation was later reported in [41]. More recently, [16] developed an adaptive position controller that compensated for unknown constant disturbances while producing globally asymptotically decaying position tracking errors. This controller, however, required exact knowledge of the spacecraft parameters. Recently, a nonlinear adaptive relative positioning controller for MSFF was developed in [13] despite parametric uncertainty in the spacecraft dynamics. As an endnote, the problem of pulse-type nonlinear control design for MSFF constitutes an open research problem which is being investigated.

References

[1] J. Ahmed, V. T. Coppola, and D. S. Bernstein, "Adaptive Asymptotic Tracking of Spacecraft Attitude Motion with Inertia Matrix Identification," *Journal of Guidance, Control, and Dynamics*, Vol. 21, No. 5, Sept.-Oct. 1998, pp. 684–691.

[2] J. Ahmed and D. S. Bernstein, "Globally Convergent Adaptive Control of Spacecraft Angular Velocity Without Inertia Modeling," *Proceedings of the American Control Conference*, San Diego, CA, June 1999, pp. 1540–1544.

[3] W. W. Anderson and C. R. Keckler, "An Integrated Power/Attitude Control System (IPACS) for Space Application," *Proceedings of the 5th IFAC Symposium on Automatic Control in Space*, 1973.

[4] F. Bauer et al., "Satellite Formation Flying using an Innovative Autonomous Control System (AUTOCON) Environment," *Proceedings of the AIAA Guidance, Navigation, and Control Conference*, New Orleans, LA, 1997, pp. 657–666.

[5] J. D. Bosković, S. M. Li, and R. K. Mehra, "Globally Stable Adaptive Tracking Control Design for Spacecraft under Input Saturation," *Proceedings of the IEEE Conference on Decision and Control*, Phoenix, AZ, Dec. 1999, pp. 1952–1957.

[6] T. Burg, D. Dawson, J. Hu, and M. de Queiroz, "An Adaptive Partial State Feedback Controller for RLED Robot Manipulators," *IEEE Transactions on Automatic Control*, Vol. 41, No. 7, July 1996, pp. 1024–1031.

[7] V. A. Chobotov (ed.), *Orbital Mechanics*, Washington, DC: AIAA, 1996, pp. 31–33.

[8] W. H. Clohessy and R. S. Wiltshire, "Terminal Guidance System for Satellite Rendezvous," *Journal of Aerospace Science*, Vol. 27, No. 9, 1960, pp. 653–658.

[9] B. T. Costic, M. S. de Queiroz, D. M. Dawson, and Y. Fang, "Energy Management and Attitude Control Strategies using Flywheels," *Proceedings of the IEEE Conference on Decision and Control*, Orlando, FL, Dec. 2001, pp. 3435–3440.

[10] B. T. Costic, D. M. Dawson, M. S. de Queiroz, and V. Kapila, "A Quaternion-Based Adaptive Attitude Tracking Controller Without Velocity Measurements," *AIAA Journal of Guidance, Control, and Dynamics*, Vol. 24, No. 6, Nov. 2001, pp. 1214–1222.

[11] M. Dalsmo and O. Egeland, "State Feedback $\mathcal{H}_\infty$-Suboptimal Control of a Rigid Spacecraft," *IEEE Transactions on Automatic Control*, Vol. 42, No. 8, Aug. 1997, pp. 1186–1189.

[12] D. M. Dawson, J. Hu, and T. C. Burg, *Nonlinear Control of Electric Machinery*, New York, NY: Marcel Dekker, 1998, pp. 1–19.

[13] M. S. de Queiroz, V. Kapila, and Q. Yan, "Adaptive Nonlinear Control of Multiple Spacecraft Formation Flying," *AIAA Journal of Guidance, Control, and Dynamics*, Vol. 23, No. 3, May-June 2000, pp. 385–390.

[14] J. L. Fausz and D. J. Richie, "Flywheel Simultaneous Attitude Control and Energy Storage Using a VSCMG Configuration," *Proceedings of the IEEE Conference on Control Applications*, Anchorage, AK, Sept. 2000, pp. 991–995.

[15] J. R. Guinn, "Autonomous Navigation for the New Millenium Program Earth Orbiter 1 Mission," *Proceedings of the AIAA Guidance, Navigation, and Control Conference*, New Orleans, LA, 1997, pp. 612–617.

[16] F. Y. Hadaegh, W. M. Lu, and P. C. Wang, "Adaptive Control of Formation Flying Spacecraft for Interferometry," *Proceedings of the IFAC Conference on Large Scale Systems*, Rio Patras, Greece, 1998, pp. 97–102.

[17] C. D. Hall, "High-Speed Flywheels for Integrated Energy Storage and Attitude Control," *Proceedings of the American Control Conference*, Albuquerque, NM, June 1997, pp. 1894–1898.

[18] P. C. Hughes, *Spacecraft Attitude Dynamics*, New York, NY: Wiley, 1994.

[19] T. R. Kane, P. W. Likins, and D. A. Levinson, *Spacecraft Dynamics*, New York, NY: McGraw-Hill, 1983.

[20] V. Kapila, A.G. Sparks, J. Buffington, and Q. Yan, "Spacecraft Formation Flying: Dynamics and Control," *Proceedings of the American Control Conference*, San Diego, CA, 1999, pp. 4137–4141.

[21] M. Krstić, I. Kanellakopoulos, and P. Kokotovic, *Nonlinear and Adaptive Control Design*, New York, NY: Wiley, 1995.

[22] M. Krstić and P. Tsiotras, "Inverse Optimal Stabilization of a Rigid Spacecraft," *IEEE Tranactions on Automatic Control*, Vol. 44, No. 5, May 1999, pp. 1042–1049.

[23] J. B. Kuipers, *Quaternions and Rotation Sequences*, Princeton, NJ: Princeton University Press, 1999.

[24] K. Lau, "The New Millenium Formation Flying Optical Interferometer," *Proceedings of the AIAA Guidance, Navigation, and Control Conference*, New Orleans, LA, 1997, pp. 650–656.

[25] C. L. Leonard, W. M. Hollister, and E. V. Bergmann, "Orbital Formationkeeping with Differential Drag," *Journal of Guidance, Control, and Dynamics*, Vol. 12, No. 1, 1989, pp. 108–113.

[26] F. Lizarralde and J. T. Wen, "Attitude Control Without Angular Velocity Measurement: A Passivity Approach," *IEEE Transactions on Automatic Control*, Vol. 41, No. 3, Mar. 1996, pp. 468–472.

[27] G. Meyer, "Design and Global Analysis of Spacecraft Attitude Control Systems," *NASA Technical Report R-361*, Mar. 1971.

[28] Y. Nakamura, *Advanced Robotics Redundancy and Optimization*, Reading, MA: Addison-Wesley, 1991.

[29] D. C. Redding, N. J. Adams, and E. T. Kubiak, "Linear-Quadratic Stationkeeping for the STS Orbiter," *Journal of Guidance, Control, and Dynamics*, Vol. 12, No. 2, 1989, pp. 248–255.

[30] A. Robertson, T. Corazzini, and J. P. How, "Formation Sensing and Control Technologies for a Separated Spacecraft Interferometer," *Proceedings of the American Control Conference*, Philadelphia, PA, 1998, pp. 1574–1579.

[31] S. Salcudean, "A Globally Convergent Angular Velocity Observer for Rigid Body Motion," *IEEE Transactions on Automatic Control*, Vol. 36, No. 12, Dec. 1991, pp. 1493–1497.

[32] H. Schaub, M. R. Akella, and J. L. Junkins, "Adaptive Control of Nonlinear Attitude Motions Realizing Linear Closed-Loop Dynamics," *Proceedings of the American Control Conference*, San Diego, CA, June 1999, pp. 1563–1567.

[33] J. Schilling and R. Spores, "Comparison of Propulsion Options for TechSat 21 Mission," *Air Force Research Laboratory-Formation Flying and Micro-Propulsion Workshop*, Lancaster, CA, 1998.

[34] M. D. Shuster, "A Survey of Attitude Representations," *J. Astronautical Sciences*, Vol. 41, No. 4, 1993, pp. 439–517.

[35] J. -J. E. Slotine and W. Li, *Applied Nonlinear Control*, Englewood Cliffs, NJ: Prentice-Hall, 1991, pp. 122–126.

[36] M. Spong and M. Vidyasagar, *Robot Dynamics and Control*, New York, NY: John Wiley, 1989.

[37] P. Tsiotras, "Further Passivity Results for the Attitude Control Problem," *IEEE Transactions on Automatic Control*, Vol. 43, No. 11, Nov. 1998, pp. 1597–1600.

[38] P. Tsiotras, H. Shen, and C. Hall, "Satellite Attitude Control and Power Tracking with Energy/Momentum Wheels," *Journal of Guidance, Control, and Dynamics*, Vol. 24, No. 1, Jan.-Feb. 2001, pp. 23–34.

[39] R. H. Vassar and R. B. Sherwood, "Formationkeeping for a Pair of Satellites in a Circular Orbit," *Journal of Guidance, Control, and Dynamics*, Vol. 8, No. 2, 1985, pp. 235–242.

[40] P. K. C. Wang and F. Y. Hadaegh, "Coordination and Control of Multiple Microspacecraft Moving in Formation," *Journal of Astronautical Sciences*, Vol. 44, No. 3, 1996, pp. 315–355.

[41] P. K. C. Wang, F. Y. Hadaegh, and K. Lau, "Synchronized Formation Rotation and Attitude Control of Multiple Free-Flying Spacecraft," *Journal of Guidance, Control, and Dynamics*, Vol. 22, No. 1, 1999, pp. 1582–1589.

[42] J. T. Wen and K. Kreutz-Delgado, "The Attitude Control Problem," *IEEE Transactions on Automatic Control*, Vol. 36, No. 10, Oct. 1991, pp. 1148–1156.

[43] H. Wong, M. S. de Queiroz, and V. Kapila, "Adaptive Tracking Control Using Synthesized Velocity from Attitude Measurements," *Proceedings of the American Control Conference*, Chicago, IL, June 2000, pp. 1572–1576.

[44] J. S. C. Yuan, "Closed-Loop Manipulator Control Using Quaternion Feedback," *IEEE Transactions on Robotics and Automation*, Vol. 4, No. 4, Aug. 1988, pp. 434–440.

6

Underactuated Systems

6.1 Introduction

The engineering systems described in the previous chapters are fully actuated (the number of control inputs (actuators) equal the number of degrees of freedom). However, because of actuator failures or various construction constraints some applications are underactuated (the degrees of freedom exceed the number of control inputs). Underactuated systems present challenging control problems since the control design must typically exploit some coupling between the unactuated states and the actuated states to achieve the control objective. In the subsequent sections, the particular control issues related to the underactuated nature of several engineering applications are examined.

The first underactuated application examined in this chapter is an overhead crane system. The control of overhead crane systems has been a heavily investigated problem due to both the theoretical challenges and the practical importance. Specifically, precise payload positioning by an overhead crane is difficult because the payload can exhibit a pendulum-like swinging motion. These payload swings can result in several performance and safety concerns including (i) payload damage, (ii) damage to the surrounding environment or personnel, and (iii) large internal forces that can result in reduced payload-carrying capacity or premature failure of stressed parts. Motivated by the desire to achieve fast and precise payload positioning while mitigating the above performance and safety concerns, in the first section of this chapter, a linear controller is proven to yield asymptotic regulation of the underactuated payload and 2 degree-of-freedom (2-DOF) gantry dynamics of an overhead crane. Specifically, a linear feedback loop at the gantry creates an artificial spring/damper system that absorbs the payload energy. Utilizing LaSalle's Invariant Set Theorem, the linear controller is proven to asymptotically regulate the overhead crane dynamics. Motivated by the heuristic concept that improved coupling between the gantry and the payload dynamics will provide a mechanism for improved transient response, two nonlinear energy-based coupling control laws are then designed by incorporating additional nonlinear feedback terms with the linear controller. Experimental results illustrate that the increased coupling of the nonlinear controllers results in improved transient response (e.g., reduced overshoot and faster settling time) over the linear control law.

Over the last decade, there has been considerable interest in designing controllers for vertical take-off and landing (VTOL) aircraft. The VTOL control problem is complicated by the fact that the underactuated dynamics are nonlinear, nonminimum-phase, and subject to nonholonomic (nonintegrable) constraints. Moreover, it has been acknowledged that standard techniques to decouple the rolling moment and the lateral acceleration, such as static input-output linearization approaches, fail to produce satisfactory performance, since input-output linearization often results in the unstable roll dynamics being unobservable (see [25]). In the second section of this chapter, a reference trajectory generator and a global invertible transformation are developed to rewrite the VTOL kinematics in a similar form as Brockett's Nonholonomic Integrator [9]. Based on the structure of the transformed dynamics, a dynamic oscillator-based control strategy is developed. The approach uses a series of transformations to manipulate the VTOL dynamics into a suitable form which allows a Lyapunov-based controller to be designed. The resulting position and attitude tracking and regulation errors of the VTOL aircraft are proven to be globally uniformly ultimately bounded (GUUB) to a neighborhood that can be made arbitrarily small. Extensions are also provided to illustrate how the open-loop error systems of an automotive steering problem and an underactuated surface vessel problem can be expressed in a similar form to Brockett's Nonholonomic Integrator, allowing similar control designs to be developed.

Over the past ten years, the attitude control of rigid body systems has become an active area of research. Among its many applications are the attitude control of rigid aircraft and spacecraft[1] systems (the interested reader is referred to [54] for a literature review of the many different types of applications). Rigid spacecraft applications are often required to perform highly accurate slewing and/or pointing maneuvers that force the spacecraft to rotate along a relatively large amplitude trajectory. These performance requirements mandate that the control design be predicated on the use of the nonlinear spacecraft model [1]. This nonlinear model is typically represented by Euler's dynamic equation, which is used to describe the time evolution of the angular velocity vector, and the kinematic equation, which relates the time derivatives of the orientation angles to the angular velocity vector. In the third section of this chapter, the satellite kinematics are formulated in terms of the constrained unit quaternion as a means to express the spacecraft attitude without singularities. Based on the form of the underactuated satellite kinematics, a similar (quaternion-based) dynamic oscillator control structure is also developed to achieve uniformly ultimately bounded (UUB) tracking and regulation provided a sufficient condition is satisfied based on the initial conditions of the system. As an extension to the design, an integrator backstepping technique is used to incorporate the dynamic model of an axisymmetric satellite.

6.2 Overhead Crane Systems

In this section, a linear controller and two nonlinear coupling control laws are developed for an overhead crane system with a 2-DOF gantry. By utilizing a Lyapunov-based stability analysis along with LaSalle's Invariance Set Theorem, asymptotic regulation of the gantry and payload position is proven for the controllers.

6.2.1   System Model

The dynamic model for the underactuated overhead crane system given in Figure 6.1 is assumed to have the following form [42]

$$M(q)\ddot{q} + V_m(q, \dot{q})\dot{q} + G(q) = u. \tag{6.1}$$

In (6.1), $M(q) \in \mathbb{R}^{4 \times 4}$, $V_m(q, \dot{q}) \in \mathbb{R}^{4 \times 4}$, and $G(q) \in \mathbb{R}^4$, represent the inertia, centripetal-Coriolis, and gravity terms (for details regarding the components of these matrices see Definition B.3 of Appendix B), respectively, $\dot{q}(t)$, $\ddot{q}(t) \in \mathbb{R}^4$, represent the first and second time derivatives of

---

[1] As in previous chapters, the words spacecraft and satellite are used interchangeably.

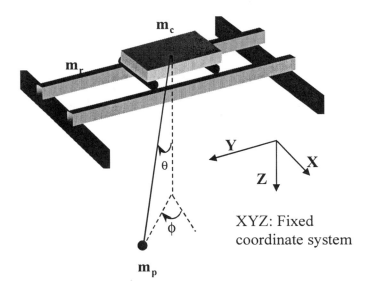

FIGURE 6.1. 3-DOF overhead crane system.

$q(t) \in \mathbb{R}^4$ that is defined as follows

$$q = \begin{bmatrix} x & y & \theta & \phi \end{bmatrix}^T \tag{6.2}$$

where $x(t)$, $y(t) \in \mathbb{R}$ denote the Cartesian coordinates of the gantry with respect to a fixed inertial reference frame, $\theta(t) \in \mathbb{R}$ denotes the payload angle with respect to the vertical, $\phi(t) \in \mathbb{R}$ denotes the projection of the payload angle along the X-coordinate axis, and $u(t) \in \mathbb{R}^4$ is defined as

$$u = [F_x \quad F_y \quad 0 \quad 0]^T \tag{6.3}$$

where $F_x(t), F_y(t) \in \mathbb{R}$ represent the control force inputs acting on the cart and rail, respectively. The dynamic model given in (6.1) exhibits the following properties that are used in the subsequent control development and stability analysis.

Property 6.1: Symmetric and Positive-Definite Inertia Matrix

The symmetric and positive-definite inertia matrix $M(q)$ satisfies the following inequalities

$$k_1 \|\xi\|^2 \leq \xi^T M(q)\xi \leq k_2 \|\xi\|^2 \qquad \forall \xi \in \mathbb{R}^4 \tag{6.4}$$

where k_1, $k_2 \in \mathbb{R}$ are positive bounding constants, and $\|\cdot\|$ denotes the standard Euclidean norm.

Property 6.2: Skew-Symmetry

The inertia and Coriolis matrices satisfy the following skew-symmetric relationship

$$\xi^T \left(\frac{1}{2} \dot{M} - V_m \right) \xi = 0 \quad \forall \xi \in \mathbb{R}^4 \tag{6.5}$$

where $\dot{M}(q)$ denotes the time derivative of the inertia matrix.

In a similar manner as in [10, 37], the dynamic model given in (6.1) is assumed to have the following characteristics.

Assumption 6.1: The payload and the gantry are connected by a mass-less rigid link.

Assumption 6.2: The angular position and velocity of the payload and the planar position and velocity of the gantry are measurable.

Assumption 6.3: The gantry mass and the length of the connecting rod are known.

Assumption 6.4: The connection between the payload link and the gantry is frictionless and does not rotate about the connecting link (i.e., the payload does not rotate about the link axis).

Assumption 6.5: The angular position of the payload mass is restricted according to following inequalities

$$-\pi < \theta(t) < \pi \tag{6.6}$$

where $\theta(t)$ is measured from the vertical position (see Figure 6.1).

Remark 6.1 *The model given by (6.1) could be modified to include other dynamic effects associated with the gantry dynamics (e.g., gantry friction, viscous damping coefficients, mass moment of inertia of the gantry and rail motors). However, these additional dynamic effects were not included in the model since these effects can be directly cancelled by the controller.*

6.2.2 Open-Loop Error System

To express the dynamic model in a form that facilitates the subsequent control development and stability analysis, both sides of (6.1) are premultiplied by $M^{-1}(q)$ to obtain the following expression

$$\ddot{q} = M^{-1} \left(u - V_m \dot{q} - G \right) \tag{6.7}$$

where $M^{-1}(q) \in \mathbb{R}^{4 \times 4}$ is guaranteed to exist since the determinant of $M(q)$, denoted by $\det(M)$, is given by the following positive function

$$\det(M) = I^2 (m_p + m_c)(m_p + m_r + m_c)$$

$$+ m_p L^2 I \left[(m_p + m_r + m_c) m_c (1 + \sin^2 \theta) \right.$$

$$+ (m_p + m_c) m_p \sin^2 \theta + m_r m_p (\sin^2 \phi + 2 \sin^2 \theta \cos^2 \phi) \Big]$$

$$+ m_p^2 L^4 \sin^2 \theta \left((m_r m_p \sin^2 \theta \cos^2 \phi) \right.$$

$$+ m_c (m_r + m_c + m_p \sin^2 \theta) \Big) .$$

$$(6.8)$$

In (6.8), the parameters $m_p, m_r, m_c \in \mathbb{R}$ represent the payload mass, rail mass, and cart mass, respectively, $I \in \mathbb{R}$ denotes the moment of inertia of the payload, $L \in \mathbb{R}$ represents the length of the crane rod, and $g \in \mathbb{R}$ represents the gravity constant. After performing some algebraic manipulation, the first two rows of (6.7) can be expressed as follows

$$\ddot{x} = \frac{1}{\det(M)} (p_{11} F_x + p_{12} F_y + w_1) \tag{6.9}$$

$$\ddot{y} = \frac{1}{\det(M)} (p_{12} F_x + p_{22} F_y + w_2) \tag{6.10}$$

where the measurable terms $p_{11}(q)$, $p_{12}(q)$, $p_{22}(q)$, $w_1(q, \dot{q})$, $w_2(q, \dot{q}) \in \mathbb{R}$ are defined as follows

$$p_{11} = m_p^2 L^2 I \left(\sin^2 \phi + 2 \sin^2 \theta \cos^2 \phi \right) + m_p I^2 + m_c m_p L^4 \sin^2 \theta$$

$$+ m_c m_p L^2 I (1 + \sin^2 \theta) + m_c I^2 + m_p^3 L^4 \cos^2 \phi \sin^4 \theta$$

$$(6.11)$$

$$p_{12} = -m_p^3 L^4 \sin \phi \cos \phi \sin^4 \theta - m_p^2 L^2 I \sin \phi \cos \phi \left(\sin^2 \theta - \cos^2 \theta \right) \quad (6.12)$$

$$p_{22} = m_p^3 L^4 \sin^4 \theta \sin^2 \phi + m_p^2 L^2 I \left[1 + (\sin^2 \theta - \cos^2 \theta) \sin^2 \phi \right]$$

$$+ (m_p + m_r + m_c) I^2 + (m_r + m_c) m_p^2 L^4 \sin^2 \theta \tag{6.13}$$

$$+ (m_r + m_c) m_p L^2 I (1 + \sin^2 \theta)$$

$$w_1 = m_p L \sin\theta \sin\phi \left[(m_p + m_c) I + m_p m_c L^2 \sin^2\theta\right]$$

$$\left[\dot{\phi}^2 \left(m_p L^2 \sin^2\theta + I\right) + \dot{\theta}^2 \left(m_p L^2 + I\right)\right]$$

$$-2 I m_p L \dot{\theta} \dot{\phi} \cos\theta \cos\phi \left[(m_p + m_c) I + m_p L^2 \left(m_c + m_p \sin^2\theta\right)\right]$$

$$+ m_p^2 g L^2 \sin\theta \cos\theta \sin\phi \left[(m_p + m_c) I + m_p m_c L^2 \sin^2\theta\right] \tag{6.14}$$

$$w_2 = m_p L \sin\theta \cos\phi \left[(m_p + m_r + m_c) I + (m_r + m_c) m_p L^2 \sin^2\theta\right]$$

$$\left[d^2 \left(m_p L^2 \sin^2\theta + I\right) + p^2 \left(m_p L^2 + I\right)\right]$$

$$+ 2 m_p L I \dot{\theta} \dot{\phi} \cos\theta \sin\phi \left[(m_p + m_r + m_c)\left(m_p L^2 + I\right) - m_p^2 L^2 \cos^2\theta\right]$$

$$+ m_p g L \sin\theta \left[(m_r + m_c) m_p^2 L^3 \sin^2\theta \cos\theta \cos\phi \right.$$

$$\left. + (m_r + m_c + m_p) m_p L I \cos\theta \cos\phi\right]. \tag{6.15}$$

To write the open-loop dynamics given in (6.9) and (6.10) in a more compact form for the subsequent control development, the composite vector $r(t) \in \mathbb{R}^2$ is defined as follows

$$r = \begin{bmatrix} x & y \end{bmatrix}^T. \tag{6.16}$$

After taking the second time derivative of $r(t)$ and then utilizing the expressions given in (6.9–6.15), the open-loop dynamics given in (6.9) and (6.10) can be expressed as follows

$$\ddot{r} = \begin{bmatrix} \ddot{x} & \ddot{y} \end{bmatrix}^T = \frac{1}{\det(M)}(PF + W) \tag{6.17}$$

where $P(q) \in \mathbb{R}^{2 \times 2}$ and $W(q, \dot{q}) \in \mathbb{R}^2$ are defined as follows

$$P = \begin{bmatrix} p_{11} & p_{12} \\ p_{12} & p_{22} \end{bmatrix} \quad W = \begin{bmatrix} w_1 \\ w_2 \end{bmatrix}, \tag{6.18}$$

and $F(t) \in \mathbb{R}^2$ is defined as

$$F = \begin{bmatrix} F_x & F_y \end{bmatrix}^T \tag{6.19}$$

where $p_{ij}(q)$, $w_i(q)$ for $i, j = 1, 2$ are given in (6.11–6.15). Given the expressions in (6.11–6.13), it is straightforward to prove that

$$p_{11} > 0 \quad \text{and} \quad p_{11} p_{22} - p_{12}^2 \geq m_p (m_p + m_r + m_c) I^4, \tag{6.20}$$

and hence, (6.18) and (6.20) can be used to prove that $P(q)$ is positive-definite symmetric and invertible,[2] where the inverse of $P(q)$, denoted by $P^{-1}(q)$, is also positive-definite and symmetric.

To facilitate the subsequent Lyapunov-based control design and stability analysis, the potential and kinetic energy of the overhead crane system is determined. Specifically, the total energy of the overhead crane system, denoted by $E(q, \dot{q}) \in \mathbb{R}$, is given as follows

$$E(q, \dot{q}) = \underbrace{\frac{1}{2} \dot{q}^T M(q) \dot{q}}_{\substack{\text{Kinetic} \\ \text{Energy}}} + \underbrace{m_p g L (1 - \cos(\theta))}_{\substack{\text{Potential} \\ \text{Energy}}} \geq 0. \tag{6.21}$$

After taking the time derivative of (6.21), substituting (6.1) for $M(q)\ddot{q}(t)$, and cancelling common terms, the following expression can be obtained

$$\dot{E} = \dot{r}^T F \tag{6.22}$$

where (6.2), (6.5–6.16), (6.19) and Definition B.3 of Appendix B were utilized.

6.2.3   Control Design and Analysis

The control objective of this section is to regulate the planar gantry position of the overhead crane to a constant desired position, denoted by $r_d \in \mathbb{R}^2$, which is explicitly defined as

$$r_d = [x_d \quad y_d]^T \tag{6.23}$$

while simultaneously regulating the payload angle $\theta(t)$ to zero. To quantify the objective of regulating the overhead crane to a constant desired position, a gantry position error signal, denoted by $e(t) \in \mathbb{R}^2$, is defined as follows

$$e(t) = r - r_d. \tag{6.24}$$

Remark 6.2 *As in [42], the crane dynamic model given in (6.1) exploits a projection of the payload angle along the X-coordinate axis, denoted by $\phi(t)$. By injecting this artificial state, the dynamic model can be written in a manner that facilitates the development of controllers that achieve*

---

[2] The expression given in (6.18) and the fact that $P(q)$ is a leading minor of the positive-definite matrix $M^{-1}(q)$ could also be used to conclude that $P(q)$ is positive-definite, symmetric, and invertible [27].

the control objective. Unfortunately, the overall stability analysis is complicated by the fact that an additional unactuated state is injected into the system. From a physical standpoint, if the payload angle, denoted by $\theta(t)$, is regulated to zero, then from Figure 6.1, the payload is regulated to the desired location, and hence, the control objective is not defined in terms of regulating $\phi(t)$.

Linear Control Law

A linear control law for the overhead crane system can be designed as follows

$$F = -\frac{1}{k_E}\left(k_d \dot{r} + k_p e\right) \tag{6.25}$$

where k_d, k_E, $k_p \in \mathbb{R}$ are positive constant control gains. The linear control law given in (6.25) yields asymptotic regulation of the overhead crane system as stated in the following theorem.

Theorem 6.1 *The linear control law given in (6.25) ensures asymptotic regulation of the overhead crane system in the sense that*

$$\lim_{t \to \infty} \left(\begin{array}{ccc} x(t) & y(t) & \theta(t) \end{array}\right) = \left(\begin{array}{ccc} x_d & y_d & 0 \end{array}\right) \tag{6.26}$$

where x_d and y_d were defined in (6.23).

Proof: To prove Theorem 6.1, we define a nonnegative function $V(t) \in \mathbb{R}$ as follows

$$V = k_E E + \frac{1}{2} k_p e^T e. \tag{6.27}$$

The time derivative of (6.27) can be obtained as follows

$$\dot{V} = \dot{r}^T \left(k_E F + k_p e\right) \tag{6.28}$$

where (6.22) and the time derivative of (6.24) were utilized. After substituting (6.25) into (6.28) for $F(t)$ and then cancelling common terms, the following expression is obtained

$$\dot{V} = -k_d \dot{r}^T \dot{r}. \tag{6.29}$$

Based on the expressions given in (6.4), (6.21), (6.24), (6.27), and (6.29), it is clear $r(t), e(t), \dot{q}(t) \in \mathcal{L}_\infty$. Based on the fact that $r(t), e(t), \dot{q}(t) \in \mathcal{L}_\infty$, (6.2) and (6.16) can be used to show that $x(t), \dot{x}(t), y(t), \dot{y}(t), \dot{r}(t), \dot{\theta}(t), \dot{\phi}(t) \in \mathcal{L}_\infty$. Given that $e(t), \dot{r}(t) \in \mathcal{L}_\infty$, (6.25) can be used to determine that $F(t) \in \mathcal{L}_\infty$. The expressions given in (6.3) and (6.19) can now be used to determine that $F_x(t), F_y(t), u(t) \in \mathcal{L}_\infty$.

Based on the fact that all of the closed-loop signals remain bounded, LaSalle's Invariance Theorem (see Lemma A.22 of Appendix A) can now be used to prove (6.26). To this end, the set Γ is defined as the set of all points where

$$\dot{V} = 0. \tag{6.30}$$

In the set Γ, we see from (6.29) and (6.30) that

$$\dot{r}(t) = 0 \qquad \ddot{r}(t) = 0, \tag{6.31}$$

and hence, (6.16), (6.27), (6.30), and (6.31) can be used to determine that $x(t)$, $y(t)$, and $V(t)$ are constant. Furthermore, from (6.22), (6.24), and (6.31), we see that

$$\dot{E}(q, \dot{q}) = \dot{e}(t) = 0. \tag{6.32}$$

Based on (6.32), we see that $E(q, \dot{q})$ and $e(t)$ are constant, and hence, from (6.25) and (6.31), it is clear that $F(t)$ is constant. To complete the proof, stability of the system must be examined for the case when $\dot{\theta} = 0$ and when $\dot{\theta} \neq 0$. In this analysis, given in Lemma B.21 of Appendix B, the result given in (6.26) is proven under the proposition that $\dot{\theta} = 0$. Furthermore, Lemma B.21 proves the proposition that $\dot{\theta} \neq 0$ leads to contradictions, and hence, is invalid. $\square$

The stability analysis for Theorem 6.1 indicates that the control objective is met and that all signals in the dynamics and the controller remain bounded for all time except for the signal $\phi(t)$. (Note that by assumption, the payload angle, denoted by $\theta(t)$, is assumed to be bounded.) The boundedness of $\phi(t)$ is insignificant from a theoretical point of view since $\phi(t)$ only appears in the dynamics and control as arguments of bounded trigonometric functions. The stability result for the linear control law (and the subsequent coupling control laws) is not considered to be global due to Assumption 6.5. However, from a practical standpoint it seems rare that the payload of the overhead crane system would need to violate this assumption.

Heuristically, the only way for the energy from the payload motion to be dissipated is through the coupling between the payload dynamics and the gantry dynamics. That is, a linear feedback loop at the gantry creates an artificial spring/damper system which absorbs the payload energy through the natural gantry/payload coupling. However, from our experience with many control experiments on overhead crane testbeds, we have observed that a linear feedback loop at the gantry will always provide poor performance because perfect compensation for the gantry friction is not achieved (and it never will be). That is, the uncompensated gantry friction effects tend to retard the natural coupling between the gantry/payload dynamics, and hence, prevent payload energy from being dissipated by the linear feedback loop at the gantry. In the following sections, controllers are developed that improve the performance of the linear feedback loop at the gantry by incorporating additional nonlinear terms in the control law that depend on the payload dynamics. Although the subsequent controllers yield the same stability result as the linear controller, the increase in gantry/payload coupling by the additional nonlinear terms results in improved performance when compared to the simple linear gantry controller. That is, the subsequent controllers exploit additional energy-based nonlinear terms that provide increased payload swing feedback through the coupling between the gantry and the payload. The improved performance of the subsequent nonlinear coupling controllers has been demonstrated by simulation results and by the experimental results presented in Section 6.2.4.

E^2 Coupling Control Law

As stated previously, improved performance may result from increasing the coupling between the gantry and the unactuated payload. Motivated by the desire to increase the gantry/payload coupling, the following E^2 coupling control law is developed[3]

$$F = \Omega^{-1} \left(-k_d \dot{r} - k_p e - \frac{k_v}{\det(M)} W \right) \tag{6.33}$$

where $\Omega(t) \in \mathbb{R}^{2 \times 2}$ is a positive-definite invertible matrix[4] defined as follows

$$\Omega = k_E E I_2 + \frac{k_v}{\det(M)} P \tag{6.34}$$

k_E, k_p, k_d, $k_v \in \mathbb{R}$ are positive constant control gains, I_2 denotes the standard 2×2 identity matrix, and $\det(M)$, $P(q)$, and $W(q, \dot{q})$ were defined in (6.8) and (6.18). The E^2 coupling control law given in (6.33) yields asymptotic regulation of the overhead crane system as stated in the following theorem.

---

[3] The control strategy is called an E^2 coupling control law because its structure is motivated by a squared energy term and an additional squared gantry velocity term in the Lyapunov function. The structure of this controller is inspired by the previous work given in [35] for an inverted pendulum.

[4] Since P and I_2 are positive-definite symmetric matrices, and k_E, k_v, $E(q, \dot{q})$, and $\det(M(q))$ are positive scalars, it can be proven that $\Omega(t)$ is positive-definite and invertible.

Theorem 6.2 *The E^2 coupling control law given in (6.33) ensures asymptotic regulation of the overhead crane system in the sense that*

$$\lim_{t\to\infty} \begin{pmatrix} x(t) & y(t) & \theta(t) \end{pmatrix} = \begin{pmatrix} x_d & y_d & 0 \end{pmatrix} \tag{6.35}$$

where x_d and y_d were defined in (6.23).

Proof: To prove Theorem 6.2, we define a nonnegative function $V(t) \in \mathbb{R}$ as follows

$$V = \frac{1}{2}k_E E^2 + \frac{1}{2}k_p e^T e + \frac{1}{2}k_v \dot{r}^T \dot{r}. \tag{6.36}$$

After taking the time derivative of (6.36) and then substituting (6.17), (6.22), and the time derivative of (6.24) into the resulting expression for $\ddot{r}(t)$, $\dot{E}(q,\dot{q})$, and $\dot{e}(t)$, respectively, the following expression can be obtained

$$\dot{V} = \dot{r}^T \left(\Omega F + \frac{k_v}{\det(M)} W + k_p e \right) \tag{6.37}$$

where (6.34) was utilized. After substituting (6.33) into (6.37) for $F(t)$ and then cancelling common terms, (6.37) can be rewritten as follows

$$\dot{V} = -k_d \dot{r}^T \dot{r}. \tag{6.38}$$

Based on the expressions given in (6.6), (6.21), (6.24), (6.36), and (6.38), it is clear that $r(t)$, $\dot{r}(t)$, $e(t)$, $\dot{q}(t)$, $E(q,\dot{q}) \in \mathcal{L}_\infty$, and hence, (6.2) and (6.16) can be used to determine that $x(t)$, $\dot{x}(t)$, $y(t)$, $\dot{y}(t)$, $\dot{\theta}(t)$, $\dot{\phi}(t) \in \mathcal{L}_\infty$. The expressions given in (6.8), (6.11–6.15), (6.18), and (6.20) can be used to show that $\det(M(q))$, $P(q)$, $P^{-1}(q)$, $W(q,\dot{q}) \in \mathcal{L}_\infty$. Given (6.206) and the fact that $E(q,\dot{q})$, $P(q) \in \mathcal{L}_\infty$, (6.34) can be used to show that $\Omega(t) \in \mathcal{L}_\infty$. Given the following expression for the determinant of $\Omega(t)$

$$\det(\Omega) = (k_E E)^2 + k_E E \frac{k_v}{\det(M)} (p_{11} + p_{22}) + \left(\frac{k_v}{\det(M)} \right)^2 (p_{11}p_{22} - p_{12}^2) \tag{6.39}$$

the expressions given in (6.8), (6.11-6.13), (6.20), and (6.21) can be utilized to determine that

$$\det(\Omega) \geq \left(\frac{k_v}{\det(M)} \right)^2 m_p (m_p + m_r + m_c) I^4. \tag{6.40}$$

Based on the fact that $\det(M(q))$, $\Omega(t) \in \mathcal{L}_\infty$, (6.40) can be used to prove that $\Omega^{-1}(t) \in \mathcal{L}_\infty$. Given that $e(t)$, $\dot{r}(t)$, $\det(M(q))$, $W(q,\dot{q})$, $\Omega^{-1}(t) \in \mathcal{L}_\infty$, it is clear from (6.33) that $F(t) \in \mathcal{L}_\infty$. From (6.3) and (6.19), $F_x(t)$, $F_y(t)$, $u(t) \in \mathcal{L}_\infty$. Since all of the closed-loop signals remain bounded and the time derivative of (6.36) is decreasing or constant as indicated by (6.38), Lemma B.22 of Appendix B can be used to prove Theorem 6.2. $\square$

Gantry Kinetic Energy Coupling Control Law

To illustrate how additional controllers can also be derived to increase the gantry/payload coupling, the following gantry kinetic energy coupling control law is designed[5]

$$F = \frac{-k_d \dot{r} - k_p e - k_v P^{-1} W - \frac{1}{2} k_v \left(\frac{d}{dt}\left(\det(M) P^{-1}\right)\right) \dot{r}}{k_E + k_v} \tag{6.41}$$

where k_E, k_p, k_d, and $k_v \in \mathbb{R}$ are positive constant control gains, and $\det(M)$, $P(q)$, and $W(q,\dot{q})$ were defined in (6.8) and (6.18). The gantry kinetic energy coupling control law given in (6.41) yields asymptotic regulation of the overhead crane system as stated in the following theorem.

Theorem 6.3 *The gantry kinetic energy coupling control law given in (6.41) ensures asymptotic regulation of the overhead crane system in the sense that*

$$\lim_{t \to \infty} \left(\begin{array}{ccc} x(t) & y(t) & \theta(t) \end{array} \right) = \left(\begin{array}{ccc} x_d & y_d & 0 \end{array} \right) \tag{6.42}$$

where x_d and y_d were defined in (6.23).

Proof: To prove Theorem 6.3, we define a nonnegative function $V(t) \in \mathbb{R}$ as follows

$$V = k_E E + \frac{1}{2} k_v \dot{r}^T \left(\det(M) P^{-1}\right) \dot{r} + \frac{1}{2} k_p e^T e. \tag{6.43}$$

After taking the time derivative of (6.43) and then substituting (6.22) and the time derivative of (6.24) for $\dot{E}(q,\dot{q})$ and $\dot{e}(t)$, respectively, the following expression is obtained

$$\dot{V} = \dot{r}^T \left(k_E F + k_v \left(\det(M) P^{-1}\right) \ddot{r} + \frac{1}{2} k_v \left(\frac{d}{dt}\left(\det(M) P^{-1}\right)\right) \dot{r} + k_p e \right). \tag{6.44}$$

By utilizing (6.17), (6.44) can be rewritten as follows

$$\dot{V} = \dot{r}^T \left((k_E + k_v) F + k_v P^{-1} W + \frac{1}{2} k_v \left(\frac{d}{dt}\left(\det(M) P^{-1}\right)\right) \dot{r} + k_p e \right). \tag{6.45}$$

After substituting (6.41) for $F(t)$ and then simplifying the resulting expression, the following expression can be obtained

$$\dot{V} = -k_d \dot{r}^T \dot{r}. \tag{6.46}$$

---

[5] The control strategy is called a gantry kinetic energy coupling control law because the control structure is derived from an additional gantry kinetic energy-like term in the Lyapunov function.

Based on the expressions given in (6.21), (6.24), (6.43), and (6.46), it is clear that $r(t), \dot{r}(t), e(t), \dot{q}(t) \in \mathcal{L}_\infty$. Hence, from (6.2) and (6.16) it is clear that $x(t), \dot{x}(t), y(t), \dot{y}(t), \dot{r}(t), \dot{\theta}(t), \dot{\phi}(t) \in \mathcal{L}_\infty$. The expressions given in (6.8), (6.11–6.15), (6.18), and (6.20) can be used to determine that $\det(M(q))$, $P^{-1}(q), W(q, \dot{q}) \in \mathcal{L}_\infty$. Based on the fact that (see Lemma B.24 of Appendix B)

$$\frac{d}{dt} P^{-1} = -P^{-1}\left(\frac{d}{dt} P\right) P^{-1},$$

the time derivative of the product $\det(M)P^{-1}(q)$ can be determined as follows

$$\frac{d}{dt}\left(\det(M)P^{-1}\right) = \left(\frac{d}{dt}\det(M)\right)P^{-1} - \det(M)P^{-1}\left(\frac{d}{dt}P\right)P^{-1}. \quad (6.47)$$

The time derivative of the determinant of $M(q)$ given in (6.47) can be obtained from (6.8) as follows

$$\frac{d}{dt}\det(M) = m_p L^2 I \left[((m_p + m_r + m_c) m_c + (m_p + m_c) m_p) \right.$$

$$\cdot \left(2\dot{\theta}\sin\theta\cos\theta\right) + 2m_r m_p \cos\phi\dot{\phi}\sin\phi$$

$$\left. \cdot \cos\phi\left(2\dot{\theta}\sin\theta\cos\theta\cos\phi - 2\dot{\phi}\sin^2\theta\sin\phi\right)\right]$$

$$+ m_p^2 L^4 \left(2\dot{\theta}\sin\theta\cos\theta\right)\left[m_r m_p \sin^2\theta\cos^2\phi \right. \qquad (6.48)$$

$$+ m_c\left(m_r + m_c + m_p\sin^2\theta\right)\right] + m_p^2 L^4 \sin^2\theta\left[m_r m_p\right.$$

$$\cdot \left(\left(2\dot{\theta}\sin\theta\cos\theta\cos^2\phi\right) - \left(2\dot{\phi}\sin^2\theta\cos\phi\sin\phi\right)\right)$$

$$\left. + m_c m_p\left(2\dot{\theta}\sin\theta\cos\theta\right)\right].$$

The time derivative of each element of $P(q)$ given in (6.47) can be obtained from (6.11–6.13) as follows

$$\dot{p}_{11} = m_p^2 L^2 I\left(2\dot{\phi}\sin\phi\cos\phi + 4\dot{\theta}\sin\theta\cos\theta\cos^2\phi\right.$$

$$\left. - 4\dot{\phi}\sin^2\theta\sin\phi\cos\phi\right) + 2m_c m_p L^4\dot{\theta}\sin\theta\cos\theta \qquad (6.49)$$

$$+ 2m_c m_p L^2 I\dot{\theta}\sin\theta\cos\theta - 2m_p^3 L^4\dot{\phi}\sin^4\theta\sin\phi\cos\phi$$

$$+ 4m_p^3 L^4\dot{\theta}\cos^2\phi\sin^3\theta\cos\theta$$

$$\dot{p}_{12} = -m_p^3 L^4 \dot{\phi} \left(\cos^2 \phi - \sin^2 \phi\right) \sin^4 \theta$$

$$-4m_p^3 L^4 \dot{\theta} \sin \phi \cos \phi \sin^3 \theta \cos \theta$$

$$-m_p^2 L^2 I \dot{\phi} \left(\cos^2 \phi - \sin^2 \phi\right) \left(\sin^2 \theta - \cos^2 \theta\right) \tag{6.50}$$

$$-4m_p^2 L^2 I \dot{\theta} \sin \phi \cos \phi \sin \theta \cos \theta$$

$$\dot{p}_{22} = 4m_p^2 L^2 \dot{\theta} \sin \theta \cos \theta \sin^2 \phi \left(m_p L^2 \sin^2 \theta + I\right)$$

$$+2m_p^2 L^2 \dot{\phi} \sin \phi \cos \phi \left[m_p L^2 \sin^4 \theta + I \left(\sin^2 \theta - \cos^2 \theta\right)\right] \tag{6.51}$$

$$+2 \left(m_r + m_c\right) m_p L^2 \dot{\theta} \sin \theta \cos \theta \left(m_p L^2 + I\right).$$

Based on the fact that $q(t), \dot{q}(t), P^{-1}(q) \in \mathcal{L}_\infty$, it is clear that the expressions given in (6.47–6.51) can be utilized to prove that $\frac{d}{dt}\left(\det(M)P^{-1}\right) \in \mathcal{L}_\infty$. Given that $e(t), \dot{r}(t), P^{-1}(q), W(q,\dot{q}), \frac{d}{dt}\left(\det(M)P^{-1}\right) \in \mathcal{L}_\infty$, (6.41) can be used to show that $F(t) \in \mathcal{L}_\infty$. Finally, (6.3) and (6.19) can be used to show that $F_x(t), F_y(t), u(t) \in \mathcal{L}_\infty$. Since all of the closed-loop signals remain bounded and the time derivative of (6.43) is decreasing or constant as indicated by (6.46), Lemma B.23 of Appendix B can now be used to prove Theorem 6.3. $\square$

6.2.4   Experimental Setup and Results

The controllers given in (6.25), (6.33), and (6.41) were implemented on the InTeCo overhead crane testbed [29] shown in Figure 6.2. The testbed consists of two primary components: a mechanical system and a data acquisition and control system. For the mechanical system, the rail and the cart are driven by AC servo motors, and the payload is connected to a cable attached to the underside of the cart through a shaft mounted encoder. The physical parameters of the overhead crane testbed were determined as follows

$$m_p = 0.73 \text{ [kg]} \qquad m_c = 1.06 \text{ [kg]} \qquad m_r = 6.4 \text{ [kg]} \tag{6.52}$$

$$I = 0.005 \text{ [kg} \cdot \text{m}^2] \qquad L = 0.7 \text{ [m]}.$$

The data acquisition and control algorithms are implemented on a Pentium 266 MHz PC running under the RT-Linux operating system. The Matlab/Simulink environment and Real Time Linux Target [55] were used to implement the controllers. The Quanser MultiQ I/O board was used for the input/output operations. Specifically, four encoder channels of the MultiQ I/O board were used to measure the position of the cart and the payload angle, and two of the D/A channels were used to output the voltage to the DC motor. The output voltages were subject to two levels of amplification. The first level of amplification was achieved by an OP07 operational amplifier (i.e., signal conditioning), and the second level of amplification was enabled by a Techron linear power amplifier. The electrical dynamics associated with the testbed were neglected, and in an attempt to compensate for gantry friction effects, the following voltage-force relationship was implemented

$$V = k_s((F + F_s \text{sgn}(\dot{x}))$$

where k_s is a scaling constant, V denotes the voltage output to the linear amplifier, F was designed in (6.25), (6.33), and (6.41), and F_s is the coefficient of static friction. The values that were used for k_s and F_s are given below

$$k_s = \frac{1}{200} \qquad F_s = 150. \tag{6.53}$$

To determine F_s, open-loop voltages were applied to the motor with increasing amplitude until the gantry began to move. The value for F_s that results in gantry movement was recorded as the static friction force. While numerous experimental methods exist to determine the static friction coefficient, this simple approach yields suitable results to demonstrate and compare the effectiveness of the developed controllers.

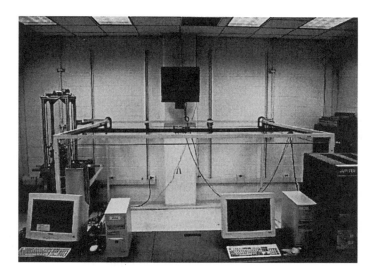

FIGURE 6.2. Experimental setup.

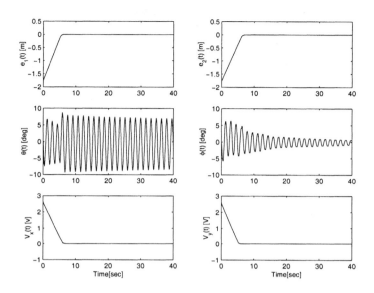

FIGURE 6.3. Results for the PD controller.

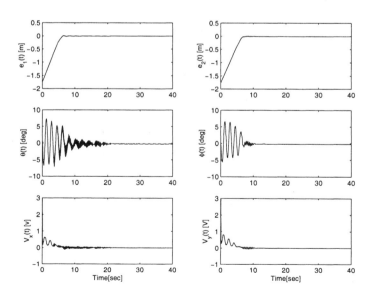

FIGURE 6.4. Results for the E^2 coupling control law.

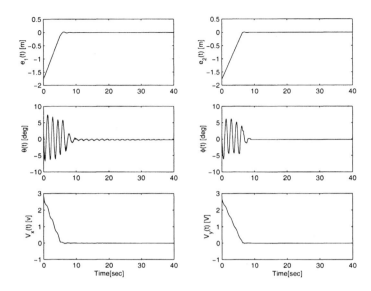

FIGURE 6.5. Results for the gantry kinetic energy coupling control law.

Each controller was implemented at a sampling frequency of 1 [kHz] and the desired gantry position was selected as follows

$$\begin{bmatrix} x_d & y_d \end{bmatrix}^T = \begin{bmatrix} 1.75 & 1.75 \end{bmatrix}^T [\text{m}]. \tag{6.54}$$

For each of the experiments, the initial conditions were set to zero and the control gains were tuned until the best performance was achieved. The resulting control gains from each controller are given in Table 6.1 where $\text{diag}\{\cdot\}$ is used to denote a diagonal matrix.[6]

The resulting gantry position error, payload angle, and the input force are shown in Figure 6.3 for the linear control law, Figure 6.4 for the E^2 coupling control law, and Figure 6.5 for the gantry kinetic energy coupling control law. A summary of the performance of the controllers given in (6.25), (6.33), and (6.41) is provided in Table 6.2. The settling time is defined as the interval between the starting time and the time when the angle $\theta(t)$ remained within ± 0.5 [deg] of the equilibrium position and the response of $x(t)$ and $y(t)$ remained within 5% of the final values.

---

[6] The stability analysis required that the gains k_d, k_p, k_E, k_v be defined as scalars; however, k_d, k_p, k_E, k_v were defined as matrices during the experiment. Although theoretical justification is not provided, experimental experience indicates that this modification usually improves the tracking performance in real-time implementations.

TABLE 6.1. Control gains

	Linear Control Law	E^2 Coupling Control Law	Gantry Kinetic Energy Coupling Control Law
k_d	diag $\{0.5, 0.5\}$	diag $\{1, 1\}$	diag $\{0.6, 0.2\}$
k_p	diag $\{3, 3\}$	diag $\{5, 5\}$	diag $\{3.2, 3\}$
k_E	diag $\{1, 1\}$	diag $\{0.01, 0.01\}$	diag $\{0.8, 0.87\}$
k_v	Not Applicable	diag $\{0.29, 0.66\}$	diag $\{0.2, 0.13\}$

TABLE 6.2. Performance comparison

	Linear Control Law	E^2 Coupling Control Law	Gantry Kinetic Energy Coupling Control Law
X % Overshoot	0%	0.56%	1.01%
Y % Overshoot	0%	0.36%	1.00%
Settling Time ($|\theta| \leq 0.5$ [deg])	> 40 [sec]	19.9 [sec]	9.3 [sec]

The coupling control laws exhibit superior settling time when compared to the linear controller based on the results given in Table 6.2 and Figures 6.3–6.5. One reason for this superior performance is that the only payload/gantry coupling that exists for the linear controller is the natural coupling. When the gantry approaches the desired setpoint, friction damps the gantry to the extent that no overshoot can be observed. Hence, the gantry stops and the payload swings freely. In contrast, the payload/gantry coupling for the coupling control laws is enhanced, resulting in faster damping of the payload swing.

6.3 VTOL Systems

In this section, a nonlinear tracking controller is developed for the nonminimum-phase underactuated planar VTOL aircraft problem. A Lyapunov-based stability analysis is presented to demonstrate that the position and orientation tracking errors are globally exponentially forced to a neighborhood about zero which can be made arbitrarily small. A unified framework is developed that solves both the planar VTOL regulation and tracking problems.

6.3.1   System Model

Dynamic Model

The dynamic model for a planar VTOL aircraft can be written as follows [25]

$$\ddot{x} = -\sin(\phi)\, u_t + \varepsilon \cos(\phi)\, u_m \tag{6.55}$$

$$\ddot{y} = \cos(\phi)\, u_t + \varepsilon \sin(\phi)\, u_m - 1 \tag{6.56}$$

$$\dot{\phi} = v_1 \tag{6.57}$$

$$\dot{v}_1 = u_m. \tag{6.58}$$

For the dynamic model given in (6.55–6.58), $x(t)$, $y(t)$, $\phi(t) \in \mathbb{R}$ denote the Cartesian position and orientation respectively, of the center of mass (COM) of the aircraft, $v_1(t) \in \mathbb{R}$ denotes the angular velocity of the aircraft, $u_t(t)$, $u_m(t) \in \mathbb{R}$ represent the vertical force and the rotational torque applied to the aircraft, -1 represents the scaled gravitational acceleration, and $\varepsilon \in \mathbb{R}$ is a nonminimum-phase parameter that represents the constant coupling between the rolling moment and the lateral acceleration of the aircraft (see Figure 6.6). To simplify the subsequent control development, the following transformation is defined

$$x_1 = x - \varepsilon \sin(\phi) \tag{6.59}$$

$$y_1 = y + \varepsilon \cos(\phi) \tag{6.60}$$

where $x_1(t)$, $y_1(t) \in \mathbb{R}$ represent the "shifted" Cartesian position. After taking the second time derivatives of (6.59) and (6.60), substituting (6.55) and (6.56) into the resulting expression, and then cancelling common terms, the shifted Cartesian system dynamics can be written as follows

$$\ddot{x}_1 = -\left(u_t - \varepsilon\dot{\phi}^2\right)\sin(\phi) \tag{6.61}$$

$$\ddot{y}_1 = \left(u_t - \varepsilon\dot{\phi}^2\right)\cos(\phi) - 1. \tag{6.62}$$

Reference Model

Since the VTOL aircraft is subject to motion constraints (e.g., nonholonomic constraints), the desired trajectory must be designed under the same constraints to ensure that the trajectory is feasible. To this end, a VTOL reference model is defined based on the structure of the system dynamics given by (6.55–6.58) as follows

$$\ddot{x}_r = -\sin(\phi_r)\, u_{rt} + \varepsilon \cos(\phi_r)\, u_{rm} \tag{6.63}$$

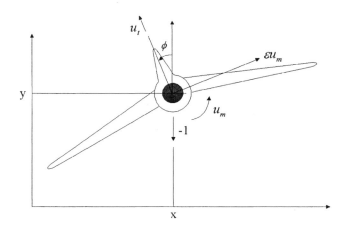

FIGURE 6.6. VTOL aircraft.

$$\ddot{y}_r = \cos\left(\phi_r\right) u_{rt} + \varepsilon \sin\left(\phi_r\right) u_{rm} - 1 \qquad (6.64)$$

$$\dot{\phi}_r = v_{r1} \qquad (6.65)$$

$$\dot{v}_{r1} = u_{rm}. \qquad (6.66)$$

In the reference model given in (6.63–6.66), $x_r\left(t\right)$, $y_r\left(t\right)$, $\phi_r\left(t\right) \in \mathbb{R}$ denote the reference Cartesian position and orientation, respectively, $v_{r1}\left(t\right) \in \mathbb{R}$ is a reference variable used to facilitate the analysis, and $u_{rt}\left(t\right)$, $u_{rm}\left(t\right) \in \mathbb{R}$ denote the reference input signals. The reference signals $u_{rt}\left(t\right)$ and $u_{rm}\left(t\right)$ are assumed to be selected such that $u_{rt}\left(t\right)$, $u_{rm}\left(t\right)$, $\dot{u}_{rt}\left(t\right)$, and $\dot{u}_{rm}\left(t\right) \in \mathcal{L}_\infty$ and $x_r\left(t\right)$, $y_r\left(t\right)$, $\phi_r\left(t\right)$, $\dot{x}_r\left(t\right)$, $\dot{y}_r\left(t\right)$, $\dot{\phi}_r\left(t\right)$, $\ddot{x}_r\left(t\right)$, $\ddot{y}_r\left(t\right)$, $\ddot{\phi}_r\left(t\right) \in \mathcal{L}_\infty$. To facilitate the subsequent error system development, a transformation similar to the one given in (6.59) and (6.60) is applied to the reference model as follows

$$x_{r1} = x_r - \varepsilon \sin\left(\phi_r\right) \qquad (6.67)$$

$$y_{r1} = y_r + \varepsilon \cos\left(\phi_r\right) \qquad (6.68)$$

where $x_{r1}\left(t\right)$, $y_{r1}\left(t\right) \in \mathbb{R}$ represent the "shifted" reference Cartesian position. After taking the second time derivatives of (6.67) and (6.68), substituting (6.63) and (6.64) into the resulting expression, and then cancelling common terms, the shifted reference Cartesian dynamics can be written as follows

$$\ddot{x}_{r1} = -\left(u_{rt} - \varepsilon\dot{\phi}_r^2\right)\sin\left(\phi_r\right) \qquad (6.69)$$

$$\ddot{y}_{r1} = \left(u_{rt} - \varepsilon\dot{\phi}_r^2\right)\cos\left(\phi_r\right) - 1. \qquad (6.70)$$

6.3.2 Control Objective

The control objective is to force the position and orientation of an aircraft performing the VTOL maneuver to track a desired trajectory. To quantify this objective, we define the position tracking error, denoted by $\tilde{x}(t)$, $\tilde{y}(t) \in \mathbb{R}$, and the orientation tracking error, denoted by $\tilde{\phi}(t) \in \mathbb{R}$, as follows

$$\tilde{x} = x - x_r \qquad \tilde{y} = y - y_r \qquad \tilde{\phi} = \phi - \phi_r. \qquad (6.71)$$

After using (6.59), (6.60), (6.67), and (6.68), the position tracking error can be expressed in terms of the shifted position coordinates as follows

$$\tilde{x} = \tilde{x}_1 + \varepsilon \left[\sin(\phi) - \sin(\phi_r)\right] \qquad \tilde{y} = \tilde{y}_1 - \varepsilon \left[\cos(\phi) - \cos(\phi_r)\right] \qquad (6.72)$$

where $\tilde{x}_1(t)$, $\tilde{y}_1(t) \in \mathbb{R}$ denote shifted Cartesian tracking error signals that are explicitly defined as follows

$$\tilde{x}_1 = x_1 - x_{r1} \qquad \tilde{y}_1 = y_1 - y_{r1}. \qquad (6.73)$$

As a direct consequence of the Mean Value Theorem (see Lemma A.1 of Appendix A), it can be shown that

$$|\sin(\phi) - \sin(\phi_r)| \le \left|\tilde{\phi}\right| \qquad |\cos(\phi) - \cos(\phi_r)| \le \left|\tilde{\phi}\right|; \qquad (6.74)$$

hence, (6.72) and (6.74) can be used to prove that

$$|\tilde{x}| \le |\tilde{x}_1| + \varepsilon \left|\tilde{\phi}\right| \qquad |\tilde{y}| \le |\tilde{y}_1| + \varepsilon \left|\tilde{\phi}\right|. \qquad (6.75)$$

From (6.75), we can prove that if the shifted position tracking errors and the orientation tracking error are driven to zero then the actual aircraft position tracking errors will be forced to zero. Likewise, if the shifted position tracking errors and the orientation tracking error are driven to a neighborhood about zero, then the actual aircraft position tracking errors will driven to a similar (but slightly larger) neighborhood. Based on these facts, the shifted position tracking errors are used in the subsequent control design and stability analysis to reduce the complexity of the design. To further simplify the control design and stability analysis, a filtered tracking error signal, denoted by $r(t) \in \mathbb{R}^2$, is also defined as follows

$$r = \begin{bmatrix} r_x \\ r_y \end{bmatrix} = \begin{bmatrix} \dot{\tilde{x}}_1 + \mu \tilde{x}_1 \\ \dot{\tilde{y}}_1 + \mu \tilde{y}_1 \end{bmatrix} \qquad (6.76)$$

where $\mu \in \mathbb{R}$ represents a constant positive control gain.

6.3.3   Open-Loop Error System

Since the VTOL aircraft system is subject to nonholonomic constraints, we are motivated to transform the error system so that it has a similar structure to Brockett's Nonholonomic Integrator [9]. By performing this transformation greater insight can be gained into the control development. To this end, the following global invertible transformation is defined

$$
\begin{bmatrix} w \\ z_1 \\ z_2 \end{bmatrix} = \begin{bmatrix} \tilde{\phi}\sin\phi + 2\cos\phi & -\tilde{\phi}\cos\phi + 2\sin\phi & 0 \\ 0 & 0 & 1 \\ -\sin\phi & \cos\phi & 0 \end{bmatrix} \begin{bmatrix} r_x \\ r_y \\ \tilde{\phi} \end{bmatrix} \tag{6.77}
$$

where $w(t) \in \mathbb{R}$ and $z(t) = \begin{bmatrix} z_1(t) & z_2(t) \end{bmatrix}^T \in \mathbb{R}^2$ are transformed tracking error variables. After taking the time derivative of (6.77) and using (6.57), (6.58), (6.61), (6.62), (6.65), (6.66), and (6.69–6.71), the following open-loop tracking error system can be developed

$$
\dot{w} = u^T J^T z + f \tag{6.78}
$$

$$
\dot{z} = u \tag{6.79}
$$

$$
\dot{u}_1 = -u_{rm} + u_m. \tag{6.80}
$$

In (6.78), $f(\phi, z, \dot{\tilde{x}}_1, \dot{\tilde{y}}_1) \in \mathbb{R}$ is defined as follows

$$
\begin{aligned}
f = \ & 2\left(-\ddot{x}_{r1}\cos(\phi) - \sin(\phi) - \ddot{y}_{r1}\sin(\phi) + v_{r1}z_2\right) \\
& +2\mu\left(\dot{\tilde{x}}_1\cos(\phi) + \dot{\tilde{y}}_1\sin(\phi)\right),
\end{aligned} \tag{6.81}
$$

$J \in \mathbb{R}^{2\times2}$ is a skew-symmetric matrix defined as follows

$$
J = \begin{bmatrix} 0 & -1 \\ 1 & 0 \end{bmatrix}, \tag{6.82}
$$

and $u(t) = \begin{bmatrix} u_1(t) & u_2(t) \end{bmatrix}^T \in \mathbb{R}^2$ denotes a transformed control signal. Specifically, $u(t)$ is related to the control signals $u_t(t)$ and $v_1(t)$ given in (6.55–6.58) through the following globally invertible transformation

$$
u = T^{-1}\begin{bmatrix} u_t \\ v_1 \end{bmatrix} - \Pi \qquad \begin{bmatrix} u_t \\ v_1 \end{bmatrix} = T(u + \Pi) \tag{6.83}
$$

where $T(r_x, r_y, \phi) \in \mathbb{R}^{2\times2}$ and $\Pi(\phi, v_1, \dot{\tilde{x}}_1, \dot{\tilde{y}}_1) \in \mathbb{R}^2$ are defined as follows

$$
T = \begin{bmatrix} r_x\cos(\phi) + r_y\sin(\phi) & 1 \\ 1 & 0 \end{bmatrix} \tag{6.84}
$$

$$\Pi = \begin{bmatrix} v_{r1} \\ \varepsilon v_1^2 + \cos(\phi) - \ddot{x}_{r1}\sin(\phi) + \ddot{y}_{r1}\cos(\phi) \\ -\mu\left(\dot{\tilde{y}}_1\cos(\phi) - \dot{\tilde{x}}_1\sin(\phi)\right) \end{bmatrix}. \tag{6.85}$$

6.3.4 Control Development

Based on the subsequent stability analysis and the structure of the open-loop error system given in (6.78–6.80), the control signal $u_d(t) \in \mathbb{R}^2$ is designed as follows

$$u_d(t) = \begin{bmatrix} u_{d1} & u_2 \end{bmatrix}^T = u_a - k_2 z. \tag{6.86}$$

For the controller given in (6.86), $u_a(t) \in \mathbb{R}^2$ is defined as

$$u_a = \left(\frac{k_1 w + f}{\delta_d^2}\right) J z_d + \Omega_1 z_d \tag{6.87}$$

where $z_d(t) \in \mathbb{R}^2$ is generated by the following oscillator-like differential equation

$$\dot{z}_d = \frac{\dot{\delta}_d}{\delta_d} z_d + \left(\frac{k_1 w + f}{\delta_d^2} + w\Omega_1\right) J z_d \qquad z_d^T(0) z_d(0) = \delta_d^2(0) \tag{6.88}$$

and $k_1, k_2 \in \mathbb{R}$ denote positive constant control gains. The terms $\Omega_1(t)$, $\delta_d(t) \in \mathbb{R}$ given in (6.87) and (6.88) are defined as follows

$$\Omega_1 = k_2 + \frac{\dot{\delta}_d}{\delta_d} + \frac{k_1 w^2 + w f}{\delta_d^2} \tag{6.89}$$

$$\delta_d = \gamma_0 \exp(-\gamma_1 t) + \varepsilon_1 \tag{6.90}$$

where γ_0, γ_1, $\varepsilon_1 \in \mathbb{R}$ are positive constant control gains, and $f(\cdot)$ was defined in (6.81). The control input $u_m(t)$ of (6.80) is designed as follows

$$u_m = k_3 \eta + \tilde{z}_1 - w z_2 + \dot{u}_{d1} + u_{rm} \tag{6.91}$$

where $k_3 \in \mathbb{R}$ is a positive constant control gain. The error signal $\eta(t) \in \mathbb{R}$ introduced in (6.91) is defined as follows

$$\eta = u_{d1} - u_1, \tag{6.92}$$

and the error signal $\tilde{z}_1(t)$ of (6.91) is an element of the vector $\tilde{z}(t) = \begin{bmatrix} \tilde{z}_1(t) & \tilde{z}_2(t) \end{bmatrix} \in \mathbb{R}^2$ that is defined as follows

$$\tilde{z} = z_d - z. \tag{6.93}$$

After substituting the time derivative of $u_a(t)$ defined in (6.87) into the time derivative of (6.86), the expression for $\dot{u}_{d1}(t)$ introduced in (6.91) can be determined as follows

$$
\dot{u}_{d1} = \; -\left(\frac{k_1\dot{w} + \dot{f}}{\delta_d^2}\right) z_{d2} + 2\left(\frac{(k_1w + f)\,\dot{\delta}_d}{\delta_d^3}\right) z_{d2}
$$

$$
+\dot{\Omega}_1 z_{d1} + \Omega_1 \dot{z}_{d1} - \left(\frac{k_1w + f}{\delta_d^2}\right) \dot{z}_{d2} - k_2\dot{z}_1
$$

(6.94)

where the time derivatives of $\Omega_1(t)$ and $f(t)$ are explicitly given by the following expressions

$$
\dot{\Omega}_1 = \frac{\ddot{\delta}_d}{\delta_d} - \frac{\dot{\delta}_d^2}{\delta_d^2} + \frac{(2k_1w + f)\,\dot{w} + w\dot{f}}{\delta_d^2} - 2\frac{(k_1w^2 + wf)\,\dot{\delta}_d}{\delta_d^3}
$$

(6.95)

$$
\dot{f} = \; 2\left[-\dot{\phi}\left(u_{rt} - \varepsilon\dot{\phi}_r^2\right) \cos\left(\tilde{\phi}\right) \right.
$$
$$
- \sin\left(\tilde{\phi}\right)\left[\dot{u}_{rt} - 2\varepsilon\dot{\phi}_r\ddot{\phi}_r + \mu\left(u_{rt} - \varepsilon\dot{\phi}_r^2\right)\right]
$$
$$
\left. +\mu\dot{\phi}\left(-\tilde{x}_1 \sin\left(\phi\right) + \tilde{y}_1 \cos\left(\phi\right)\right) + v_{r1}\dot{z}_2 + u_{rm}z_2 \right].
$$

(6.96)

Remark 6.3 *Motivation for the structure of (6.88) is obtained by taking the time derivative of $z_d^T(t)z_d(t)$ as follows*

$$
\frac{d}{dt}\left(z_d^T z_d\right) = 2z_d^T \dot{z}_d
$$
$$
= 2z_d^T \left(\frac{\dot{\delta}_d}{\delta_d} z_d + \left(\frac{k_1w + f}{\delta_d^2} + w\Omega_1\right) Jz_d\right)
$$

(6.97)

where (6.88) has been utilized. By exploiting the skew-symmetry of J, the expression given in (6.97) can be rewritten as follows

$$
\frac{d}{dt}\left(z_d^T z_d\right) = 2\frac{\dot{\delta}_d}{\delta_d} z_d^T z_d.
$$

(6.98)

As a result of the selection of the initial conditions given in (6.88), the following solution to the differential expression given in (6.98) can be obtained

$$
z_d^T z_d = \|z_d\|^2 = \delta_d^2.
$$

(6.99)

The relationship given by (6.99) will be used during the subsequent error system development and stability analysis.

6.3.5   Closed-Loop Error System

To facilitate the closed-loop error system development for $w(t)$, the control input $u_{d1}(t)$ is injected by adding and subtracting the product $u_{d1}(t)z_2(t)$ to the right side of the open-loop dynamic expression for $w(t)$ given in (6.78) to obtain the following expression

$$\dot{w} = \begin{bmatrix} u_{d1} & u_2 \end{bmatrix} J^T z - \eta z_2 + f \qquad (6.100)$$

where (6.92) was used. After substituting (6.86) for $\begin{bmatrix} u_{d1}(t) & u_2(t) \end{bmatrix}$ and then adding and subtracting the product $u_a^T(t) J z_d(t)$ to the resulting expression, the dynamics for $w(t)$ can be written as follows

$$\dot{w} = -u_a^T J z_d + u_a^T J \tilde{z} - \eta z_2 + f \qquad (6.101)$$

where (6.93) and the fact that $J^T = -J$ were utilized. By substituting (6.87) for only the first occurrence of $u_a(t)$ in (6.101), utilizing the equality given by (6.99), exploiting the skew-symmetry of J defined in (6.82), and the fact that $J^T J = I_2$ (I_2 denotes the standard 2×2 identity matrix), the final expression for the closed-loop error system for $w(t)$ can be obtained as follows

$$\dot{w} = -k_1 w + u_a^T J \tilde{z} - \eta z_2. \qquad (6.102)$$

To determine the closed-loop error system for $\tilde{z}(t)$, we take the time derivative of (6.93), substitute (6.88) for $\dot{z}_d(t)$, and substitute (6.79) for $\dot{z}(t)$ to obtain the following expression

$$\dot{\tilde{z}} = \frac{\dot{\delta}_d}{\delta_d} z_d + \left(\frac{k_1 w + f}{\delta_d^2} + w\Omega_1 \right) J z_d - \begin{bmatrix} u_{d1} & u_2 \end{bmatrix}^T + \begin{bmatrix} \eta & 0 \end{bmatrix}^T \qquad (6.103)$$

where the control input $u_{d1}(t)$ was injected by adding and subtracting the vector $\begin{bmatrix} u_{d1}(t) & 0 \end{bmatrix}^T$ to the right side of (6.103) and then using (6.92). After substituting (6.86) for the vector $\begin{bmatrix} u_{d1}(t) & u_2(t) \end{bmatrix}^T$ and then substituting (6.87) for $u_a(t)$ into the resulting expression, (6.103) can be rewritten as follows

$$\dot{\tilde{z}} = \frac{\dot{\delta}_d}{\delta_d} z_d + w\Omega_1 J z_d - \Omega_1 z_d + k_2 z + \begin{bmatrix} \eta & 0 \end{bmatrix}^T. \qquad (6.104)$$

After substituting (6.89) for only the second occurrence of $\Omega_1(t)$ in (6.104), using the fact that $JJ = -I_2$, and then cancelling common terms, (6.104) can be rewritten as follows

$$\dot{\tilde{z}} = -k_2 \tilde{z} + w J \left[\left(\frac{k_1 w + f}{\delta_d^2} \right) J z_d + \Omega_1 z_d \right] + \begin{bmatrix} \eta & 0 \end{bmatrix}^T \qquad (6.105)$$

where (6.93) was utilized. Finally, since the bracketed term in (6.105) is equal to $u_a(t)$ defined in (6.87), the final expression for the closed-loop error system for $\tilde{z}(t)$ can be obtained as follows

$$\dot{\tilde{z}} = -k_2\tilde{z} + wJu_a + \begin{bmatrix} \eta & 0 \end{bmatrix}^T. \qquad (6.106)$$

The following closed-loop error system for $\eta(t)$ is obtained by taking the time derivative of (6.92) and then substituting (6.80) and (6.91) into the resulting expression for $\dot{u}_1(t)$ and $u_m(t)$, respectively

$$\dot{\eta} = -k_3\eta + wz_2 - \tilde{z}_1. \qquad (6.107)$$

6.3.6   Stability Analysis

The controller given in (6.86–6.91) forces the tracking errors for the VTOL aircraft to exponentially converge to an arbitrarily small neighborhood about the origin as stated in the following theorem.

Theorem 6.4 *The VTOL aircraft controller given in (6.86–6.91) ensures that the position/orientation tracking error signals defined in (6.71) are GUUB in the sense that*

$$|\tilde{x}(t)|, |\tilde{y}(t)| \leq \beta_3 \exp(-\min\{k_1, k_2, k_3, \gamma_1, \mu\}\, t) + \varepsilon_3 \qquad (6.108)$$

$$\left|\tilde{\phi}(t)\right| \leq \beta_1 \exp\left(-\min\{k_1, k_2, k_3, \gamma_1\}\, t\right) + \varepsilon_1 \qquad (6.109)$$

where γ_0, μ were introduced in (6.90) and (6.76), respectively. The positive constants β_3, $\varepsilon_3 \in \mathbb{R}$ given in (6.108) are defined as follows

$$\beta_3 = \max\{|\tilde{x}_1(0)|, |\tilde{y}_1(0)|\} + \frac{\beta_1(\beta_1 + 2\varepsilon_1 + 2)}{|\mu - \min\{k_1, k_2, k_3, \gamma_1\}|} + \varepsilon\beta_1 \qquad (6.110)$$

$$\varepsilon_3 = \varepsilon_1 \left(\frac{(1 + \varepsilon_1)}{\mu} + \varepsilon\right)$$

where ε was introduced in (6.55) and (6.56), γ_1, ε_1 were introduced in (6.90), and the positive constant $\beta_1 \in \mathbb{R}$ is defined as follows

$$\beta_1 = \|\Psi(0)\| + \gamma_0. \qquad (6.111)$$

In (6.111), the vector $\Psi(t) \in \mathbb{R}^4$ is defined as follows

$$\Psi = \begin{bmatrix} w & \eta & \tilde{z}^T \end{bmatrix}^T. \qquad (6.112)$$

Proof: To prove Theorem 6.4, we define a nonnegative function $V(t) \in \mathbb{R}$ as follows

$$V = \frac{1}{2}w^2 + \frac{1}{2}\eta^2 + \frac{1}{2}\tilde{z}^T\tilde{z}. \tag{6.113}$$

After taking the time derivative of (6.113) and using the closed-loop error systems given in (6.102), (6.106), and (6.107), the following expression can be obtained

$$\dot{V} = w\left(-k_1 w + u_a^T J\tilde{z} - \eta z_2\right) + \eta\left(-k_3\eta + wz_2 - \tilde{z}_1\right) \tag{6.114}$$

$$+\tilde{z}^T\left(-k_2\tilde{z} + wJu_a + \begin{bmatrix} \eta & 0 \end{bmatrix}^T\right).$$

After utilizing the fact that $J^T = -J$, cancelling common terms, and utilizing (6.113), the expression given in (6.114) can be rewritten as follows

$$\dot{V} \le -2\min\{k_1, k_2, k_3\} V. \tag{6.115}$$

By invoking Lemma A.10 of Appendix A, the differential inequality given in (6.115) can be solved as follows

$$V(t) \le \exp(-2\min\{k_1, k_2, k_3\} t)V(0). \tag{6.116}$$

Based on (6.113), the expression given in (6.116) can be rewritten as

$$\|\Psi(t)\| \le \|\Psi(0)\| \exp(-\min\{k_1, k_2, k_3\} t), \tag{6.117}$$

where $\Psi(t)$ was defined in (6.112).

The expressions given in (6.112) and (6.117) can be used to prove that $w(t), \eta(t), \tilde{z}(t) \in \mathcal{L}_\infty$. After utilizing (6.93), (6.99), and the fact that $\tilde{z}(t)$, $\delta_d(t) \in \mathcal{L}_\infty$, we can conclude that $z(t), z_d(t) \in \mathcal{L}_\infty$. From the fact that $z(t), w(t) \in \mathcal{L}_\infty$, the following inverse transformation of (6.77)

$$\begin{bmatrix} r_x \\ r_y \\ \tilde{\phi} \end{bmatrix} = \frac{1}{2}\begin{bmatrix} \cos\phi & 0 & \left(\tilde{\phi}\cos\phi - 2\sin\phi\right) \\ \sin\phi & 0 & \left(\tilde{\phi}\sin\phi + 2\cos\phi\right) \\ 0 & 2 & 0 \end{bmatrix}\begin{bmatrix} w \\ z_1 \\ z_2 \end{bmatrix} \tag{6.118}$$

can be used to prove that $r_x(t), r_y(t), \tilde{\phi}(t) \in \mathcal{L}_\infty$. Based on the fact that $r_x(t), r_y(t), \tilde{\phi}(t) \in \mathcal{L}_\infty$ and that the reference trajectory is selected so that $x_{r1}(t), y_{r1}(t), \phi_r(t), \dot{x}_{r1}(t), \dot{y}_{r1}(t), \dot{\phi}_r(t), \ddot{x}_{r1}(t), \ddot{y}_{r1}(t) \in \mathcal{L}_\infty$, the expressions given in (6.71), (6.73), and (6.76) and Lemma A.13 of Appendix A can be used to prove that $\dot{\tilde{x}}_1(t), \dot{\tilde{y}}_1(t), \tilde{x}_1(t), \tilde{y}_1(t), \dot{x}_1(t), \dot{y}_1(t), x_1(t)$, $y_1(t), \phi(t) \in \mathcal{L}_\infty$. Using the fact that $r_x(t), r_y(t), \phi(t), \dot{x}(t), \dot{y}(t) \in \mathcal{L}_\infty$,

(6.81) and (6.84) can be used to prove that $f(\cdot), T(\cdot) \in \mathcal{L}_\infty$. Based on these facts, (6.85–6.90) and (6.92) can be used to show that $u_{d1}(t)$, $u_a(t)$, $\dot{z}_d(t)$, $\Omega_1(t)$, $u_1(t)$, $u_2(t)$, $\Pi(\cdot) \in \mathcal{L}_\infty$. From (6.83), we can now conclude that $u_t(t)$, $v_1(t) \in \mathcal{L}_\infty$. Based on the previous boundedness results, (6.94–6.96) can now be used to prove that $\dot{u}_{d1}(t) \in \mathcal{L}_\infty$. Based on the facts that $\eta(t)$, $\tilde{z}(t)$, $w(t)$, $z(t)$, $\dot{u}_{d1}(t)$, $u_{rm}(t) \in \mathcal{L}_\infty$, the expression given in (6.91) can be used to prove that $u_m(t) \in \mathcal{L}_\infty$.

To prove (6.109), we first show that $z(t)$ of (6.77) is GUUB by applying the triangle inequality to (6.93) to obtain the following bound

$$\|z\| \leq \|\tilde{z}\| + \|z_d\| \leq \beta_1 \exp\left(-\min\{k_1, k_2, k_3, \gamma_1\} t\right) + \varepsilon_1 \qquad (6.119)$$

where (6.90), (6.99), (6.112), and (6.117) have been utilized, β_1 is defined in (6.111), and γ_0, γ_1, ε_1 were introduced in (6.90). Based on (6.118) and the bound given in (6.119), the result given in (6.109) can now be obtained. To prove (6.108), the transformation given in (6.118) and the bounds given in (6.117) and (6.119) can be used to obtain the following inequalities

$$
\begin{aligned}
|r_x| &\leq \left|\frac{1}{2}\cos\phi\right| |w| + \left|\frac{1}{2}\cos\phi\right| \left|\tilde{\phi}\right| |z_2| + |\sin\phi| \, |z_2| \\
&\leq |w| + |z_1| \, |z_2| + |z_2| \\
&\leq \beta_2 \exp(-\min\{k_1, k_2, k_3, \gamma_1\} t) + \varepsilon_2
\end{aligned}
\qquad (6.120)
$$

where the positive constants β_2, $\varepsilon_2 \in \mathbb{R}$ are defined as follows

$$\beta_2 = \left(2\beta_1 + 2\beta_1\varepsilon_1 + \beta_1^2\right) \qquad \varepsilon_2 = \varepsilon_1 + \varepsilon_1^2. \qquad (6.121)$$

The same upper bound given in (6.120) can be developed for $|r_y(t)|$ from (6.118). Based on the inequalities given in (6.120), an upper bound can now be developed for $\tilde{x}_1(t)$. Specifically, the expression given in (6.76) can be integrated as follows (see Lemma A.19 of Appendix A)

$$\tilde{x}_1(t) = \tilde{x}_1(0)\exp(-\mu t) + \int_0^t \exp(-\mu(t-\sigma))r_x(\sigma)\, d\sigma. \qquad (6.122)$$

By using (6.120), the integral expression given in (6.122) can be upper bounded as follows

$$
\begin{aligned}
|\tilde{x}_1(t)| \;\leq\; & |\tilde{x}_1(0)| \exp(-\mu t) + \exp(-\mu t) \int_0^t \exp(\mu \sigma)\varepsilon_2 d\sigma \\
& + \exp(-\mu t) \int_0^t \left(\beta_2 \exp((\mu - \min\{k_1,k_2,k_3,\gamma_1\})\sigma)\right) d\sigma
\end{aligned}
$$

$$
\begin{aligned}
\leq\; & |\tilde{x}_1(0)| \exp(-\mu t) + \frac{\varepsilon_2}{\mu} - \frac{\varepsilon_2}{\mu}\exp(-\mu t) \\
& + \frac{\beta_2 \exp(-\min\{k_1,k_2,k_3,\gamma_1\} t) - \beta_2 \exp(-\mu t)}{|\mu - \min\{k_1,k_2,k_3,\gamma_1\}|}
\end{aligned}
$$

$$
\leq\; |\tilde{x}_1(0)| \exp(-\mu t) + \frac{\varepsilon_2}{\mu} + \frac{\beta_2 \exp(-\min\{k_1,k_2,k_3,\gamma_1\} t)}{|\mu - \min\{k_1,k_2,k_3,\gamma_1\}|}.
$$
(6.123)

Likewise, the following upper bound can be developed for $|\tilde{y}_1(t)|$

$$
|\tilde{y}_1(t)| \leq |\tilde{y}_1(0)| \exp(-\mu t) + \frac{\varepsilon_2}{\mu} + \frac{\beta_2 \exp(-\min\{k_1,k_2,k_3,\gamma_1\} t)}{|\mu - \min\{k_1,k_2,k_3,\gamma_1\}|}.
$$
(6.124)

The result given in (6.108) can now be obtained by utilizing (6.75), (6.123), and (6.124). $\square$

Remark 6.4 *From the definition of ε_3 given in (6.110), it is clear that the tracking error variables, $\tilde{x}(t)$, $\tilde{y}(t)$, and $\tilde{\phi}(t)$ can be made arbitrarily small by reducing the design parameters $\varepsilon, \varepsilon_1$, and/or by increasing μ. Moreover, the rate of convergence of the errors to this arbitrarily small neighborhood around zero can be controlled through the design parameters k_1, k_2, k_3, γ_1, and μ as is evident from (6.108) and (6.109).*

Remark 6.5 *The occurrence of the term $(\mu - \min\{k_1,k_2,k_3,\gamma_1\})$ in the denominator of (6.123) and (6.124) seems to indicate a potential singularity in the control law. However, in the event that $\mu = \min\{k_1,k_2,k_3,\gamma_1\}$, the solution of the linear differential equations results in repeated roots. Specifically, if $\mu = \min\{k_1,k_2,k_3,\gamma_1\}$, then (6.123) and (6.124) are now given by the following expressions*

$$
|\tilde{x}_1(t)| \leq |\tilde{x}_1(0)| \exp(-\mu t) + \left(2\beta_1 + 2\beta_1\varepsilon_1 + \beta_1^2\right) t \exp(-\mu t) + \frac{\varepsilon_2}{\mu} \quad (6.125)
$$

$$
|\tilde{y}_1(t)| \leq |\tilde{y}_1(0)| \exp(-\mu t) + \left(2\beta_1 + 2\beta_1\varepsilon_1 + \beta_1^2\right) t \exp(-\mu t) + \frac{\varepsilon_2}{\mu}. \quad (6.126)
$$

Remark 6.6 *Since no restrictions were placed on the desired trajectory, it is straightforward that the tracking control development can also be applied to solve the regulation problem.*

6.3.7   Simulation Results

The controller given in (6.86–6.91) was simulated for a VTOL aircraft that is modeled by the expressions given in (6.55–6.58) where the coupling parameter ε was selected to be 0.1. The reference trajectory was generated according to (6.63–6.66) where $u_{rt}(t)$ and $u_{rm}(t)$ were selected as follows

$$
\begin{aligned}
u_{rt} &= \left(1 - \exp\left(-0.1t^2\right)\right)\cos\left(0.2t\right) \\
u_{rm} &= \left(1 - \exp\left(-0.1t^2\right)\right)\sin\left(0.2t\right).
\end{aligned}
$$

The actual and reference position and orientation and the respective first time derivatives were initialized to zero, and the auxiliary signal $z_d(t)$ was initialized as

$$
z_d(0) = \left[\begin{array}{cc} 0.0 & 1.04 \end{array}\right]^T.
$$

The control gains that resulted in the best performance were selected as follows

$$
\begin{aligned}
&k_1 = 10.0 \quad k_2 = 300.0 \quad k_3 = 10.0 \quad \gamma_0 = 1.0 \\
&\gamma_1 = 0.01 \quad \mu = 0.1 \quad\quad \varepsilon_1 = 0.04.
\end{aligned}
$$

The tracking control inputs are illustrated in Figures 6.7 and 6.8 and the resulting position/orientation tracking errors are shown in Figures 6.9–6.11.

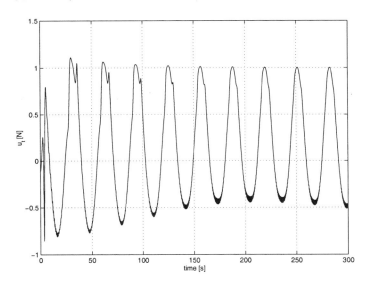

FIGURE 6.7. Control input $u_t(t)$.

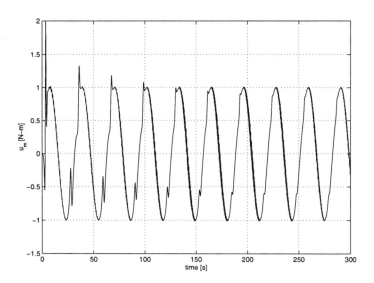

FIGURE 6.8. Control input $u_m(t)$.

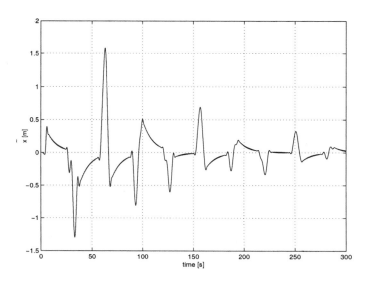

FIGURE 6.9. Position tracking error $\tilde{x}(t)$.

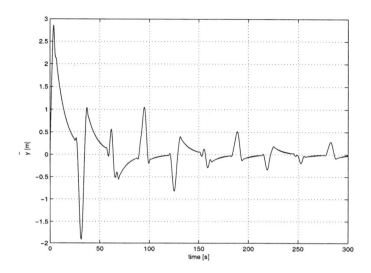

FIGURE 6.10. Position tracking error $\tilde{y}(t)$.

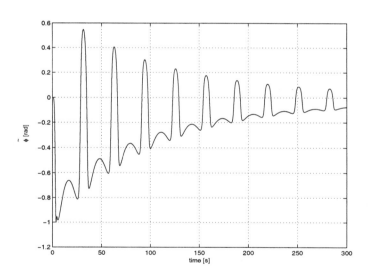

FIGURE 6.11. Orientation tracking error.

6.3.8 Automotive Steering Extension

Model Development

The kinematic equations of motion of the COM for a front wheel driven and steered vehicle can be written as follows (see Figure 6.12)

$$\dot{q} = S(q)v \qquad (6.127)$$

where $\dot{q}(t) \in \mathbb{R}^3$ represents the time derivative of the Cartesian position and orientation of the vehicle COM defined by $q(t) \in \mathbb{R}^3$ as follows

$$q(t) = \begin{bmatrix} x(t) & y(t) & \phi(t) \end{bmatrix}^T, \qquad (6.128)$$

the transformation matrix $S(q) \in \mathbb{R}^{3 \times 3}$ is defined as follows

$$S(q) = \begin{bmatrix} \cos(\phi) & -\sin(\phi) & 0 \\ \sin(\phi) & \cos(\phi) & 0 \\ 0 & 0 & 1 \end{bmatrix}, \qquad (6.129)$$

and the velocity vector $v(t) \in \mathbb{R}^3$ is defined as

$$v = \begin{bmatrix} v_1 & v_2 & v_3 \end{bmatrix}^T. \qquad (6.130)$$

For the vector given in (6.128), $x(t)$, $y(t)$, $\phi(t) \in \mathbb{R}$ denote the measurable position/orientation of the COM of the vehicle with respect to a fixed inertial reference frame. For the vector given in (6.130), $v_1(t)$, $v_2(t)$, and $v_3(t) \in \mathbb{R}$ denote the vehicle's measurable longitudinal, lateral, and yaw velocities, respectively. Under the assumptions that (i) the body-fixed coordinate axis coincides with the center of gravity (CG), (ii) the mass distribution is homogeneous, (iii) the heave, pitch, and roll modes can be neglected, and (iv) the half tread of the vehicle is small (i.e., a bicycle model), the vehicle dynamic model can be expressed in the following form [26]

$$\begin{aligned} m\dot{v}_1 &= 2(F_{xf} + F_{xr}) + mv_2v_3 - F_D(v_1) \qquad (6.131) \\ m\dot{v}_2 &= 2(F_{yf} + F_{yr}) - mv_2v_3 \\ I\dot{v}_3 &= 2(l_f F_{yf} - l_r F_{yr}). \end{aligned}$$

For the vehicle dynamic model given in (6.131), $l_f, l_r \in \mathbb{R}$ denote the constant distances from the COM to front and rear axle, respectively, $F_D(t) \in \mathbb{R}$ represents the known aerodynamic drag of the vehicle, $F_{xf}(t)$, $F_{yf}(t) \in \mathbb{R}$ denote the known front wheel forces acting perpendicular and parallel to the rear wheel axis, respectively, and $F_{xr}(t)$, $F_{yr}(t) \in \mathbb{R}$ denote the known rear wheel forces acting perpendicular and along the rear wheel axis, respectively (see Figure 6.12). In the subsequent development, $F_{xf}(t)$ and $F_{yf}(t)$ are assumed to be the control inputs (see [49] for a discussion regarding the relationship between $F_{xf}(t)$, $F_{yf}(t)$ and the actual angular wheel velocity and the steering angle control inputs).

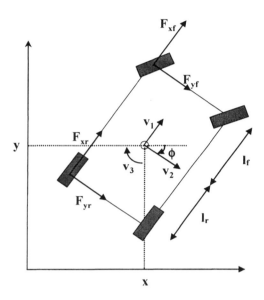

FIGURE 6.12. Inertial and body-fixed coordinate frames.

Given the structure of the dynamic model in (6.131) and the assumption of exact model knowledge, feedback linearization techniques can be applied to simplify the subsequent analysis. Specifically, by designing the input signals $F_{xf}(t)$ and $F_{yf}(t)$ in the following manner

$$F_{xf} = \frac{1}{2}F_D(v_1) - F_{xr} + \frac{m}{2}u_t \qquad (6.132)$$

$$F_{yf} = \frac{m}{2l_f}u_m - F_{yr},$$

the dynamics given in (6.131) can be partially feedback linearized as follows

$$\dot{v}_1 = u_t + v_2v_3 \qquad (6.133)$$

$$\dot{v}_2 = \frac{u_m}{l_f} - v_1v_3$$

$$\dot{v}_3 = \left(\frac{m}{I}\right)\left(u_m - 2\frac{(l_f + l_r)}{m}F_{yr}\right)$$

where $u_t(t)$ and $u_m(t) \in \mathbb{R}$ are subsequently designed control inputs. By taking the time derivative of (6.127) and utilizing (6.129), (6.130), and (6.133), the vehicle dynamic model can be expressed in terms of the inertial reference frame as follows

$$\ddot{x} = u_t \cos(\phi) - \varepsilon \sin(\phi) u_m \tag{6.134}$$

$$\ddot{y} = u_t \sin(\phi) + \varepsilon \cos(\phi) u_m \tag{6.135}$$

$$\dot{\phi} = v_3 \tag{6.136}$$

$$\dot{v}_3 = \frac{m}{I}\left(u_m - \frac{1}{\varepsilon}g\right) \tag{6.137}$$

where $g(t), \varepsilon \in \mathbb{R}$ are defined as follows

$$g = 2\frac{(l_f + l_r)}{m}\varepsilon F_{yr} \qquad \varepsilon = \frac{1}{l_f}. \tag{6.138}$$

To facilitate the development of the open-loop error system in a similar form as the VTOL problem, the "shifted" Cartesian position for the automobile is defined as follows

$$x_1 = x - \left(\frac{I}{m}\right)\varepsilon \cos(\phi) \tag{6.139}$$

$$y_1 = y - \left(\frac{I}{m}\right)\varepsilon \sin(\phi). \tag{6.140}$$

After taking the second time derivatives of (6.139) and (6.140), substituting (6.134–6.137) into the resulting equations, and then cancelling common terms, the shifted Cartesian system dynamics can be written as follows

$$\ddot{x}_1 = \left(u_t + \frac{I}{m}\varepsilon\dot{\phi}^2\right)\cos(\phi) - g\sin(\phi) \tag{6.141}$$

$$\ddot{y}_1 = \left(u_t + \frac{I}{m}\varepsilon\dot{\phi}^2\right)\sin(\phi) + g\cos(\phi). \tag{6.142}$$

Reference Model Development

Based on the structure of the system dynamics given by (6.134–6.137), the reference model for the automotive steering problem is defined in the following manner

$$\ddot{x}_r = u_{rt}\cos(\phi_r) - \varepsilon\sin(\phi_r)u_{rm} \tag{6.143}$$

$$\ddot{y}_r = u_{rt}\sin(\phi_r) + \varepsilon\cos(\phi_r)u_{rm} \tag{6.144}$$

$$\dot{\phi}_r = v_{3r} \tag{6.145}$$

$$\dot{v}_{3r} = \frac{m u_{rm}}{I} \qquad (6.146)$$

where $x_r(t)$, $y_r(t)$, $\phi_r(t) \in \mathbb{R}$ denote the reference Cartesian position and orientation, respectively, $v_{3r}(t) \in \mathbb{R}$ denotes the reference yaw velocity, and $u_{rt}(t)$, $u_{rm}(t) \in \mathbb{R}$ denote the reference input signals that are assumed to be selected such that $u_{rt}(t)$, $u_{rm}(t)$, $\dot{u}_{rt}(t)$, $\dot{u}_{rm}(t) \in \mathcal{L}_{\infty}$ and $x_r(t)$, $y_r(t)$, $\phi_r(t)$, $\dot{x}_r(t)$, $\dot{y}_r(t)$, $\dot{\phi}_r(t)$, $\ddot{x}_r(t)$, $\ddot{y}_r(t)$, $\ddot{\phi}_r(t) \in \mathcal{L}_{\infty}$. A transformation similar to (6.139) and (6.140) is applied to the reference system as follows

$$x_{r1} = x_r - \frac{I}{m}\varepsilon\cos(\phi_r) \qquad (6.147)$$

$$y_{r1} = y_r - \frac{I}{m}\varepsilon\sin(\phi_r) \qquad (6.148)$$

where $x_{r1}(t)$, $y_{r1}(t) \in \mathbb{R}$ represented the shifted reference Cartesian position. After taking the second time derivatives of (6.147) and (6.148), utilizing (6.143) and (6.144), and then cancelling common terms, the shifted reference Cartesian dynamics can be rewritten as follows

$$\ddot{x}_{r1} = \left(u_{rt} + \frac{I}{m}\varepsilon\dot{\phi}_r^2\right)\cos(\phi_r) \qquad (6.149)$$

$$\ddot{y}_{r1} = \left(u_{rt} + \frac{I}{m}\varepsilon\dot{\phi}_r^2\right)\sin(\phi_r). \qquad (6.150)$$

Remark 6.7 *The inputs to the reference generator, denoted by $u_{rt}(t)$ and $u_{rm}(t)$, are analogous to the throttle and steering torque inputs, respectively, for a vehicle with a conventional steering system. These reference inputs can be selected appropriately to generate the position and orientation reference signals.*

Open-Loop Error System

To write the open-loop error system in a similar form as the VTOL problem, the transformation given in (6.77) is modified as follows

$$\begin{bmatrix} w \\ z_1 \\ z_2 \end{bmatrix} = \begin{bmatrix} \tilde{\phi}\cos\phi - 2\sin\phi & \tilde{\phi}\sin\phi + 2\cos\phi & 0 \\ 0 & 0 & -1 \\ \cos\phi & \sin\phi & 0 \end{bmatrix} \begin{bmatrix} r_x \\ r_y \\ \tilde{\phi} \end{bmatrix} \qquad (6.151)$$

where $r_x(t)$, $r_y(t)$, and $\tilde{\phi}(t)$ were defined in (6.71) and (6.76). After taking the time derivative of (6.151) and using similar algebraic manipulation as described for the development of the VTOL open-loop error system, the open-loop error dynamics for the automotive steering problem can be expressed as follows

$$\dot{w} = u^T J^T z + f + 2g \qquad (6.152)$$

$$\dot{z} = u \tag{6.153}$$

$$\dot{u}_1 = \left(\frac{m}{I}\right)\left(-u_m + u_{rm} + \frac{1}{\varepsilon}g\right). \tag{6.154}$$

For the open-loop error system of (6.152), J denotes the skew-symmetric matrix defined in (6.82), $g(t)$ is defined in (6.138), and $f(\phi, z, \tilde{x}_1, \tilde{y}_1) \in \mathbb{R}$ is defined as follows

$$f = 2\left[\ddot{x}_{r1}\sin(\phi) - \ddot{y}_{r1}\cos(\phi) + v_{3r}z_2\right] - 2\mu\left(\dot{\tilde{x}}_1\sin(\phi) - \dot{\tilde{y}}_1\cos(\phi)\right) \tag{6.155}$$

where μ was introduced in (6.76) and $\tilde{x}_1(t)$, $\tilde{y}_1(t)$ were defined in (6.73). The control signal $u(t) = \begin{bmatrix} u_1(t) & u_2(t) \end{bmatrix}^T \in \mathbb{R}^2$ given in (6.152) and (6.153) is related to the control signals $u_t(t)$ and $v_3(t)$ through the following globally invertible transformation

$$u = T^{-1}\begin{bmatrix} u_t \\ v_3 \end{bmatrix} - \Pi \qquad \begin{bmatrix} u_t \\ v_3 \end{bmatrix} = T(u + \Pi) \tag{6.156}$$

where $T(r_x, r_y, \phi) \in \mathbb{R}^{2\times2}$ and $\Pi(\phi, v_3, \dot{\tilde{x}}_1, \dot{\tilde{y}}_1) \in \mathbb{R}^2$ are defined as follows

$$T = \begin{bmatrix} -r_x\sin(\phi) + r_y\cos(\phi) & 1 \\ -1 & 0 \end{bmatrix} \tag{6.157}$$

$$\Pi = \begin{bmatrix} -v_{3r} \\ -\varepsilon\dfrac{I}{m}v_3^2 + \ddot{x}_{r1}\cos(\phi) + \ddot{y}_{r1}\sin(\phi) \\ -\mu\left(\dot{\tilde{x}}_1\cos(\phi) + \dot{\tilde{y}}_1\sin(\phi)\right) \end{bmatrix}. \tag{6.158}$$

Based on the similarities in the structure of the open-loop tracking error dynamics given in (6.78–6.85) with the dynamics given in (6.152–6.158), the control structure given in (6.86–6.91) can be easily modified to achieve GUUB tracking and regulation for the automotive steering problem (see [49]).

6.3.9   Surface Vessel Extension

Model Formulation

The kinematic model for an underactuated surface vessel can be written as follows

$$\dot{q} = S(q)v \tag{6.159}$$

where $q(t)$ is defined in (6.128) and $x(t)$, $y(t)$, $\phi(t) \in \mathbb{R}$ now denote the measurable position/orientation of the COM of the surface vessel with respect to a fixed inertial reference frame, $S(q)$ is defined in (6.129), and $v(t)$ is defined in (6.130) where $v_1(t)$, $v_2(t)$, and $v_3(t) \in \mathbb{R}$ now denote the measurable surge, sway, and yaw velocities of the surface vessel, respectively (see Figure 6.13). Under the assumptions that (i) the body-fixed coordinate axis coincides with the center of gravity (CG), (ii) the mass distribution is homogeneous, (iii) the hydrodynamic damping terms of order higher than one are negligible, (iv) changes in the inertia are negligible, and (v) the heave, pitch, and roll modes can be neglected, the dynamic model for a neutrally buoyant surface vessel with two axes of symmetry can be expressed in the following form [23]

$$M\dot{v} + D(v)v = \tau_0 \tag{6.160}$$

where $\dot{v}(t)$ denotes the time derivative of $v(t)$ given in (6.159).

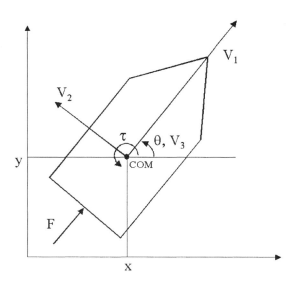

FIGURE 6.13. Actuator diagram for an underactuated surface vessel.

In the dynamic model given in (6.160), $M \in \mathbb{R}^{3\times3}$ represents the following constant diagonal positive-definite inertia matrix

$$M = \begin{bmatrix} m & 0 & 0 \\ 0 & m & 0 \\ 0 & 0 & I_o \end{bmatrix} \tag{6.161}$$

where $m, I_o \in \mathbb{R}$ represent the mass and inertia of the surface vessel, respectively, $D(v) \in \mathbb{R}^{3\times3}$ represents the Centripetal-Coriolis and hydrodynamic damping effects as follows

$$D(v) = \begin{bmatrix} -X_{v1} & 0 & -mv_2 \\ 0 & -Y_{v2} & mv_1 - Y_{v3} \\ 0 & -N_{v2} & -N_{v3} \end{bmatrix}. \tag{6.162}$$

In (6.162), X_{v1}, Y_{v2}, Y_{v3}, N_{v2}, and $N_{v3} \in \mathbb{R}$ denote constant damping coefficients, and $\tau_0(t) \in \mathbb{R}^3$ denotes the following force-torque control input vector

$$\tau_0(t) = \begin{bmatrix} F & 0 & \tau \end{bmatrix}^T \tag{6.163}$$

where $F(t) \in \mathbb{R}$ denotes a control force that is applied to produce a forward thrust, and $\tau(t) \in \mathbb{R}$ denotes a torque that is applied about the center of gravity. As in the automotive steering example, the open-loop dynamics for the surface vessel can be partially feedback linearized as follows

$$\begin{bmatrix} \dot{v}_1 \\ \dot{v}_2 \\ \dot{v}_3 \end{bmatrix} = \begin{bmatrix} F_1 + v_2 v_3 \\ \dfrac{1}{m}(Y_{v2}v_2 + Y_{v3}v_3) - v_1 v_3 \\ \tau_1 \end{bmatrix} \tag{6.164}$$

by designing $F(t)$ and $\tau(t)$ as follows

$$F = -X_{v1}v_1 + mF_1 \tag{6.165}$$

$$\tau = -N_{v2}v_2 - N_{v3}v_3 + I_o\tau_1 \tag{6.166}$$

where (6.159–6.163) were utilized and $F_1(t)$, $\tau_1(t) \in \mathbb{R}$ denote subsequently designed control inputs. By taking the time derivative of (6.159) and using (6.129), (6.130), and (6.164), the surface vessel dynamic model can be expressed in terms of the inertial reference frame as follows

$$\begin{bmatrix} \ddot{x} \\ \ddot{y} \\ \ddot{\phi} \end{bmatrix} = \begin{bmatrix} F_1 \cos\phi - \dfrac{Y_{v2}}{m}(\dot{y}\cos\phi - \dot{x}\sin\phi)\sin\phi - \dfrac{Y_{v3}}{m}v_3\sin\phi \\ F_1 \sin\phi + \dfrac{Y_{v2}}{m}(\dot{y}\cos\phi - \dot{x}\sin\phi)\cos\phi + \dfrac{Y_{v3}}{m}v_3\cos\phi \\ v_3 \end{bmatrix}. \tag{6.167}$$

Based on the structure of the surface vessel model given by (6.167), a reference model is defined as follows

$$
\begin{bmatrix} \ddot{x}_r \\ \ddot{y}_r \\ \dot{\phi}_r \end{bmatrix} = \begin{bmatrix} F_{1r}\cos\phi_r - \dfrac{\sin\phi_r}{m}\left(Y_{v2}\left(\dot{y}_r\cos\phi_r - \dot{x}_r\sin\phi_r\right) + Y_{v3}v_{3r}\right) \\ F_{1r}\sin\phi_r + \dfrac{\cos\phi_r}{m}\left(Y_{v2}\left(\dot{y}_r\cos\phi_r - \dot{x}_r\sin\phi_r\right) + Y_{v3}v_{3r}\right) \\ v_{3r} \end{bmatrix}
$$

$$(6.168)$$

where $x_r(t)$, $y_r(t)$, $\phi_r(t) \in \mathbb{R}$ denote the reference Cartesian position and orientation, respectively, and $v_{3r}(t)$, $F_{1r}(t) \in \mathbb{R}$ denote the reference yaw velocity and thrust force, respectively, that are assumed to be selected such that $F_{1r}(t) \in \mathcal{L}_\infty$, $v_{3r}(t)$, $\dot{v}_{3r}(t) \in \mathcal{L}_\infty$, and $x_r(t)$, $y_r(t)$, $\phi_r(t)$, $\dot{x}_r(t)$, $\dot{y}_r(t)$, $\dot{\phi}_r(t)$, $\ddot{x}_r(t)$, $\ddot{y}_r(t)$, $\ddot{\phi}_r(t) \in \mathcal{L}_\infty$.

Open-Loop Error System

To write the open-loop error system in a similar form as the VTOL aircraft and automobile steering problems, the transformation given in (6.77) is modified as follows

$$
\begin{bmatrix} w \\ z_1 \\ z_2 \end{bmatrix} = \begin{bmatrix} -\tilde{\phi}\cos\phi + 2\sin\phi & -\tilde{\phi}\sin\phi - 2\cos\phi & 2\dfrac{Y_{v3}}{m} \\ 0 & 0 & 1 \\ \cos\phi & \sin\phi & 0 \end{bmatrix} \begin{bmatrix} r_x \\ r_y \\ \tilde{\phi} \end{bmatrix}
$$

$$(6.169)$$

where $r_x(t)$, $r_y(t)$, and $\tilde{\phi}(t)$ were defined in (6.71) and (6.76). After taking the time derivative of (6.169) and using (6.129), (6.130), (6.159), (6.164), (6.167), and (6.168), the resulting expression for the open-loop tracking error dynamics can be determined as follows

$$\dot{w} = u^T J^T z + f \qquad (6.170)$$
$$\dot{z} = u$$
$$\dot{u}_1 = -\dot{v}_{3r} + \tau_1.$$

For the open-loop error system given in (6.170), J denotes the skew-symmetric matrix defined in (6.82), $f(\phi, v_2, z, \dot{\tilde{x}}, \dot{\tilde{y}}) \in \mathbb{R}$ is defined as follows

$$
\begin{aligned}
f ={}& 2\left(v_{3r}z_2 - F_{1r}\sin z_1 + \mu\left(\sin\phi\,\dot{\tilde{x}} - \cos\phi\,\dot{\tilde{y}}\right)\right) \qquad (6.171) \\
&+ \frac{2}{m}Y_{v2}\left((\dot{y}_r\cos\phi_r - \dot{x}_r\sin\phi_r)\cos z_1 - v_2\right) + \frac{2Y_{v3}v_{3r}}{m}\left(\cos z_1 - 1\right)
\end{aligned}
$$

where μ was introduced in (6.76) and $u(t) = \begin{bmatrix} u_1(t) & u_2(t) \end{bmatrix}^T \in \mathbb{R}^2$ is related to the control signals $F_1(t)$ and $v_3(t)$. Specifically, $u(t)$ is related to

$F_1(t)$ and $v_3(t)$ through the following globally invertible transformation

$$u = T^{-1} \begin{bmatrix} F_1 \\ v_3 \end{bmatrix} - \Pi \qquad \begin{bmatrix} F_1 \\ v_3 \end{bmatrix} = T(u + \Pi) \qquad (6.172)$$

where $T(r_x, r_y, \phi) \in \mathbb{R}^{2 \times 2}$ and $\Pi(z, \phi, v_1) \in \mathbb{R}^2$ are defined as follows

$$T = \begin{bmatrix} r_x \sin\phi - r_y \cos\phi & 1 \\ 1 & 0 \end{bmatrix} \qquad (6.173)$$

$$\Pi = \begin{bmatrix} v_{3r} \\ F_{1r} \cos z_1 + \dfrac{1}{m} Y_{v2} (\dot{y}_r \cos\phi_r - \dot{x}_r \sin\phi_r) \sin z_1 \\ -\mu(v_1 - \dot{x}_r \cos\phi - \dot{y}_r \sin\phi) + \dfrac{Y_{v3}}{m} v_{3r} \sin z_1 \end{bmatrix}. \qquad (6.174)$$

Based on the similarities in the structure of the open-loop tracking error dynamics given in (6.170–6.174) with the open-loop dynamics given for the VTOL and the automotive steering problems, the control structure given in (6.86–6.91) can be easily modified to achieve GUUB tracking and regulation for the underactuated surface vessel (see [6]).

6.4 Satellite Systems

In this section, a model-based controller is designed for the attitude tracking and regulation problems for a rigid underactuated satellite. To address the underactuated nature of the satellite systems examined in this section, a dynamic oscillator-based design is utilized that is inspired by the development for the VTOL aircraft controller in the previous section. In contrast to the dynamic oscillator designed for the planar VTOL problem, the control design in this section is based on the unit quaternion parameterization. Provided the initial errors satisfy a sufficient condition, a kinematic control design is proven to exponentially force the tracking and regulation errors to an arbitrarily small neighborhood about the origin (i.e., UUB tracking and regulation). By employing integrator backstepping control techniques, an axisymmetric satellite extension is provided that also yields UUB tracking and regulation with an exponential rate of convergence. Simulation results are provided to illustrate the performance of the axisymmetric satellite tracking and regulation controllers under various initial conditions.

6.4.1   System Model

Let $\mathcal{F}$ and $\mathcal{I}$ be orthogonal coordinate frames attached to the center of mass of a satellite and to an inertial reference frame, respectively. The kinematic model for a rigid satellite expressed in the unit quaternion parameterization is given as follows [28, 30]

$$\dot{q} = B(q)\omega \qquad (6.175)$$

where $q(t) =[\ q_o(t)\quad q_v^T(t)\ ]^T \in \mathbb{R}^4$ with $q_0(t) \in \mathbb{R}$ and where $q_v(t) \in \mathbb{R}^3$ represents the measurable unit quaternion that describes the orientation of $\mathcal{F}$ with respect to $\mathcal{I}$, and $\omega(t) =[\ \omega_1(t)\quad \omega_2(t)\quad \omega_3(t)\ ]^T \in \mathbb{R}^3$ denotes the relative angular velocity between $\mathcal{F}$ and $\mathcal{I}$. For the underactuated satellite, $\omega_1(t)$ is assumed to be a measurable exogenous bounded signal and $\omega_2(t)$ and $\omega_3(t)$ are assumed to be the control inputs. In the kinematic model given in (6.175), the Jacobian-type matrix $B(q) \in \mathbb{R}^{4\times3}$ is defined as follows

$$B(q) = \frac{1}{2}\left[\begin{array}{c} -q_v^T \\ q_o I_3 + q_v^\times \end{array}\right] \qquad (6.176)$$

where the notation $\zeta^\times$, $\forall\zeta =[\ \zeta_1\quad \zeta_2\quad \zeta_3\ ]^T$ denotes the following skew-symmetric matrix

$$\zeta^\times = \left[\begin{array}{ccc} 0 & -\zeta_3 & \zeta_2 \\ \zeta_3 & 0 & -\zeta_1 \\ -\zeta_2 & \zeta_1 & 0 \end{array}\right]. \qquad (6.177)$$

The unit quaternion and the Jacobian-type matrix $B(q)$ defined in (6.176) are subject to the following constraints

$$q^T q = 1 \qquad B^T B = I_3 \qquad (6.178)$$

where I_3 denotes the 3×3 identity matrix. Based on (6.178), the kinematic model given in (6.175) can be rewritten as follows

$$\omega = B^T(q)\dot{q}. \qquad (6.179)$$

6.4.2   Control Objective

The satellite attitude tracking control objective is to force the orientation of $\mathcal{F}$ to a desired orientation, denoted by $\mathcal{F}_d$, where the orientation of $\mathcal{F}_d$ with respect to $\mathcal{I}$ is specified by a desired unit quaternion $q_d(t) =[\ q_{od}(t)\quad q_{vd}^T(t)\ ]^T$ with $q_{od}(t) \in \mathbb{R}$ and $q_{vd}(t) \in \mathbb{R}^3$. The time derivative of $q_d(t)$ is related to the desired angular velocity of $\mathcal{F}_d$, denoted by $\bar{\omega}_d(t) = [\ \bar{\omega}_{d1}(t)\quad \bar{\omega}_{d2}(t)\quad \bar{\omega}_{d3}(t)\ ]^T \in \mathbb{R}^3$, through the following expression

$$\dot{q}_d = B(q_d)\bar{\omega}_d. \qquad (6.180)$$

To quantify the mismatch between the actual and desired satellite attitudes, a rotation matrix $\tilde{R}(e) \in SO(3)$ that brings $\mathcal{F}_d$ onto $\mathcal{F}$ is defined as follows

$$\tilde{R} = RR_d^T = \left(e_0^2 - e_v^T e_v\right) I_3 + 2e_v e_v^T - 2e_0 e_v^\times. \tag{6.181}$$

For the rotation matrix given in (6.181), $R(q) \in SO(3)$ denotes the rotation matrix that brings $\mathcal{I}$ onto $\mathcal{F}$ as follows

$$R = \left(q_0^2 - q_v^T q_v\right) I_3 + 2q_v q_v^T - 2q_0 q_v^\times, \tag{6.182}$$

$R_d(q_d) \in SO(3)$ denotes the rotation matrix that brings $\mathcal{I}$ onto $\mathcal{F}_d$ as follows

$$R_d = \left(q_{0d}^2 - q_d^T q_d\right) I_3 + 2q_d q_d^T - 2q_{0d} q_d^\times, \tag{6.183}$$

and the quaternion tracking error $e(t) = [\begin{array}{cc} e_0(t) & e_v^T(t) \end{array}]^T$ with $e_0(t) \in \mathbb{R}$ and $e_v(t) = [\begin{array}{ccc} e_{v1}(t) & e_{v2}(t) & e_{v3}(t) \end{array}]^T \in \mathbb{R}^3$ is defined as

$$e = \begin{bmatrix} e_0 \\ e_v \end{bmatrix} = \begin{bmatrix} q_0 q_{0d} + q_v^T q_{vd} \\ q_{0d} q_v - q_0 q_{vd} + q_v^\times q_{vd} \end{bmatrix}. \tag{6.184}$$

Based on (6.178) and (6.184), it is clear that the unit quaternion tracking error satisfies the following constraint

$$e^T e = e_0^2 + e_v^T e_v = 1 \tag{6.185}$$

where

$$0 \leq |e_0(t)| \leq 1 \qquad 0 \leq \|e_v(t)\| \leq 1. \tag{6.186}$$

Based on the previous definitions, the satellite attitude tracking control objective can be stated as follows

$$\lim_{t \to \infty} \tilde{R}(e) = I_3. \tag{6.187}$$

The control objective can also be written in terms of the unit quaternion tracking error. Specifically, from (6.185), it can be proven that

$$\text{if} \quad \lim_{t \to \infty} \|e_v(t)\| = 0, \quad \text{then} \quad \lim_{t \to \infty} |e_0(t)| = 1; \tag{6.188}$$

hence, (6.181) and (6.188) can be used to prove that

$$\text{if} \quad \lim_{t \to \infty} \|e_v(t)\| = 0 \quad \text{then} \quad \lim_{t \to \infty} \tilde{R}(e) = I_3. \tag{6.189}$$

Based on (6.188) and (6.189), the subsequent control design is motivated by the desire to force $e_v(t)$ to zero (or to some arbitrarily small neighborhood). The control objective is predicated on the assumption that $q(t)$ and $\dot{q}(t)$ are measurable signals.

Remark 6.8 *The tracking control objective is formulated under the assumption that the desired trajectory input, denoted by $q_d(t)$, is provided by a trajectory planning module that ensures that $q_{0d}(t)$, $q_{vd}(t)$, and the respective first two time derivatives are bounded for all time (note that $q_{0d}(t)$ can be calculated via (6.180)). However, the trajectory planning module can also generate $\bar{\omega}_d(t)$ from the following expression*

$$\bar{\omega}_d = B^T(q_d)\dot{q}_d \tag{6.190}$$

where (6.178) was utilized.

6.4.3   Open-Loop Error System

The open-loop tracking error dynamics can be developed by taking the time derivative of (6.184) and utilizing (6.179), (6.180), and (6.190) as follows

$$\dot{e}_0 = -\frac{1}{2}e_v^T\tilde{\omega} \tag{6.191}$$

$$\dot{e}_v = \frac{1}{2}\left(e_v^\times + e_0 I_3\right)\tilde{\omega} \tag{6.192}$$

where the angular velocity tracking error $\tilde{\omega}(t) \in \mathbb{R}^3$ is defined as follows

$$\tilde{\omega} = \omega - \omega_d \tag{6.193}$$

where the auxiliary term $\omega_d(t) = \begin{bmatrix} \omega_{d1}(t) & \omega_{d2}(t) & \omega_{d3}(t) \end{bmatrix}^T \in \mathbb{R}^3$ is defined as follows

$$\omega_d = \tilde{R}\bar{\omega}_d. \tag{6.194}$$

To facilitate the subsequent control design and stability analysis, the open-loop error system given in (6.192) can be rewritten as follows

$$\dot{x} = \frac{1}{2}\left(z^T J \begin{bmatrix} \omega_2 & \omega_3 \end{bmatrix}^T + f\right) \tag{6.195}$$

$$\dot{z} = \frac{1}{2}\left(Jz\tilde{\omega}_1 + \Omega_e\left(\begin{bmatrix} \omega_2 & \omega_3 \end{bmatrix}^T - \begin{bmatrix} \omega_{d2} & \omega_{d3} \end{bmatrix}^T\right)\right). \tag{6.196}$$

In the error system given in (6.195) and (6.196), the error variables $x(t) \in \mathbb{R}$, $z(t) \in \mathbb{R}^2$ are related to $e_v(t)$ of (6.184) as follows

$$x = e_{v1} \qquad z = \begin{bmatrix} e_{v2} & e_{v3} \end{bmatrix}^T, \tag{6.197}$$

J denotes the skew-symmetric matrix given in (6.82), $f(e_0, e_2, e_3, \tilde{\omega}_1) \in \mathbb{R}$ is defined as follows

$$f = e_0\tilde{\omega}_1 - e_2\omega_{d3} + e_3\omega_{d2} \tag{6.198}$$

where $\omega_{d2}(t)$, $\omega_{d3}(t)$ were defined in (6.194), and $\Omega_e(e_0, e_1) \in \mathbb{R}^{2 \times 2}$ is defined as

$$\Omega_e = \begin{bmatrix} e_0 & -e_1 \\ e_1 & e_0 \end{bmatrix}. \tag{6.199}$$

6.4.4 Kinematic Control

Based on the structure of the open-loop error system given in (6.195) and (6.196) and the subsequent stability analysis, the kinematic control input signals $\omega_2(t)$ and $\omega_3(t)$ given in (6.175) are designed as follows

$$\begin{bmatrix} \omega_2(t) & \omega_3(t) \end{bmatrix}^T = -k_p e_0 z + \begin{bmatrix} \omega_{d2} & \omega_{d3} \end{bmatrix}^T + u_a. \tag{6.200}$$

For the kinematic controller given in (6.200), $k_p \in \mathbb{R}$ is a constant control gain, and $u_a(t) \in \mathbb{R}^2$ is defined as follows

$$u_a = \Pi_0 J z_d + \Pi_1 z_d. \tag{6.201}$$

For the control signal given in (6.201), $z_d(t) \in \mathbb{R}^2$ is defined by the following dynamic oscillator-like relationship

$$\dot{z}_d = \frac{\dot{\delta}_d}{\delta_d} z_d + \Lambda J z_d \qquad z_d^T(0) z_d(0) = \delta_d^2(0) \tag{6.202}$$

and $\Pi_0(t)$, $\Pi_1(t)$, $\Lambda(t)$, $\delta_d(t) \in \mathbb{R}$ are defined as follows

$$\Pi_0 = \frac{k_a x + e_0 \tilde{\omega}_1}{\delta_d^2} \tag{6.203}$$

$$\Pi_1 = \frac{2\dot{\delta}_d}{e_0 \delta_d} + k_p e_0 \tag{6.204}$$

$$\Lambda = \frac{1}{2}(k_p e_0 x + e_0 \Pi_0 + \tilde{\omega}_1) \tag{6.205}$$

$$\delta_d = \gamma_0 \exp(-\gamma_1 t) + \varepsilon_1 \tag{6.206}$$

where k_a, γ_0, γ_1, $\varepsilon_1 \in \mathbb{R}$ are positive constant design parameters.

Remark 6.9 *Note that based on the structure of the dynamic oscillator given in (6.202) with $\delta_d(t)$ given in (6.206), the development given in (6.97) and (6.98) can be utilized to obtain the result given (6.99) (i.e., $\|z_d(t)\| = \delta_d(t)$).*

6.4.5   Closed-Loop Error System

To facilitate the closed-loop error system development for $x(t)$, we substitute (6.200) into (6.195) for $\omega_2(t)$ and $\omega_3(t)$, cancel common terms, then add and subtract the product $z_d^T(t)Ju_a(t)$ to the resulting expression as follows

$$\dot{x} = \frac{1}{2}\left(e_0\tilde{\omega}_1 + z_d^T Ju_a - \tilde{z}^T Ju_a\right) \tag{6.207}$$

where $\tilde{z}(t) \in \mathbb{R}^2$ is defined as

$$\tilde{z} = z_d - z. \tag{6.208}$$

After substituting (6.201) for only the first occurrence of $u_a(t)$ in (6.207) and then utilizing (6.203), the equality given by (6.99), the skew-symmetry of J defined in (6.82), and the fact that $JJ = -I_2$ (note that I_2 denotes the standard 2×2 identity matrix), the closed-loop error system for $x(t)$ can be developed as follows

$$\dot{x} = -\frac{1}{2}k_a x - \frac{1}{2}\tilde{z}^T Ju_a. \tag{6.209}$$

To determine the closed-loop error system for $\tilde{z}(t)$, we take the time derivative of (6.208), substitute (6.202) for $\dot{z}_d(t)$, and then substitute (6.196) for $\dot{z}(t)$ to obtain the following expression

$$\dot{\tilde{z}} = \frac{\dot{\delta}_d}{\delta_d}z_d + \Lambda J z_d - \frac{1}{2}Jz\tilde{\omega}_1 - \frac{1}{2}\Omega_e\begin{bmatrix} \omega_2 - \omega_{d2} \\ \omega_3 - \omega_{d3} \end{bmatrix}. \tag{6.210}$$

After substituting (6.200) into (6.210) for $\omega_2(t)$ and $\omega_3(t)$, and then cancelling common terms, the closed-loop error system for $\tilde{z}(t)$ can be determined as follows

$$\dot{\tilde{z}} = \frac{\dot{\delta}_d}{\delta_d}z_d + \Lambda J z_d + \frac{1}{2}k_p e_0^2 z - \frac{1}{2}Jz\left(\tilde{\omega}_1 + k_p e_0 x\right) - \frac{1}{2}\Omega_e u_a \tag{6.211}$$

where (6.199) was utilized. After substituting (6.201) into (6.211) for $u_a(t)$, utilizing (6.199) and (6.203–6.205), cancelling common terms and exploiting the skew-symmetry of J, the following expression for the closed-loop error dynamics for $\tilde{z}(t)$ can be determined

$$\dot{\tilde{z}} = \frac{1}{2}\left(-k_p e_0^2 \tilde{z} + J\tilde{z}\left(\tilde{\omega}_1 + k_p e_0 x\right) + x Ju_a\right). \tag{6.212}$$

6.4.6   Stability Analysis

Based on the controller given in (6.200–6.206), the tracking errors for the underactuated satellite exponentially converge to an arbitrarily small neighborhood about the origin as stated in the following theorem.

Theorem 6.5 *Given the closed-loop systems in (6.209) and (6.212), the kinematic controller of (6.200–6.206) ensures UUB tracking and regulation in the sense that*

$$||e_v(t)|| \leq \lambda_1 \exp(-\lambda_2 t) + \lambda_3 \varepsilon_1 \qquad (6.213)$$

provided that the initial errors satisfy the following sufficient condition

$$||e_v(0)|| \leq \frac{\delta - 8(\gamma_0 + \varepsilon_1)}{9} \qquad (6.214)$$

where λ_1, λ_2, $\lambda_3 \in \mathbb{R}$ denote positive constants, and $\delta \in \mathbb{R}$ is some positive constant selected as follows

$$8(\gamma_0 + \varepsilon_1) < \delta < 1 \qquad (6.215)$$

where the control gains γ_0, ε_1 are given in (6.206).

Proof: To prove Theorem 6.5, we define a nonnegative function $V(t) \in \mathbb{R}$ as follows

$$V = x^2 + \tilde{z}^T \tilde{z}. \qquad (6.216)$$

After taking the time derivative of (6.216) and utilizing the closed-loop error systems given in (6.209) and (6.212), the following expression can be obtained

$$\dot{V} = x\left(-k_a x - \tilde{z}^T J u_a\right) + \tilde{z}^T\left(-k_p e_0^2 \tilde{z} + J\tilde{z}(\tilde{\omega}_1 + k_p e_0 x) + x J u_a\right). \qquad (6.217)$$

After cancelling common terms and exploiting the skew-symmetry of J, (6.217) can be simplified as follows

$$\begin{aligned}
\dot{V} &= -k_a x^2 - k_p e_0^2 \tilde{z}^T \tilde{z} \qquad (6.218)\\
&\leq -\min(k_a, k_p e_0^2) V
\end{aligned}$$

where (6.216) was utilized. The right-hand side of (6.218) can be further upper bounded as follows

$$\dot{V} \leq -\beta V \quad \text{if} \quad |e_0(t)| \geq \sqrt{1 - \delta^2} > 0 \qquad (6.219)$$

where $\beta \in \mathbb{R}$ is a positive bounding constant, and δ is a positive constant selected according to (6.215). The solution to the differential inequality given in (6.219) can be obtained as follows

$$V(t) \leq \exp(-\beta t)V(0) \quad \text{if} \quad |e_0(t)| \geq \sqrt{1 - \delta^2} > 0. \qquad (6.220)$$

By utilizing (6.216), the inequality given in (6.220) can be rewritten as

$$||\zeta(t)|| \leq ||\zeta(0)|| \exp(-\frac{\beta}{2}t) \quad \text{if} \quad |e_0(t)| \geq \sqrt{1 - \delta^2} > 0 \qquad (6.221)$$

where $\zeta(t) \in \mathbb{R}^3$ is defined as

$$\zeta = \begin{bmatrix} x & \tilde{z}^T \end{bmatrix}^T. \tag{6.222}$$

After applying the triangle inequality to (6.208) and to the norm of $e_v(t)$, the following inequalities can be formulated

$$\|z\| \leq \|\tilde{z}\| + \|z_d\| \leq \|\tilde{z}\| + \gamma_0 + \varepsilon_1 \tag{6.223}$$

$$\|e_v\| \leq |x| + 2 (\|\tilde{z}\| + \gamma_0 + \varepsilon_1) \tag{6.224}$$

where (6.197) and (6.206) were utilized. By utilizing (6.185), the sufficient condition given in (6.219–6.221) can be rewritten as follows

$$\|e_v\| = \sqrt{1 - e_0^2} \leq \delta. \tag{6.225}$$

After using (6.224), the sufficient condition given in (6.225) can be expressed in terms of the Lyapunov variables given in (6.216) as follows

$$|x(t)| + 2 \|\tilde{z}(t)\| \leq \delta - 2 (\gamma_0 + \varepsilon_1). \tag{6.226}$$

After utilizing (6.197), the sufficient condition given in (6.226) can be rewritten as follows

$$\|\zeta(t)\| \leq \frac{\delta - 2 (\gamma_0 + \varepsilon_1)}{3}. \tag{6.227}$$

Based on (6.221) and (6.227), we can conclude that

$$\|\zeta(t)\| \leq \|\zeta(0)\| \quad \text{if} \quad \|\zeta(t)\| \leq \frac{\delta - 2 (\gamma_0 + \varepsilon_1)}{3}. \tag{6.228}$$

The inequalities described by (6.228) imply that $\|\zeta(t)\|$ can be upper bounded by $\|\zeta(0)\|$ provided $\|\zeta(t)\|$ is upper bounded by the second inequality of (6.228). Since $\|\zeta(0)\|$ is a constant that can be independently restricted (i.e., it is a measure of the initial error between the desired and system orientations), the following upper bound for $\|\zeta(0)\|$ can be developed so that both inequalities given in (6.228) are satisfied

$$\|\zeta(t)\| \leq \|\zeta(0)\| \quad \text{if} \quad \|\zeta(0)\| \leq \frac{\delta - 2 (\gamma_0 + \varepsilon_1)}{3}. \tag{6.229}$$

By using (6.221), (6.222), and (6.229), the following inequalities can now be obtained

$$\|x\|, \; \|\tilde{z}\| \leq \|\zeta(0)\| \exp \left(-\frac{\beta}{2} t \right) \quad \text{if} \quad \|\zeta(0)\| \leq \frac{\delta - 2 (\gamma_0 + \varepsilon_1)}{3}. \tag{6.230}$$

From (6.206), (6.223), and (6.230), the following bound for $z(t)$ can be developed

$$\|z\| \leq \|\zeta(0)\| \exp\left(-\frac{\beta}{2}t\right) + \gamma_0 \exp(-\gamma_1 t) + \varepsilon_1$$
$$\text{if}\quad \|\zeta(0)\| \leq \frac{\delta - 2(\gamma_0 + \varepsilon_1)}{3}. \tag{6.231}$$

After utilizing (6.206), (6.208), (6.222), and (6.229), the following expression can be obtained

$$\|\zeta(0)\| \leq 3\|e_v(0)\| + 2(\gamma_0 + \varepsilon_1) \tag{6.232}$$

The inequality given in (6.232) can now be used to rewrite the sufficient condition given in (6.229–6.231) as in (6.214). Therefore, provided the condition given in (6.214) in satisfied, it is straightforward to prove that $x(t)$, $\tilde{z}(t)$, $z(t)$, $z_d(t) \in \mathcal{L}_\infty$ from (6.99), (6.230), and (6.231). Since $x(t)$, $z(t) \in \mathcal{L}_\infty$ and the reference trajectory is selected such that $q_{0d}(t)$, $q_{vd}(t) \in \mathcal{L}_\infty$, (6.184) and (6.197) can be used to prove that $q_v(t)$, $q_0(t) \in \mathcal{L}_\infty$. Based on the facts that $q_{0d}(t)$, $q_{vd}(t)$, $q_0(t)$, $q_v(t) \in \mathcal{L}_\infty$, (6.181) and (6.184) can be used to prove that $e(t)$, $\tilde{R}(e) \in \mathcal{L}_\infty$. Based on the assumption that $\omega_d(t)$, $\omega_1(t) \in \mathcal{L}_\infty$ and the fact that $\tilde{R}(e) \in \mathcal{L}_\infty$, (6.193) and (6.194) can be used to prove that $\tilde{\omega}_1(t) \in \mathcal{L}_\infty$. The preceding boundedness assertions and (6.198) can now be used to prove that $f(t) \in \mathcal{L}_\infty$. Based on (6.185) and (6.225), it can also be proven that

$$|e_0(t)| = \sqrt{1 - \|e_v(t)\|^2} \geq \sqrt{1 - \delta^2} > 0 \tag{6.233}$$

provided the conditions given in (6.214) and (6.215) are satisfied. Given (6.233), we can now utilize (6.200–6.206) to prove that $\omega_2(t)$, $\omega_3(t)$, $u_a(t)$, $\dot{z}_d(t)$, $\Pi_0(\cdot)$, $\Pi_1(\cdot) \in \mathcal{L}_\infty$. Based on the fact that all of the signals are bounded under closed-loop operation, (6.230) and (6.231) can be used to prove the result given in (6.213), provided the conditions given in (6.214) and (6.215) are satisfied. $\square$

Remark 6.10 *The initial condition magnitude restriction given in (6.214) serves as a restriction on the desired unit quaternion, denoted by $q_d(t)$, as indicated by the definition of $e_v(t)$ given in (6.184). That is, given the initial state of the system $q(0)$, $q_d(0)$ must be selected such that $\|e_v(0)\|$ satisfies the sufficient condition given in (6.214).*

6.4.7 Axisymmetric Satellite Extension

The controller given in (6.200–6.206) was designed based on the assumption that the satellite velocities can be directly used as the control input (i.e., the controller is based on the satellite kinematics). In practice, the satellite has dynamic effects that will result in a mismatch between the desired velocity control signals and the actual satellite velocity. To address the velocity mismatch, the satellite dynamics can be incorporated through integrator backstepping techniques. Specifically, in this extension $q(t)$ and $\dot{q}(t)$ are assumed to be measurable and that the underactuated rigid satellite is modeled by (6.175) and by the following dynamic model [28]

$$M\dot{\omega} = -\omega^{\times} M\omega + \tau \tag{6.234}$$

where the notation $\zeta^{\times}$ is given in (6.177). For the dynamic model given in (6.234), $M \in \mathbb{R}^{3\times3}$ denotes a diagonal matrix of the principal moments of inertia which is explicitly defined as follows

$$M = \text{diag}\{m_1, m_2, m_3\} \tag{6.235}$$

where diag$\{\cdot\}$ is used to denote the formation of diagonal matrix, m_1, m_2, $m_3 \in \mathbb{R}$ are positive constants, and the torque control input vector denoted by $\tau(t) \in \mathbb{R}^3$ is explicitly defined as

$$\tau = [0 \quad \tau_2 \quad \tau_3]^T \tag{6.236}$$

where $\tau_2(t)$, $\tau_3(t) \in \mathbb{R}$ are control torques applied in the direction of the last two principal inertia axes (i.e., the first principal inertia axis is underactuated). To facilitate the control design, the dynamics given in (6.234) are rewritten in the following advantageous form

$$\dot{\omega}_1 = c_1\omega_2\omega_3 \tag{6.237}$$
$$\dot{\omega}_2 = c_2\omega_1\omega_3 + \bar{\tau}_2 \tag{6.238}$$
$$\dot{\omega}_3 = c_3\omega_1\omega_2 + \bar{\tau}_3 \tag{6.239}$$

where the control signals, denoted by $\bar{\tau}_2(t)$, $\bar{\tau}_3(t) \in \mathbb{R}$, are defined as follows

$$\bar{\tau}_2 = \frac{\tau_2}{m_2} \qquad \bar{\tau}_3 = \frac{\tau_3}{m_3}$$

while c_1, c_2, $c_3 \in \mathbb{R}$ are defined as

$$c_1 = \frac{m_2 - m_3}{m_1} \qquad c_2 = \frac{m_3 - m_1}{m_2} \qquad c_3 = \frac{m_1 - m_2}{m_3} \, .$$

The axisymmetric satellite dynamics can now be obtained by setting $c_1 = 0$ in (6.237), and hence, from (6.237) it is clear that

$$\dot{\omega}_1 = 0. \tag{6.240}$$

To compensate for the mismatch between the actual velocity and the desired velocity, integrator backstepping techniques can be used to design the control torque input $\bar{\tau}(t) = [\bar{\tau}_2(t), \quad \bar{\tau}_3(t)]^T \in \mathbb{R}^2$ as follows

$$\bar{\tau} = k_e \left(\begin{bmatrix} \omega_{c2} \\ \omega_{c3} \end{bmatrix} - \begin{bmatrix} \omega_2 \\ \omega_3 \end{bmatrix} \right) - \begin{bmatrix} c_2\omega_1\omega_3 \\ c_3\omega_1\omega_2 \end{bmatrix} + \begin{bmatrix} \dot{\omega}_{c2} \\ \dot{\omega}_{c3} \end{bmatrix} + \Omega_e^T \tilde{z} - xJ^T z$$

(6.241)

where $k_e \in \mathbb{R}$ is a positive control gain, $\omega_{c2}(t), \omega_{c3}(t) \in \mathbb{R}$ represent the kinematic control inputs (i.e., the desired velocity) defined as follows

$$\begin{bmatrix} \omega_{c2}(t) & \omega_{c3}(t) \end{bmatrix}^T = -k_p e_0 z + \begin{bmatrix} \omega_{d2} & \omega_{d3} \end{bmatrix}^T + u_a$$

(6.242)

where $e_0(t)$, $\omega_{d2}(t)$, $\omega_{d3}(t)$, $z(t)$, k_p, and $u_a(t)$ were defined in (6.184), (6.194), (6.197), (6.200), and (6.201), respectively, and $\dot{\omega}_{c2}(t)$, $\dot{\omega}_{c3}(t)$ can be obtained by differentiating (6.242). In contrast to the result given in [53], the axisymmetric extension given in this section does not require the restrictions that $\omega_1 = 0$ for the stabilization problem (i.e., ω_1 can be any bounded constant value) or that $\tilde{R}\bar{\omega}_{d1}(t) = \omega_1(0)$ for the tracking problem (i.e., we only need to ensure that $\tilde{R}\bar{\omega}_{d1}(t)$ and its first time derivative are bounded for all time, and that $\omega_1(0)$ is bounded).

6.4.8   Simulation Results

In this section, numerical simulations for the underactuated axisymmetric satellite given in (6.175) and (6.237–6.239) are provided to illustrate the performance of the controller given in (6.197–6.206), (6.241), and (6.242). Specifically, tracking and regulation simulations are provided for different initial condition configurations. For each simulation, the inertia parameters for the axisymmetric satellite given in (6.235) were selected as follows

$$m_1 = 20 \text{ [kgm}^2] \quad m_2 = 15 \text{ [kgm}^2] \quad m_3 = 15 \text{ [kgm}^2].$$

(6.243)

Setpoint Controller

To illustrate the regulation performance of the controller given in (6.197–6.206), (6.241), and (6.242), the initial attitude of the satellite was set to the following values

$$q_0(0) = \sqrt{0.1} \qquad q_v(0) = \begin{bmatrix} 0 & -\sqrt{0.45} & -\sqrt{0.45} \end{bmatrix}^T$$

(6.244)

such that (6.178) is satisfied, $\omega_1(t)$ was selected as 0 [rads^{-1}], and the oscillator signal $z_d(t)$ was initialized as follows

$$z_d(0) = \begin{bmatrix} 2.001 & 0 \end{bmatrix}^T.$$

(6.245)

After selecting the control gains in the following manner

$$k_p = 1.0 \qquad k_a = 1.0 \qquad \gamma_0 = 2.0 \qquad \gamma_1 = 0.2$$
$$\varepsilon_1 = 0.001 \quad k_1 = 10.0 \quad k_2 = 10.0,$$

(6.246)

the resulting satellite attitude regulation error and control torque inputs are depicted in Figure 6.14 and Figure 6.15, respectively. A second simulation was performed for the regulation problem using the same initial conditions and control gains, with the exception that $\omega_1(t) = 0.005$ [rads^{-1}]. The resulting regulation error and control torque input are given in Figure 6.16 and Figure 6.17, respectively.

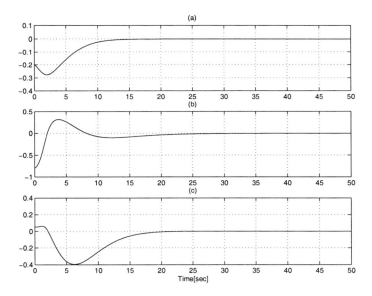

FIGURE 6.14. Orientation error: (a) $e_{v1}(t)$, (b) $e_{v2}(t)$, and (c) $e_{v3}(t)$.

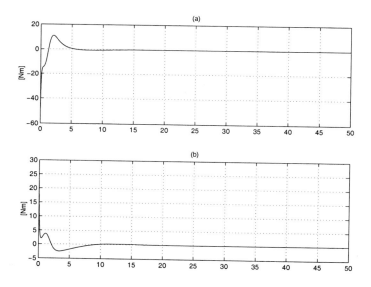

FIGURE 6.15. Applied torque: (a) $\tau_2(t)$ and (b) $\tau_3(t)$.

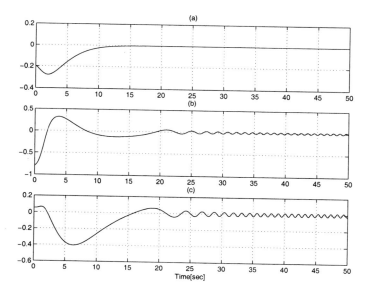

FIGURE 6.16. Orientation error: (a) $e_{v1}(t)$, (b) $e_{v2}(t)$, and c) $e_{v3}(t)$.

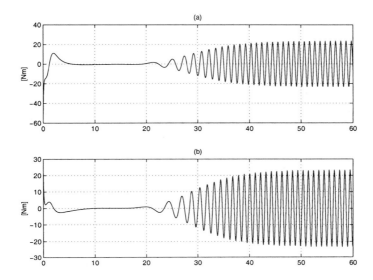

FIGURE 6.17. Applied torque: (a) $\tau_2(t)$ and (b) $\tau_3(t)$.

Tracking Controller

For the satellite attitude tracking simulation, the desired trajectory was generated via (6.190) where $\bar{\omega}_d(t)$ was used as the reference input. Specifically, the reference trajectory input $\bar{\omega}_d(t)$ was selected in the following manner

$$\bar{\omega}_d(t) = \begin{bmatrix} 0 & \sin(t) & \cos(t) \end{bmatrix}^T \tag{6.247}$$

where the initial desired attitude of the satellite was selected as follows

$$q_{0d}(0) = 1 \qquad q_v(0) = \begin{bmatrix} 0 & 0 & 0 \end{bmatrix}^T \tag{6.248}$$

such that (6.178) is satisfied. The initial satellite attitude was set to the following values

$$q_0(0) = \sqrt{0.1} \qquad q_v(0) = \begin{bmatrix} 0 & \sqrt{0.45} & \sqrt{0.45} \end{bmatrix}^T, \tag{6.249}$$

$\omega_1(t)$ was selected as 0 $[\text{rads}^{-1}]$, and $z_d(t)$ was initialized as follows

$$z_d(0) = \begin{bmatrix} 2.07 & 0 \end{bmatrix}^T \tag{6.250}$$

such that (6.202) is satisfied. After selecting the control gains as follows

$$\begin{array}{llll} k_p = 10.0 & k_a = 0.5 & \gamma_0 = 2.0 & \gamma_1 = 0.5 \\ \varepsilon_1 = 0.01 & k_1 = 20.0 & k_2 = 20.0, & \end{array} \tag{6.251}$$

the resulting satellite attitude tracking error and control torque inputs are depicted in Figure 6.18 and Figure 6.19, respectively. As shown in Figure 6.18, the satellite attitude tracking error is confined to a neighborhood around zero. From (6.213), it is clear that the tracking error neighborhood can be reduced by setting ε_1 to a smaller value.

To demonstrate that any bounded desired trajectory can be tracked within an arbitrarily small neighborhood, another simulation was conducted where the initial desired and actual angular velocities were selected as $\bar{\omega}_{d1}(t) \neq \omega_1(0)$. Specifically, the desired satellite trajectory was selected as follows

$$q_{0d} = \sqrt{0.1}$$

$$q_{vd}(t) = \begin{bmatrix} \sqrt{0.9}\sin(0.1t) & \sqrt{0.45}\cos(0.1t) & \sqrt{0.45}\cos(0.1t) \end{bmatrix}^T$$
$$(6.252)$$

such that (6.178) is satisfied. The initial attitude of the satellite was set to the following values

$$q_0(0) = \sqrt{0.1} \qquad q_v(0) = \begin{bmatrix} 0 & \sqrt{0.45} & \sqrt{0.45} \end{bmatrix}^T . \qquad (6.253)$$

The initial value for $\omega_1(t)$ was arbitrarily chosen to be 0.01 [rad^{-1}], and $z_d(t)$ was initialized as follows

$$z_d(0) = \begin{bmatrix} 2.07 & 0 \end{bmatrix}^T . \qquad (6.254)$$

After selecting the control gains as follows

$$\begin{aligned} k_p &= 10.0 & k_a &= 0.5 & \gamma_0 &= 2.0 & \gamma_1 &= 0.5 \\ \varepsilon_1 &= 0.07 & k_1 &= 20.0 & k_2 &= 20.0, \end{aligned} \qquad (6.255)$$

the resulting satellite attitude tracking error and control torque inputs are depicted in Figure 6.20 and Figure 6.21. In comparison to Figure 6.19, larger torque values are shown in Figure 6.21 because $\bar{\omega}_{d1}(t) \neq \omega_1(0)$.

6.5   Background and Further Reading

Motivated by both practical applications and the fact that overhead cranes represent classical underactuated control problems, several researchers have developed various controllers for overhead crane systems. For example, Yu, Lewis and Huang [58] used a time-scale separation approach to control an overhead crane system. However, an approximate linearized model of the crane was utilized to facilitate the construction of the error systems. In

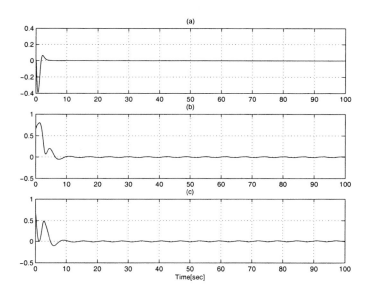

FIGURE 6.18. Orientation tracking error: (a) $e_{v1}(t)$, (b) $e_{v2}(t)$, and (c) $e_{v3}(t)$.

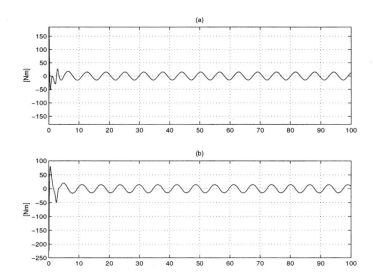

FIGURE 6.19. Applied torque: (a) $\tau_2(t)$ and (b) $\tau_3(t)$.

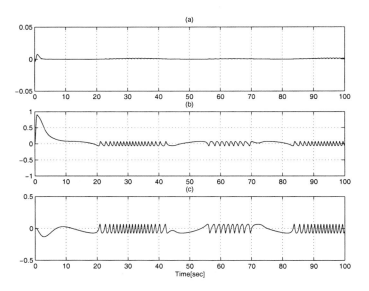

FIGURE 6.20. Orientation tracking error: (a) $e_{v1}(t)$, (b) $e_{v2}(t)$, and (c) $e_{v3}(t)$.

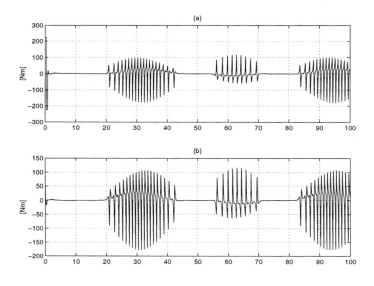

FIGURE 6.21. Applied torque: (a) $\tau_2(t)$ and (b) $\tau_3(t)$.

[56], Yoshida and Kawabe proposed a saturating control law based on a guaranteed cost control method for a linearized version of the crane system dynamics. Martindale et al. [37] used an approximate crane model to develop exact model knowledge and adaptive controllers, while Butler, Honderd, and Van Amerongen [11] exploited a modal decomposition technique to develop an adaptive controller. In [13], Chung and Hauser designed a nonlinear controller for regulating the swinging energy of the payload.

Several researchers have also examined the control problem for overhead crane systems with additional degrees of freedom. Specifically, Moustafa and Ebeid [42] derived the nonlinear dynamic model for an overhead crane and then utilized a standard linear feedback controller based on a linearized state space model. In [43], Noakes and Jansen developed a generalized input shaping approach for the linearized crane dynamics that exploited a notch filtering technique to control the motion of the bridge/trolley of an overhead crane system. More recently, Lee [33] developed a nonlinear model for overhead cranes based on a new 2-DOF swing angle definition. Based on this nonlinear model, Lee then developed an anti-swing control law for the decoupled linearized dynamics. In [47], Sakawa and Sano derived a nonlinear model for a crane system, which was subsequently linearized to facilitate the development of a control scheme that first transferred the load to a position near the equilibrium point using an open-loop controller and then used a linear feedback controller to stabilize the payload about the equilibrium point.

One of the limiting factors associated with the above overhead crane control designs is that the system nonlinearities are often excluded from the closed-loop error system design and stability analysis. To overcome this drawback, several researchers have investigated control approaches that account for the nonlinear dynamics of overhead cranes and similar systems. For example in [52], Teel used saturation functions to develop an output feedback controller that achieves a robust semi-global stability result for the ball-and-beam control problem. In [10], Burg et al. transformed the nonlinear crane dynamics into a structure that resembled the ball-and-beam problem and then adopted the research efforts of [52] to achieve asymptotic positioning from a large set of initial conditions. More recently, Fantoni, Lozano, and Spong [21] and Lozano, Fantoni, and Block [35] proposed passivity-based controllers for the inverted pendulum and the pendubot (i.e., an inverted pendulum-like robot with an unactuated second link) based on the paradigm of driving the underactuated system to a homoclinic orbit using an energy-based nonlinear controller and then switching to a linear controller to stabilize the system around its unstable equilibrium point. Using similar stability analysis techniques, Collado, Lozano, and Fantoni [14] proposed a PD controller for the overhead crane problem. In [31], Kiss, Levine, and Mullhaupt developed a PD controller for a vertical crane-winch system that only requires the measurement of the winch angle and its derivative rather than a cable angle measurement. Passivity-based interconnection and damping assignment control techniques are used in [24, 46] to stabilize underactuated mechanical systems. Specifically, the desired inertia matrix is parameterized such that energy shaping can be exploited to stabilize underactuated mechanical systems such as the ball-and-beam problem and the inverted pendulum. In [57], Yoshida developed a nonlinear energy-based controller to damp out the pendulum oscillations despite amplitude constraints on the trolly position. Recently, Fang et al. [20] developed several energy-based controllers for overhead crane systems in which additional nonlinear terms were injected into the controller to increase the coupling between the gantry position and the payload position to provide for improved transient response.

Motivated by the tactical advantages that VTOL aircraft provide, several researchers have investigated control methods that can address the nonlinear nonminimum-phase underactuated dynamics. Specifically, Hauser, Sastry, and Meyer [25] proposed an approximate input-output linearization approach. However, the controller was initially designed by assuming that the coupling between the thrust and the rolling moments can be neglected. Unfortunately, this assumption leads to magnitude restrictions on the coupling constant. Furthermore, the approach also requires that the roll of the VTOL aircraft be restricted to the interval $(-90°, +90°)$. In [36], Martin, Devasia, and Paden proposed a tracking controller based on differential flatness. Specifically, the authors exploited the fact that the output at a point fixed with respect to the aircraft body (Huygens center of oscillation) can be used to transform the system into a flat input-state system. This result did not impose any restrictions on the coupling coefficients or on the roll of the VTOL aircraft. However, this controller does not produce a solution for tracking desired trajectories. In [22], Fliess et al. developed a differential geometric approach for investigating system equivalence that is proven to reduce the dimension of complex systems. The approach was applied to several underactuated examples including an inverted pendulum and VTOL aircraft. In [34], Lin, Zhang, and Brandt designed a set-point controller using optimal control techniques. However, the control law required similar restrictions as those given in [25] (in addition, the control design methodology assumed that a part of the control input can be bounded as a state disturbance). In [38], McClamroch and Kolmanovsky proposed hybrid switching strategies for two different setpoint control problems. In [3], dynamic inversion and robust control techniques were used to deal with the nonminimum-phase dynamics. However, this type of approach imposed restrictions on the desired trajectory. In [4], output tracking and maneuver regulating controllers were proposed to illustrate the advantages of a maneuver regulating controller over an output tracking approach. In [5], nontrivial extensions of [3] and [4] were presented for the conventional take-off and landing problem. In [44, 45], Oishi and Tomlin proposed a fusion of approximate linearization and switching techniques to develop tracking controllers in a "safe" envelope. However, some of the aforementioned restrictions were not completely avoided. Sira-Ramirez recently proposed an approximate solution to the general reference trajectory tracking problem in [50].

Several solutions to the attitude control problem have been presented in the literature since the early 1970s (see [39]). In [54], the authors presented a general attitude control design framework that includes PD, model-based, and adaptive setpoint controllers. Adaptive tracking control schemes based on three-parameter kinematic representations were presented in [48, 51] to compensate for the unknown satellite inertia matrix. In [1], an adaptive attitude tracking controller based on the unit quaternion was proposed that identified the inertia matrix via periodic command signals. The work of [1] was later applied to the angular velocity tracking problem in [2]. An $\mathcal{H}_\infty$-suboptimal state feedback controller was developed for the quaternion representation in [18]. In [32], the authors designed an inverse optimal control law for attitude regulation using the backstepping method for a three-parameter representation. Recently, the authors of [8] presented a variable structure tracking controller using quaternions in the presence of inertia uncertainties and external disturbances. In [16], Costic et al. presented an adaptive control solution to the quaternion-based attitude tracking control problem that eliminated angular velocity measurements and compensated for parametric uncertainty.

Recently, there has been some interest in designing controllers for the underactuated rigid satellite tracking/regulation problem (see the Notes Section of Chapter 5 for a review of research targeting the fully actuated satellite problem). In [17], Crouch provided necessary and sufficient conditions for controllability of a rigid body in the case of one, two, or three independent actuators. In [12], Byrnes and Isidori demonstrated that a rigid satellite with only two controls cannot be asymptotically stabilized via continuous-state feedback since it does not satisfy Brockett's necessary condition [9] for smooth feedback stabilizability. In [40], Morin et al.

developed a smooth time-varying stabilizing controller by using averaging theory. A continuous time-varying/time-periodic switching controller was proposed by Coron and Kerai in [15]. Using averaging theory and Lyapunov control design techniques, Morin and Samson [41] developed a continuous time-varying controller that locally exponentially stabilized the attitude of a rigid spacecraft. In [53], Tsiotras and Luo proposed a saturated, tracking/stabilizing controller for the kinematic control of an underactuated axisymmetric spacecraft. However, the spin rate on the unactuated axis is required to be zero. Recently in [7], Behal et al. developed a tracking/stabilizing quaternion-based controller for the kinematic control of an underactuated axisymmetric satellite that eliminated the requirement for a zero spin rate of the unactuated axis as required in [53].

References

[1] J. Ahmed, V. Coppola, and D. Bernstein, "Adaptive Asymptotic Tracking of Spacecraft Attitude Motion with Inertia Matrix Identification," *Journal of Guidance, Control, and Dynamics*, Vol. 21, No. 5, Sept.-Oct. 1998, pp. 684–691.

[2] J. Ahmed and D. Bernstein, "Globally Convergent Adaptive Control of Spacecraft Angular Velocity Without Inertia Modeling," *Proceedings of the American Control Conference*, San Diego, CA, June 1999, pp. 1540–1544.

[3] S. Al-Hiddabi and N. McClamroch, "Output Tracking for Nonlinear Nonminimum-Phase VTOL Aircraft," *Proceedings of the IEEE Conference on Decision and Control*, Tampa, FL, Dec. 1998, pp. 4573–4578.

[4] S. Al-Hiddabi, J. Shen, and N. McClamroch, "A Study of Flight Maneuvers for the PVTOL Aircraft Model," *Proceedings of the American Control Conference*, San Diego, CA, June 1999, pp. 2727–2731.

[5] S. Al-Hiddabi and N. McClamroch, "Trajectory Tracking Control and Maneuver Regulation Control for the CTOL Aircraft Model," *Proceedings of the IEEE Conference on Decision and Control*, Phoenix, AZ, Dec. 1999, pp. 1958–1963.

[6] A. Behal, W. Dixon, D. Dawson, and Y. Fang, "Tracking and Regulation Control of an Underactuated Surface Vessel with Nonintegrable

Dynamics," *Proceedings of the IEEE Conference on Decision and Control*, Sydney, Australia, Dec. 2000, pp. 2150–2155.

[7] A. Behal, D. M. Dawson, E. Zergeroglu, and Y. Fang, "Nonlinear Tracking Control of an Underactuated Spacecraft," *AIAA Journal of Guidance, Control and Dynamics*, Vol.. 25, No. 5, Sept.–Oct. 2002, pp. 979–985.

[8] J. Bosković, S. Li, and R. Mehra, "Globally Stable Adaptive Tracking Control Design for Spacecraft under Input Saturation," *Proceedings of the IEEE Conference on Decision and Control*, Phoenix, AZ, Dec. 1999, pp. 1952–1957.

[9] R. Brockett, "Asymptotic Stability and Feedback Stabilization," in *Differential Geometric Control Theory* (R. Brockett, R. Millman, and H. Sussmann, eds.), Boston, MA: Birkhauser, 1983.

[10] T. Burg, D. Dawson, C. Rahn, and W. Rhodes, "Nonlinear Control of an Overhead Crane via the Saturating Control Approach of Teel," *Proceedings of the IEEE International Conference on Robotics and Automation*, 1996, pp. 3155–3160.

[11] H. Butler, G. Honderd, and J. Van Amerongen, "Model Reference Adaptive Control of a Gantry Crane Scale Model," *IEEE Control Systems Magazine*, Jan. 1991, pp. 57–62.

[12] C. Byrnes and A. Isidori, "On the Attitude Stabilization of Rigid Spacecraft," *Automatica*, Vol. 27, 1991, pp. 87–95.

[13] C. Chung and J. Hauser, "Nonlinear Control of a Swinging Pendulum," *Automatica*, Vol. 31, No. 6, 1995, pp. 851–862.

[14] J. Collado, R. Lozano, and I. Fantoni, "Control of Convey-crane Based on Passivity," *Proceedings of the American Control Conference*, 2000, pp. 1260–1264.

[15] J. Coron and E. Kerai, "Explicit Feedbacks Stabilizing the Attitude of a Rigid Spacecraft with Two Control Torques," *Automatica*, Vol. 32, 1996, pp. 669–677.

[16] B. Costic, D. Dawson, M. de Queiroz, and V. Kapila, "A Quaternion-Based Adaptive Attitude Tracking Controller Without Velocity Measurements," *Journal of Guidance, Control, and, Dynamics*, Vol. 24, No. 6, Nov. 2001, pp. 1214–1222.

[17] P. Crouch, "Spacecraft Attitude Control and Stabilization: Applications of Geometric Control Theory to Rigid Body Models," *IEEE Transactions on Automatic Control*, Vol. 29, No. 4, 1984, pp. 321–331.

[18] M. Dalsmo and O. Egeland, "State Feedback $\mathcal{H}_\infty$-Suboptimal Control of a Rigid Spacecraft," *IEEE Transactions on Automatic Control*, Vol. 42, No. 8, Aug. 1997, pp. 1186–1189.

[19] W. Dixon, D. Dawson, E. Zergeroglu, and F. Zhang, "Robust Tracking and Regulation Control for Mobile Robots," *International Journal of Robust and Nonlinear Control* (Special Issue on Control of Underactuated Nonlinear Systems), Vol. 10, 2000, pp. 199–216.

[20] Y. Fang, W. E. Dixon, D. M. Dawson, and E. Zergeroglu, "Nonlinear Coupling Control Laws for a 3-DOF Overhead Crane System," *IEEE Transactions on Mechatronics*, to appear.

[21] I. Fantoni, R. Lozano, and M. W. Spong, "Energy Based Control of the Pendubot," *IEEE Transactions on Automatic Control*, Vol. 45, No. 4, 2000, pp. 725–729.

[22] M. Fliess, J. Levine, P. Martin, and P. Rouchon, "A Lie-Backlund Approach to Equivalence and Flatness of Nonlinear Systems," *IEEE Transactions on Automatic Control*, Vol. 44, No. 5, May 1999.

[23] T. I. Fossen, *Guidance and Control of Ocean Vehicles*, Chichester, UK: John Wiley, 1994.

[24] F. Gomez-Estern, R. Ortega, F. R. Rubio, J. Acacil, "Stabilization of a Class of Underactuated Mechanical Systems via Total Energy Shaping," *Proceedings of the IEEE Conference on Decision and Control*, 2001, pp. 1137–1143.

[25] J. Hauser, S. Sastry, and G. Meyer, "Nonlinear Control Design for Slightly Nonminimum-Phase Systems: Application to V/STOL Aircraft," *Automatica*, Vol. 28, No. 4, 1992, pp. 665–679.

[26] J. Hendrikx, T. Meijlink, and R. Kriens, "Application of Optimal Control Theory to Inverse Simulation of Car Handling," *Vehicle System Dynamics*, Vol. 26, 1996, pp. 449–461.

[27] R. Horn and C. Johnson, *Matrix Analysis*, Cambridge, UK: Cambridge University Press, 1985.

[28] P. Hughes, *Spacecraft Attitude Dynamics*, New York, NY: Wiley, 1994.

[29] Intelligent Systems Control Ltd., "3DCrane: Installation and Commissioning," Version 1.2., 2000.

[30] T. Kane, P. Likins, and D. Levinson, *Spacecraft Dynamics*, New York, NY: McGraw-Hill, 1983.

[31] B. Kiss, J. Levine, and P. Mullhaupt, "A Simple Output Feedback PD Controller for Nonlinear Cranes," *Proceedings of the IEEE Conference on Decision and Control*, Dec. 2000, pp. 5097–5101.

[32] M. Krstić and P. Tsiotras, "Inverse Optimal Stabilization of a Rigid Spacecraft," *IEEE Transactions on Automatic Control*, Vol. 44, No. 5, May 1999, pp. 1042–1049.

[33] H. Lee, "Modeling and Control of a Three-Dimensional Overhead Cranes," *ASME Journal of Dynamic Systems, Measurement, and Control*, Vol. 120, 1998, pp. 471–476.

[34] F. Lin, W. Zhang, and R. Brandt, "Robust Hovering Control of a PVTOL Aircraft," *IEEE Transactions on Control Systems Technology*, Vol. 7, No. 3, 1999, pp. 343–351.

[35] R. Lozano, I. Fantoni, and D. J. Block, "Stabilization of the Inverted Pendulum Around Its Homoclinic Orbit," *Systems and Controls Letters*, Vol. 40, No. 3, 2000, pp. 197–204.

[36] P. Martin, S. Devasia, and B. Paden, "A Different Look at Output Tracking: Control of a VTOL Aircraft," *Automatica*, Vol. 32, No. 1, 1996, pp. 101–107.

[37] S. C. Martindale, D. M. Dawson, J. Zhu, and C. Rahn, "Approximate Nonlinear Control for a Two Degree of Freedom Overhead Crane: Theory and Experimentation," *Proceedings of the American Control Conference*, 1995, pp. 301–305.

[38] N. McClamroch and I. Kolmanovsky, "A Hybrid Switched Mode Control Approach for V/STOL Flight Control Problems," *Proceedings of the IEEE Conference on Decision and Control*, Kobe, Japan, Dec. 1996, pp. 2648–2653.

[39] G. Meyer, "Design and Global Analysis of Spacecraft Attitude Control Systems," *NASA Technology Report R-361*, Mar. 1971.

[40] P. Morin, C. Samson, J. Pomet, and Z. Jiang, "Time-Varying Feedback Stabilization of the Attitude of a Rigid Spacecraft with Two Controls," *Systems and Controls Letters*, Vol. 25, 1995, pp. 375–385.

[41] P. Morin and C. Samson, "Time-Varying Exponential Stabilization of a Rigid Spacecraft with Two Control Torques," *IEEE Transactions on Automatic Control*, Vol. 42, No. 4, Apr. 1997, pp. 528–534.

[42] K. A. F. Moustafa and A. M. Ebeid, "Nonlinear Modeling and Control of Overhead Crane Load Sway," *ASME Journal of Dynamic Systems, Measurement, and Control*, Vol. 110, 1988, pp. 266–271.

[43] M. W. Noakes and J. F. Jansen, "Generalized Inputs for Damped-Vibration Control of Suspended Payloads," *Robotics and Autonomous Systems*, Vol. 10, 1992, pp. 199–205.

[44] M. Oishi and C. Tomlin, "Switched Nonlinear Control of a VSTOL Aircraft," *Proceedings of the IEEE Conference on Decision and Control*, Phoenix, AZ, Dec. 1999, pp. 2685–2690.

[45] M. Oishi and C. Tomlin, "Switching in Nonminimum-Phase Systems: Applications to VSTOL Aircraft," *Proceedings of the American Control Conference*, Chicago, IL, June 2000, pp. 487–491.

[46] R. Ortega, M. W. Spong, and F. Gomez-Estern, "Stabilization of Underactuated Mechanical Systems via Interconnection and Damping Assignment," *IEEE Transactions on Automatic Control*, Vol. 47, No. 8, Aug. 2002, pp. 1218–1233.

[47] Y. Sakawa and H. Sano, "Nonlinear Model and Linear Robust Control of Overhead Traveling Cranes," *Nonlinear Analysis*, Vol. 30, No. 4, 1997, pp. 2197–2207.

[48] H. Schaub, M. Akella, and J. Junkins, "Adaptive Control of Nonlinear Attitude Motions Realizing Linear Closed-Loop Dynamics," *Proceedings of the American Control Conference*, San Diego, CA, June 1999, pp. 1563–1567.

[49] P. Setlur, D. Dawson, J. Wagner, and Y. Fang, "Nonlinear Tracking Controller Design for Steer-by-Wire Automotive Systems," *Proceedings of the American Control Conference*, Anchorage, AK, 2002, pp. 280–285.

[50] H. Sira-Ramirez, "Nonminimum-Phase Output Reference Trajectory Tracking for a PVTOL Aircraft," *Proceedings of the IEEE Conference on Control Applications*, Anchorage, AK, Sept. 2000, pp. 838–843.

[51] J. -J. E. Slotine and W. Li, *Applied Nonlinear Control*, Englewood Cliffs, NJ: Prentice-Hall, 1991.

[52] A. R. Teel, "Semi-global Stabilization of the 'Ball and Beam' Using 'Output' Feedback," *Proceedings of the American Control Conference*, 1993, pp. 2577–2581.

[53] P. Tsiotras and J. Luo, "Control of Underactuated Spacecraft with Bounded Inputs," *Automatica*, Vol. 36, No. 8, 2000, pp. 1153–1169.

[54] J. Wen and K. Kreutz-Delgado, "The Attitude Control Problem," *IEEE Transactions on Automatic Control*, Vol. 36, No. 10, Oct. 1991, pp. 1148–1162.

[55] Z. Yao, N. P. Costescu, S. P. Nagarkatti, and D. M. Dawson, "Real-Time Linux Target: A MATLAB-Based Graphical Control Environment," *Proceedings of the IEEE Conference on Control Applications*, Anchorage, AK, 2000, pp. 173–178.

[56] K. Yoshida and H. Kawabe, "A Design of Saturating Control with a Guaranteed Cost and Its Application to the Crane Control System," *IEEE Transactions on Automatic Control*, Vol. 37, No. 1, 1992, pp. 121–127.

[57] K. Yoshida, "Nonlinear Controller Design for a Crane System with State Constraints," *Proceedings of the American Control Conference*, 1998, pp. 1277–1283.

[58] J. Yu, F. L. Lewis, and T. Huang, "Nonlinear Feedback Control of a Gantry Crane," *Proceedings of the American Control Conference*, Seattle, WA, 1995, pp. 4310–4315.

Appendix A
Mathematical Background

In this appendix, several fundamental mathematical tools are presented in the form of definitions and lemmas that supplement the control development and closed-loop stability analyses presented in the previous chapters. The proofs of most of the following lemmas are omitted, but can be found in the cited references.

Definition A.1 *[9]*

Consider a function $f(t) : \mathbb{R}_+ \to \mathbb{R}$. Let the 2-norm (denoted by $\|\cdot\|_2$) of a scalar function $f(t)$ be defined as

$$\|f(t)\|_2 = \sqrt{\int_0^\infty f^2(\tau)\,d\tau}. \qquad (A.1)$$

If $\|f(t)\|_2 < \infty$, then we say that the function $f(t)$ belongs to the subspace $\mathcal{L}_2$ of the space of all possible functions (i.e., $f(t) \in \mathcal{L}_2$). Let the ∞-norm (denoted by $\|\cdot\|_\infty$) of $f(t)$ be defined as

$$\|f(t)\|_\infty = \sup_t |f(t)|. \qquad (A.2)$$

If $\|f(t)\|_\infty < \infty$, then we say that the function $f(t)$ belongs to the subspace $\mathcal{L}_\infty$ of the space of all possible functions (i.e., $f(t) \in \mathcal{L}_\infty$).

Definition A.2 *[9]*

The induced 2-norm of matrix $A(t) \in \mathbb{R}^{n \times n}$ is defined as follows

$$\|A(t)\|_{i2} = \sqrt{\lambda_{\max} \{A^T(t) A(t)\}}. \qquad (A.3)$$

Lemma A.1 *[4]*

Given a function $f : \mathbb{R}^n \to \mathbb{R}$ that is continuously differentiable on an open set $S \subset \mathbb{R}^n$ and given points $(x_{10}, ..., x_{n0})$ and $(x_1, ..., x_n)$ in S that are joined by a straight line that lies entirely in $\mathbb{R}^n$, then there exists a point $(\xi_1, ..., \xi_n)$ on the line between the endpoints, such that

$$f(x_1, ..., x_n) = f(x_{10}, ..., x_{n0}) + \sum_{j=1}^{n} \frac{\partial}{\partial x_j} f(\xi_1, ..., \xi_n)(x_j - x_{j0}). \qquad (A.4)$$

This lemma is often referred to as the Mean Value Theorem.

Lemma A.2 *[6]*

Given a function $f : \mathbb{R}^n \times \mathbb{R}^m \to \mathbb{R}^n$ that is continuously differentiable at every point (x, y) on an open set $S \subset \mathbb{R}^n \times \mathbb{R}^m$, if there is a point (x_0, y_0) on S where

$$f(x_0, y_0) = 0 \qquad (A.5)$$

and

$$\frac{\partial f}{\partial x}(x_0, y_0) \neq 0, \qquad (A.6)$$

then there are neighborhoods $U \subset \mathbb{R}^n$ and $V \subset \mathbb{R}^m$ of x_0 and y_0, respectively, such that for all $y \in V$ the expression in (A.5) has a unique solution $x \in U$. This unique solution can be written as $x = g(y)$ where g is continuously differentiable at $y = y_0$. This lemma is often referred to as the Implicit Function Theorem.

Lemma A.3 *[3]*

Given $a, b, c \in \mathbb{R}^n$, any of the following cyclic permutations leaves the scalar triple product invariant

$$a \cdot (b \times c) = b \cdot (c \times a) = c \cdot (a \times b) \qquad (A.7)$$

and the following interchange of the inner and vector product

$$a \cdot (b \times c) = (a \times b) \cdot c \qquad (A.8)$$

leaves the scalar triple product invariant where the notation $a \cdot b$ represents the dot product of a and b and the notation $a \times b$ represents the cross product of a and b.

Lemma A.4 *[3]*

Given $a, b, c \in \mathbb{R}^n$, the vector triple products satisfy the following expressions

$$a \times (b \times c) = (a \cdot c)\, b - (a \cdot b)\, c \tag{A.9}$$

$$(a \times b) \times c\, (a \cdot c)\, b - (b \cdot c)\, a \tag{A.10}$$

where the notation $a \cdot b$ represents the dot product of a and b and the notation $a \times b$ represents the cross product of a and b.

Lemma A.5 *[3]*

Given $a, b \in \mathbb{R}^n$, the vector product satisfies the following skew-symmetric property

$$a \times b = -b \times a \tag{A.11}$$

where the notation $a \times b$ represents the cross product of a and b.

Lemma A.6 *[3]*

Given $a = \begin{bmatrix} a_1 & a_2 & a_3 \end{bmatrix}^T \in \mathbb{R}^3$ and $a^\times \in \mathbb{R}^{3 \times 3}$ which is defined as follows

$$a^\times = \begin{bmatrix} 0 & -a_3 & a_2 \\ a_3 & 0 & -a_1 \\ -a_2 & a_1 & 0 \end{bmatrix} \tag{A.12}$$

then the product $a^T a^\times$ satisfies the following property

$$a^T a^\times = \begin{bmatrix} 0 & 0 & 0 \end{bmatrix}^T. \tag{A.13}$$

Lemma A.7 *[8] (Theorems 9–11)*

Given the symmetric matrix $A \in \mathbb{R}^{n \times n}$ and the diagonal matrix $D \in \mathbb{R}^{n \times n}$, then A is orthogonally similar to D and the diagonal elements of D are necessarily the eigenvalues of A.

Lemma A.8 *[9]*

If $w(t) : \mathbb{R}_+ \to \mathbb{R}$ is persistently exciting and $w(t)$, $\dot{w}(t) \in \mathcal{L}_\infty$, then the stable minimum-phase rational transfer function $\hat{H}(w)$ is also persistently exciting.

Lemma A.9 *[10]*

If $\dot{f}(t) \triangleq \frac{d}{dt} f(t)$ is bounded for $t \in [0, \infty)$, then $f(t)$ is uniformly continuous for $t \in [0, \infty)$.

Lemma A.10 *[1]*

Let $V(t)$ be a nonnegative scalar function of time on $[0, \infty)$ which satisfies the differential inequality

$$\dot{V}(t) \leq -\gamma V(t) \tag{A.14}$$

where γ is a positive constant. Given (A.14), then

$$V(t) \leq V(0) \exp(-\gamma t) \quad \forall t \in [0, \infty) \tag{A.15}$$

where $\exp(\cdot)$ denotes the base of the natural logarithm.

Lemma A.11

Given a nonnegative function denoted by $V(t) \in \mathbb{R}$ as follows

$$V = \frac{1}{2} x^2 \tag{A.16}$$

with the following time derivative

$$\dot{V} = -k_1 x^2, \tag{A.17}$$

then $x(t) \in \mathbb{R}$ is square integrable (i.e., $x(t) \in \mathcal{L}_2$).

Proof: To prove Lemma A.11, we integrate both sides of (A.17) as follows

$$-\int_0^\infty \dot{V}(t) dt = k_1 \int_0^\infty x^2(t) dt. \tag{A.18}$$

After evaluating the left side of (A.18), we can conclude that

$$k_1 \int_0^\infty x^2(t) dt = V(0) - V(\infty) \leq V(0) < \infty \tag{A.19}$$

where we used the fact that $V(0) \geq V(\infty) \geq 0$ (see (A.16) and (A.17)). Since the inequality given in (A.19) can be rewritten as follows

$$\sqrt{\int_0^\infty x^2(t) dt} \leq \sqrt{\frac{V(0)}{k_1}} < \infty \tag{A.20}$$

we can use Definition A.1 to conclude that $x(t) \in \mathcal{L}_2$. □

Lemma A.12 *[5]*

Let $A \in \mathbb{R}^{n \times n}$ be a real symmetric positive-definite matrix; therefore, all of the eigenvalues of A are real and positive. Let $\lambda_{\min}\{A\}$ and $\lambda_{\max}\{A\}$ denote the minimum and maximum eigenvalues of A, respectively, then for $\forall x \in \mathbb{R}^n$

$$\lambda_{\min}\{A\} \|x\|^2 \leq x^T A x \leq \lambda_{\max}\{A\} \|x\|^2 \tag{A.21}$$

where $\|\cdot\|$ denotes the standard Euclidean norm. This lemma is often referred to as the Rayleigh-Ritz Theorem.

Lemma A.13 *[1]*

Given a scalar function $r(t)$ and the following differential equation

$$r = \dot{e} + \alpha e \tag{A.22}$$

where $\dot{e}(t) \in \mathbb{R}$ represents the time derivative $e(t) \in \mathbb{R}$ and $\alpha \in \mathbb{R}$ is a positive constant, if $r(t) \in \mathcal{L}_\infty$, then $e(t)$ and $\dot{e}(t) \in \mathcal{L}_\infty$.

Lemma A.14 *[1]*

Given the differential equation in (A.22), if $r(t)$ is exponentially stable in the sense that

$$|r(t)| \leq \beta_0 \exp(-\beta_1 t) \tag{A.23}$$

where β_0 and $\beta_1 \in \mathbb{R}$ are positive constants, then $e(t)$ and $\dot{e}(t)$ are exponentially stable in the sense that

$$|e(t)| \leq |e(0)| \exp(-\alpha t) + \frac{\beta_0}{\alpha - \beta_1} \left(\exp(-\beta_1 t) - \exp(-\alpha t)\right) \tag{A.24}$$

and

$$|\dot{e}(t)| \leq \alpha |e(0)| \exp(-\alpha t) + \frac{\alpha \beta_0}{\alpha - \beta_1} \left(\exp(-\beta_1 t) - \exp(-\alpha t)\right) \tag{A.25}$$
$$+ \beta_0 \exp(-\beta_1 t)$$

where α was defined in (A.22).

Lemma A.15 *[1]*

Given the differential equation in (A.22), if $r(t) \in \mathcal{L}_\infty$, $r(t) \in \mathcal{L}_2$, and $r(t)$ converges asymptotically in the sense that

$$\lim_{t \to \infty} r(t) = 0, \tag{A.26}$$

then $e(t)$ and $\dot{e}(t)$ converge asymptotically in the sense that

$$\lim_{t \to \infty} e(t), \dot{e}(t) = 0. \tag{A.27}$$

Lemma A.16 *[9]*

Consider a function $f(t) : \mathbb{R}_+ \to \mathbb{R}$. If $f(t) \in \mathcal{L}_\infty$, $\dot{f}(t) \in \mathcal{L}_\infty$, and $f(t) \in \mathcal{L}_2$, then

$$\lim_{t \to \infty} f(t) = 0. \tag{A.28}$$

This lemma is often referred to as Barbalat's Lemma.

Lemma A.17 *[1, 7]*

If a scalar function $N_d(x, y)$ is given by

$$N_d = \Omega(x)xy - k_n \Omega^2(x)x^2 \tag{A.29}$$

where $x, y \in \mathbb{R}$, $\Omega(x) \in \mathbb{R}$ is a function dependent only on x, and k_n is a positive constant, then $N_d(x, y)$ can be upper bounded as follows

$$N_d \leq \frac{y^2}{k_n}. \tag{A.30}$$

The bounding of $N_d(x, y)$ in the above manner is often referred to as nonlinear damping [7] since a nonlinear control function (e.g., $k_n \Omega^2(x)x^2$) can be used to "damp-out" an unmeasurable quantity (e.g., y) multiplied by a known measurable nonlinear function, (e.g., $\Omega(x)$).

Lemma A.18 *[1]*

Let $V(t)$ be a nonnegative scalar function of time on $[0, \infty)$ which satisfies the differential inequality

$$\dot{V} \leq -\gamma V + \varepsilon \tag{A.31}$$

where γ and ε are positive constants. Given (A.31), then

$$V(t) \leq V(0) \exp\left(-\gamma t\right) + \frac{\varepsilon}{\gamma} \left(1 - \exp\left(-\gamma t\right)\right) \qquad \forall t \in [0, \infty). \tag{A.32}$$

Lemma A.19 *[1]*

If the differential equation in (A.22) can be bounded as follows

$$|r(t)| \leq \sqrt{A + B \exp(-kt)} \tag{A.33}$$

where k, A, and $B \in \mathbb{R}$ and $A + B \geq 0$, then $e(t)$ given in (A.22) can be bounded as follows

$$|e(t)| \quad \leq \quad |e(0)| \exp(-\alpha t) + \frac{a}{\alpha} \left(1 - \exp(-\alpha t)\right) \tag{A.34}$$

$$+ \frac{2b}{2\alpha - k} \left(\exp\left(-\frac{1}{2}kt\right) - \exp(-\alpha t)\right)$$

where

$$a = \sqrt{A} \quad \text{and} \quad b = \sqrt{B}. \tag{A.35}$$

Lemma A.20 *[6]*

If a function $f(t) : \mathbb{R}_+ \to \mathbb{R}$ is uniformly continuous and if the integral

$$\lim_{t \to \infty} \int_0^t |f(\tau)| \, d\tau \tag{A.36}$$

exists and is finite, then

$$\lim_{t \to \infty} |f(t)| = 0. \tag{A.37}$$

This lemma is often referred to as the integral form of Barbalat's Lemma.

Lemma A.21 *[2]*

If a given differentiable function $f(t) : \mathbb{R}_+ \to \mathbb{R}$ has a finite limit as $t \to \infty$ and if $f(t)$ has a time derivative, defined as $\dot{f}(t)$, that can be written as the sum of two functions, denoted by $g_1(t)$ and $g_2(t)$, as follows

$$\dot{f}(t) = g_1 + g_2 \tag{A.38}$$

where $g_1(t)$ is a uniformly continuous function and

$$\lim_{t \to \infty} g_2(t) = 0 \tag{A.39}$$

then

$$\lim_{t \to \infty} \dot{f}(t) = 0 \qquad \lim_{t \to \infty} g_1(t) = 0. \tag{A.40}$$

This lemma is often referred to as the Extended Barbalat's Lemma.

Lemma A.22 *[6]*

Let the origin of the following autonomous system

$$\dot{x} = f(x) \tag{A.41}$$

be an equilibrium point $x(t) = 0$ where $f(\cdot) : D \to \mathbb{R}^n$ is a map from the domain $D \subset \mathbb{R}^n$ into $\mathbb{R}^n$. Consider a continuously differentiable positive definite function $V(\cdot) : D \to \mathbb{R}^n$ containing the origin $x(t) = 0$ where

$$\dot{V}(x) \leq 0 \quad \text{in } D. \tag{A.42}$$

Let Γ be defined as the set of all points where $\left\{ x \in D | \dot{V}(x) = 0 \right\}$ and suppose that no solution can stay identically in Γ other than the trivial solution $x(t) = 0$. Then the origin is globally asymptotically stable. This Lemma is a corollary to LaSalle's Invariance Theorem.

References

[1] D. M. Dawson, J. Hu, and T. C. Burg, *Nonlinear Control of Electric Machinery*, New York, NY: Marcel Dekker, 1998.

[2] W. E. Dixon, D. M. Dawson, E. Zergeroglu, and A. Behal, *Nonlinear Control of Wheeled Mobile Robots*, London, UK: Springer-Verlag, 2001.

[3] W. Fulks, *Advanced Calculus: An Introduction to Analysis*, New York, NY: John Wiley, 1978.

[4] M. D. Greenburg, *Advanced Engineering Mathematics*, Upper Saddle River, NJ: Prentice-Hall, 1998.

[5] R. Horn and C. Johnson, *Matrix Analysis*, Cambridge, UK: Cambridge University Press, 1985.

[6] H. Khalil, *Nonlinear Systems*, Upper Saddle River, NJ: Prentice-Hall, 1996.

[7] M. Krstić, I. Kanellakopoulos, and P. Kokotović, *Nonlinear and Adaptive Control Design*, New York, NY: John Wiley, 1995.

[8] S. Perlis, *Theory of Matrices*, New York, NY: Dover Publications, 1991.

[9] S. Sastry and M. Bodson, *Adaptive Control: Stability, Convergence, and Robustness*, Englewood Cliffs, NJ: Prentice-Hall, 1989.

[10] J. -J. E. Slotine and W. Li, *Applied Nonlinear Control*, Englewood Cliffs, NJ: Prentice-Hall, 1991.

Appendix B
Supplementary Lemmas and Definitions

In this appendix, supplementary lemmas and definitions are provided to support the mathematical development in the previous chapters. Proofs are provided for most of the lemmas.

B.1 Chapter 2 Lemmas

B.1.1 Convolution Operations for Torque Filtering

Lemma B.1 *The filtered control input signal $u_f(t)$ defined in (2.32) can be expressed as the the linear parameterization given in (2.34).*

Proof: To rewrite (2.32) in terms of the linear parameterization given in (2.34), the expression given in (2.10) is written in the following form [2]

$$u = \dot{h} + g \tag{B.1}$$

where

$$\dot{h} = \frac{d}{dt}(M(q)\dot{q}) \tag{B.2}$$

and

$$g = -\dot{M}(q)\dot{q} + V_m(q, \dot{q}). \tag{B.3}$$

After substituting (B.1) into (2.32), the following expression can be obtained

$$u_f = \dot{f}(t) * Y_A(q, \dot{q})\phi + Y_B(q, \dot{q})\phi + f(t) * Y_C(q, \dot{q})\phi \qquad (B.4)$$

where the standard convolution properties

$$f * \left\{\dot{h} + g\right\} = f * \dot{h} + f * g \qquad (B.5)$$

and

$$f * \dot{h} = \dot{f} * h + f(0)h - fh(0) \qquad (B.6)$$

have been used. The linear regression matrices $Y_A(q, \dot{q})\phi$, $Y_B(q, \dot{q})\phi$, and $Y_C(q, \dot{q})\phi$ in (B.4) are defined as follows

$$Y_A(q, \dot{q})\phi = M(q(t))\dot{q}(t) \qquad (B.7)$$

$$Y_B(q, \dot{q})\phi = f(0)M(q(t))\dot{q}(t) - f(t)M(q(0))\dot{q}(0)$$

$$Y_C(q, \dot{q})\phi = -\dot{M}(q(t))\dot{q}(t) + V_m(q, \dot{q}).$$

By utilizing standard Laplace Transform techniques, the expression in (B.4) can be rewritten as follows

$$u_f = \left(\bar{Y}_A(q, \dot{q}) + Y_B(q, \dot{q}) + \bar{Y}_C(q, \dot{q})\right)\phi \qquad (B.8)$$

where the regression matrices $\bar{Y}_A(q, \dot{q})$ and $\bar{Y}_C(q, \dot{q})$ are generated by the following differential expressions

$$\dot{\bar{Y}}_A(q, \dot{q}) + \gamma\bar{Y}_A(q, \dot{q}) = -\gamma^2 Y_A(q, \dot{q}) \qquad (B.9)$$
$$\dot{\bar{Y}}_C(q, \dot{q}) + \gamma\bar{Y}_C(q, \dot{q}) = \gamma Y_C(q, \dot{q})$$

where (2.33) and (2.36) were utilized. Hence, based on (B.8) and (B.9), it is straightforward to conclude that (2.32) can be rewritten as the linear parameterization given in (2.34). $\square$

B.1.2   Control Signal Bound

Lemma B.2 *The term $\chi(q, \dot{q}, t)$ defined in (2.48) can be upper bounded by the following inequality*

$$\|\chi\| \leq \zeta_1 \|z\| \qquad (B.10)$$

where $\zeta_1 \in \mathbb{R}$ is a known positive bounding constant and $z(t) \in \mathbb{R}^6$ is defined as follows

$$z(t) = \left[\begin{array}{cccc} x(t) & y(t) & e(t) & r^T(t) \end{array}\right]^T. \tag{B.11}$$

Proof: To prove that (B.10) is a valid bound for $\chi(\cdot) = \left[\begin{array}{ccc} \chi_1 & \chi_2 & \chi_3 \end{array}\right]^T$, (2.11), (2.12), (2.22), (2.38), and (2.39) are used to compute the difference given in (2.48) as follows

$$\chi_1 = -\alpha_1 m\dot{x} - mL\left[\sin(\theta+\beta)\ddot{\theta}_d + \cos(\theta+\beta)\dot{\theta}\dot{\theta}_d\right.$$

$$\left. - \cos\beta\left(\ddot{\theta}_d\sin\theta + \dot{\theta}_d^2\cos\theta\right) + \sin\beta\left(-\ddot{\theta}_d\cos\theta + \dot{\theta}_d^2\sin\theta\right)\right] \tag{B.12}$$

$$\chi_2 = -\alpha_2 m\dot{y} - mL\left[-\cos(\theta+\beta)\ddot{\theta}_d + \sin(\theta+\beta)\dot{\theta}\dot{\theta}_d\right. \tag{B.13}$$

$$\left. + \cos\beta\left(\ddot{\theta}_d\cos\theta - \dot{\theta}_d^2\sin\theta\right) - \sin\beta\left(\ddot{\theta}_d\sin\theta + \dot{\theta}_d^2\cos\theta\right)\right]$$

$$\chi_3 = mL\left(\alpha_1\dot{x}\sin(\theta+\beta) - \alpha_2\dot{y}\cos(\theta+\beta)\right). \tag{B.14}$$

After utilizing (2.18) and (2.19), exploiting several trigonometric identities, and cancelling common terms, the expressions given in (B.12–B.14) can be rewritten as follows

$$\chi_1 = \alpha_1 m(r_1 + \alpha_1 x) + mL\cos(\theta+\beta)\dot{\theta}_d e \tag{B.15}$$

$$\chi_2 = \alpha_2 m(r_2 + \alpha_2 y) + mL\sin(\theta+\beta)\dot{\theta}_d e \tag{B.16}$$

$$\chi_3 = mL\left(-\alpha_1(r_1 + \alpha_1 x)\sin(\theta+\beta) + \alpha_2(r_2 + \alpha_2 y)\cos(\theta+\beta)\right). \tag{B.17}$$

Based on the expressions given in (B.15–B.17) and the definition of $z(t)$ given in (B.11), the bound given in (B.10) can now be directly obtained. $\square$

B.1.3   Control Signal Bound

Lemma B.3 *The auxiliary signal $\Omega(q, \dot{q}, t)$ defined in (2.45) can be upper bounded by the following inequality*

$$\|\Omega\| \leq \gamma\zeta\|\psi\| \tag{B.18}$$

where γ was defined in (2.33), $\zeta \in \mathbb{R}$ is a known positive bounding constant, and $\psi(t) \in \mathbb{R}^9$ is defined as follows

$$\psi(t) = \left[\begin{array}{cc} z^T(t) & \tilde{\phi}_2^T(t) \end{array}\right]^T. \tag{B.19}$$

Proof: To prove that (B.18) is a valid bound for $\Omega\left(q, \dot{q}, t\right)$, the expression given in (2.47) is first rewritten as

$$u = Y\phi = \chi + Y_d\phi_2 + \psi_1 \tag{B.20}$$

where $\psi_1(q, \dot{q}, t) \in \mathbb{R}$ is an auxiliary function defined as follows

$$\psi_1 = -M\dot{r} - V_m r \tag{B.21}$$

and $\chi(q, \dot{q}, t)$ was defined in (2.130). After substituting the closed-loop dynamics for $r(t)$ given by (2.52) into (B.21), and cancelling common terms, the following expression is obtained

$$\psi_1 = -Y_d\tilde{\phi}_2 - \chi + K_s r - \begin{bmatrix} k_{p1}x & k_{p2}y & 0 \end{bmatrix}^T + k_n\zeta_1^2 r. \tag{B.22}$$

Since $\dot{\theta}_d\left(t\right)$, $\ddot{\theta}_d\left(t\right)$, and the trigonometric terms in (2.38) always remain bounded, an upper bound for $Y_d\left(q, t\right)\tilde{\phi}_2$ can be formulated as follows

$$\left\|Y_d\tilde{\phi}_2\right\| \leq \zeta_y \left\|\tilde{\phi}_2\right\| \tag{B.23}$$

where $\zeta_y \in \mathbb{R}$ is a positive scalar constant. By utilizing (2.49), (2.50), and the bound given in (B.23), an upper bound for (B.22) can be formulated as follows

$$\|\psi_1\| \leq \zeta_y \left\|\tilde{\phi}_2\right\| + \zeta_1 \|z\| + K_s \|r\| + k_{p1} |x| + k_{p2} |y| + k_n\zeta_1^2 \|r\|. \tag{B.24}$$

Given the definitions for $\psi(t)$ and $z(t)$ in (2.56) and (2.50), respectively, the upper bound given in (B.24) can be rewritten in the following compact form

$$\|\psi_1\| \leq \zeta_2 \|\psi\| \tag{B.25}$$

where $\zeta_2 \in \mathbb{R}$ is a positive bounding constant. After rewriting (B.20) as follows

$$Y\phi - Y_d\phi_2 = \chi + \psi_1 \tag{B.26}$$

and then utilizing (2.49) and (B.25), the following inequality can be formulated

$$\|Y\phi - Y_d\phi_2\| \leq \|\chi\| + \|\psi_1\| \leq \zeta_1 \|z\| + \zeta_2 \|\psi\| \leq \zeta \|\psi\| \tag{B.27}$$

where $\zeta \in \mathbb{R}$ is a positive bounding constant, and $\psi(t)$ and $z(t)$ were defined in (2.56) and (2.50), respectively.

Based on (2.32), (2.37), and (2.45), $\Omega\left(q, \dot{q}, t\right)$ can be expressed as follows

$$\Omega = f * (Y\phi - Y_d\phi_2) \tag{B.28}$$

where $f(t)$ was defined in (2.33). An upper bound for (B.28) can now be formulated as follows

$$\|\Omega\| \leq \gamma \|Y\phi - Y_d\phi_2\| \tag{B.29}$$

where the standard inequality property of convolution in [4] and the upper bound $\|f(t)\| \leq \gamma$ have been applied (see (2.33)). After substituting (B.27) into (B.29), the inequality given by (B.18) can be obtained. $\square$

B.1.4   Control Signal Bound

Before developing lower and upper bounds for $P^{-1}(t)$ given in (2.43), the following preliminary lemma is presented.

Lemma B.4 *Given the definition of $P^{-1}(t)$ in (2.43), it can be shown that*

$$\|P\|_{i2} P^{-1} \geq I_3 \tag{B.30}$$

where I_3 is the 3×3 identity matrix and the matrix inequality $A > B$ for $A, B \in \mathbb{R}^{p \times p}$ is a shorthand notation to denote $\xi^T A \xi > \xi^T B \xi$, $\forall \xi \in \mathbb{R}^p$.

Proof: Given the definition of $P^{-1}(t)$ in (2.43) and the fact that $P(0)$ is selected to be positive-definite and symmetric, $P^{-1}(t)$ is positive-definite and symmetric for all time. Hence, it follows that $P(t)$ is also positive-definite and symmetric for all time. Since $P(t)$ is a positive-definite and symmetric gain matrix, the Rayleigh-Ritz Theorem (see Lemma A.12 of Appendix A) can be used to prove that

$$x^T \lambda_{\max}\{P\} x \geq x^T P x \tag{B.31}$$

where $\lambda_{\max}\{\cdot\}$ denotes the maximum eigenvalue of $\{\cdot\}$. Based on Lemma A.7 and Definition A.2 of Appendix A, the following property can be proven

$$\lambda_{\max}\{P\} = \sqrt{\lambda_{\max}\{P^T P\}} = \|P\|_{i2}. \tag{B.32}$$

By utilizing (B.32), the expression given in (B.31) can be rewritten as follows

$$x^T \|P\|_{i2} x \geq x^T P x. \tag{B.33}$$

To facilitate the remainder of the proof, $y \in \mathbb{R}$ is defined as follows

$$y = \sqrt{P} x. \tag{B.34}$$

After rewriting the right-hand side of (B.33) as follows

$$x^T P x = x^T \sqrt{P} \sqrt{P} x \tag{B.35}$$

and then utilizing the expression given in (B.34), the following expression can be obtained

$$y^T \left(\sqrt{P}\right)^{-1} \sqrt{P}\sqrt{P} \left(\sqrt{P}\right)^{-1} y = y^T I_3 y. \tag{B.36}$$

In a similar manner, (B.34) can be used to rewrite the left-hand side of (B.33) as follows

$$x^T \|P\|_{i2} \, x = \|P\|_{i2} \, y^T \left(\sqrt{P}\right)^{-1} \left(\sqrt{P}\right)^{-1} y = y^T \|P\|_{i2} \, P^{-1} y. \tag{B.37}$$

The expressions given in (B.33–B.37) can now be used to prove the result given in (B.30). $\square$

Lemma B.5 *The gain forgetting factor $\lambda(t)$ and the gain matrix $P^{-1}(t)$ defined in (2.58) and (2.43), respectively, can be upper and lower bounded by the following inequalities*

$$0 < \frac{\lambda_1 k_1 \kappa}{1 + k_1 \kappa} \leq \lambda(t) \tag{B.38}$$

$$\frac{1 + k_1 \kappa}{k_1} I_3 \leq P^{-1}(t) \leq k_2 I_3 \tag{B.39}$$

where κ is defined in (2.63).

Proof: To prove Lemma B.5, the lower bound for $P^{-1}(t)$ is obtained and then the upper and lower bounds on the gain adjusted forgetting factor $\lambda(t)$ are developed, which are used to derive the upper bound for $P^{-1}(t)$.

Lower bound for $P^{-1}(t)$:
By substituting (2.58) into (2.42), the following expression can be obtained

$$\dot{P}^{-1} = -\lambda_1 P^{-1} + \left[\frac{\lambda_1}{k_1} \|P\|_{i2} \, P^{-1} + Y_{df}^T Y_{df}\right]. \tag{B.40}$$

The solution to the differential equation given in (B.40) can be determined as follows

$$P^{-1}(t) = \quad P^{-1}(0) \exp\left(-\lambda_1 t\right) + \int_0^t \exp\left(-\lambda_1 (t - \sigma)\right)$$

$$\cdot \left[\frac{\lambda_1}{k_1} \|P(t)\|_{i2} \, P^{-1}(t) + Y_{df}^T (\cdot) \, Y_{df}(\cdot)\right] d\sigma. \tag{B.41}$$

Based on the result of Lemma B.4, the following inequality can be developed

$$P^{-1}(t) \geq \quad P^{-1}(0) \exp(-\lambda_1 t) + \frac{\lambda_1}{k_1} I_3 \exp(-\lambda_1 t) \left[\int_0^t \exp(\lambda_1 \sigma) \, d\sigma \right]$$

$$+ \int_0^t \exp(-\lambda_1 (t - \sigma)) Y_{df}^T(\cdot) Y_{df}(\cdot) \, d\sigma.$$

$$(B.42)$$

After evaluating the bracketed integral given in (B.42), the resulting expression can be simplified as follows

$$P^{-1}(t) \geq \quad \left(P^{-1}(0) - \frac{1}{k_1} I_3 \right) \exp(-\lambda_1 t) + \frac{1}{k_1} I_3$$

$$(B.43)$$

$$+ \left[\int_0^t \exp(-\lambda_1 (t - \sigma)) Y_{df}^T(\cdot) Y_{df}(\cdot) \, d\sigma \right].$$

To facilitate the remaining analysis, let the time interval $[0, t]$ be divided into n different intervals where the length of the interval from t_i to t_{i+1} is denoted by δ_i (*i.e.*, length of the interval from t_0 to t_1 is δ_0, t_1 to t_2 is δ_1,...., t_n to t_{n+1} is δ_n).

For the interval $0 \leq t \leq t_1$ where $t_1 = \delta_0$: Since the bracketed term in (B.43) is always positive, the following lower bound can be developed

$$P^{-1}(t) \geq \left(P^{-1}(0) - \frac{1}{k_1} I_3 \right) \exp(-\lambda_1 t) + \frac{1}{k_1} I_3. \qquad (B.44)$$

Provided that the gain condition given in (2.64) is satisfied, the following fact can be utilized

$$\exp(-\lambda_1 t) \geq \exp(-\lambda_1 \delta_0) \quad \forall t \in [0, t_1]$$

to rewrite (B.44) as

$$P^{-1}(t) \geq \left(P^{-1}(0) - \frac{1}{k_1} I_3 \right) \exp(-\lambda_1 \delta_0) + \frac{1}{k_1} I_3. \qquad (B.45)$$

For the interval $t_1 \leq t \leq t_2$ where $t_2 = (t_1 + \delta_1)$: Provided that the gain condition given in (2.64) is satisfied, the first term on the right-hand side of (B.43) will always be positive, and hence, the following lower bound can be formulated for (B.43)

$$P^{-1}(t) \geq \frac{1}{k_1} I_3 + \left[\int_{t_0}^{t_0 + \delta_0} \exp(-\lambda_1 (t - \sigma)) Y_{df}^T(\cdot) Y_{df}(\cdot) \, d\sigma \right]. \qquad (B.46)$$

Since σ used in (B.46) satisfies the inequality $0 \le \sigma \le t$, and $t \le \delta_0 + \delta_1$ in the interval $t_1 \le t \le t_2$, it follows that $t - \sigma \le \delta_0 + \delta_1$. Since $t - \sigma \le \delta_0 + \delta_1$, the following inequality can be developed

$$\exp\left(-\lambda_1 \left(t - \sigma\right)\right) \ge \exp\left(-\lambda_1 \left(\delta_0 + \delta_1\right)\right). \tag{B.47}$$

Hence, (B.46) can be rewritten as

$$P^{-1}\left(t\right) \ge \frac{1}{k_1} I_3 + \exp\left(-\lambda_1 \left(\delta_0 + \delta_1\right)\right) \left[\int_{t_0}^{t_0 + \delta_0} Y_{df}^T\left(\cdot\right) Y_{df}\left(\cdot\right) d\sigma \right]. \tag{B.48}$$

For interval $t_n \le t \le t_{n+1}$ **where** $t_{n+1} = (t_n + \delta_n)$**:** After extending the results of (B.46) and (B.47) to the interval $t_n \le t \le t_{n+1}$, the following expression is obtained

$$P^{-1}\left(t\right) \ge \frac{1}{k_1} I_3 + \left[\int_{t_{n-1}}^{t_{n-1} + \delta_{n-1}} \exp\left(-\lambda_1 \left(t - \sigma\right)\right) Y_{df}^T\left(\cdot\right) Y_{df}\left(\cdot\right) d\sigma \right] \tag{B.49}$$

and

$$\exp\left(-\lambda_1 \left(t - \sigma\right)\right) \ge \exp\left(-\lambda_1 \left(\delta_{n-1} + \delta_n\right)\right). \tag{B.50}$$

Hence, (B.49) can be rewritten as

$$P^{-1}\left(t\right) \ge \frac{1}{k_1} I_3 + \exp\left(-\lambda_1 \left(\delta_{n-1} + \delta_n\right)\right) \left[\int_{t_{n-1}}^{t_{n-1} + \delta_{n-1}} Y_{df}^T\left(\cdot\right) Y_{df}\left(\cdot\right) d\sigma \right]. \tag{B.51}$$

If the PE condition given in (2.57) is satisfied, then (B.51) can be lower bounded as follows

$$P^{-1}\left(t\right) \ge \frac{1}{k_1} I_3 + \exp\left(-\lambda_1 \left(\delta_{n-1} + \delta_n\right)\right) \mu I_3 \ge \left(\frac{1}{k_1} + \kappa\right) I_3 \tag{B.52}$$

where κ is defined in (2.63). It should be noted that to obtain the expression given in (2.63), the interval $0 \le t \le t_1$ and the minimum from the other intervals $t_{i-1} \le t \le t_i$ were used where $i = 2, 3, ..., n$. The lower bound on $P^{-1}\left(t\right)$ can now be formulated as

$$P^{-1}\left(t\right) \ge \left(\frac{1 + k_1 \kappa}{k_1}\right) I_3. \tag{B.53}$$

Finally, by invoking Lemma B.4 the following expression is obtained

$$P\left(t\right) \le \lambda_{\max}\left\{P\left(t\right)\right\} I_3 = \|P\left(t\right)\|_{i2} I_3 \le \frac{k_1}{1 + k_1 \kappa} I_3. \tag{B.54}$$

Bounds for $\lambda(t)$:

After applying the result given in (B.54) to (2.58), lower and upper bounds for $\lambda(t)$ can be formulated as follows

$$\lambda_1 \geq \lambda(t) \geq \lambda_1 \left(\frac{k_1 \kappa}{1 + k_1 \kappa} \right). \tag{B.55}$$

Upper bound for $P^{-1}(t)$:

After applying the lower bound in (B.55) to (2.43), the following expression is obtained

$$P^{-1}(t) \leq \quad P^{-1}(0) \exp \left(-\lambda_1 \left(\frac{k_1 \kappa}{1 + k_1 \kappa} \right) t \right)$$

$$+ \int_0^t \exp \left(-\lambda_1 \left(\frac{k_1 \kappa}{1 + k_1 \kappa} \right) (t - \sigma) \right) Y_{df}^T (\cdot) Y_{df} (\cdot) \, d\sigma, \tag{B.56}$$

which can be further upper bounded by the following expression

$$P^{-1}(t) \leq \quad P^{-1}(0) + \left\| Y_{df}^T (\cdot) Y_{df} (\cdot) \right\|_{i\infty} I_3$$

$$\cdot \left[\frac{\left(1 - \exp \left(-\lambda_1 \left(\frac{k_1 \kappa}{1 + k_1 \kappa} \right) t \right) \right) (1 + k_1 \kappa)}{\lambda_1 k_1 \kappa} \right] \tag{B.57}$$

where the definition for the induced-infinity norm of a matrix is given in Definition A.1 in Appendix A. The bracketed term in (B.57) can be upper bounded as

$$\left[\frac{\left(1 - \exp \left(-\lambda_1 \left(\frac{k_1 \kappa}{1 + k_1 \kappa} \right) t \right) \right) (1 + k_1 \kappa)}{\lambda_1 k_1 \kappa} \right] \leq \frac{(1 + k_1 \kappa)}{\lambda_1 k_1 \kappa}; \tag{B.58}$$

thus, (B.57) can be further bounded as

$$P^{-1}(t) \leq P^{-1}(0) + \frac{\left\| Y_{df}^T (\cdot) Y_{df} (\cdot) \right\|_{i\infty} (1 + k_1 \kappa)}{\lambda_1 k_1 \kappa} I_3. \tag{B.59}$$

Given that $\dot{\theta}_d(t)$ and $\ddot{\theta}_d(t)$ are upper bounded by some known positive constants, it can be shown that $\|Y_d(\cdot)\|_{i\infty}$ defined in (2.38) can also be upper bounded by some known positive constant. Given the definitions in (2.37) and (2.33), the standard inequality property of convolution in [4] can be used to prove that

$$\|Y_{df}(\cdot)\|_{i\infty} \leq \gamma \|Y_d(\cdot)\|_{i\infty} \tag{B.60}$$

where the upper bound $\|f(t)\| \leq \gamma$ for (2.33) has been used. Since $\|Y_{df}(\cdot)\|_{i\infty}$ in (B.60) is bounded, the following inequality can be applied to show that $\left\|Y_{df}(\cdot)^T Y_{df}(\cdot)\right\|_{i\infty}$ is bounded:

$$\|AB\|_i \leq \|A\|_i \|B\|_i \quad \forall A \in \mathbb{R}^{n \times m}, \forall B \in \mathbb{R}^{m \times p} \tag{B.61}$$

where $\|\cdot\|_i$ denotes any induced norm of a matrix. Since $\left\|Y_{df}(\cdot)^T Y_{df}(\cdot)\right\|_{i\infty}$ is bounded and $P^{-1}(0)$ is a positive-definite symmetric constant matrix, there exists a positive bounding constant $k_2 \in \mathbb{R}$ such that

$$P^{-1}(t) \leq k_2 I_3. \tag{B.62}$$

The result given in (2.75) can now be obtained from (B.53) and (B.62). $\square$

B.1.5   Inequality Proofs

In this section, proofs for the inequalities given in Property 2.7 are provided. The following facts are exploited to facilitate these proofs

$$\left. \begin{array}{l} |\cos a - \cos b| \leq 8\,|\tanh(a - b)| \\ |\sin a - \sin b| \leq 8\,|\tanh(a - b)| \end{array} \right\} \forall a, b \in \mathbb{R},$$

$$\tag{B.63}$$

$$\|a\| = \sqrt{\sum_{i=1}^{n} |a_i|^2} \geq |a_i| \quad \forall a = [a_1, a_2, ..., a_n]^T \in \mathbb{R}^n.$$

Lemma B.6 *The following inequality is valid for the transformed ship dynamics of (2.95)*

$$\|M^*(u) - M^*(w)\|_{i\infty} \leq \zeta_m \|Tanh(u - w)\| \quad \forall u, w \in \mathbb{R}^3. \tag{B.64}$$

Proof: After substituting the definition of $M^*(\cdot)$ given in (2.96) into (2.102), the following bound can be obtained

$$\|M^*(u) - M^*(w)\|_{i\infty} \leq \max_j \left\{ \tilde{M}_j^*(u, w) \right\} \quad \forall j = 1, 2, 3 \tag{B.65}$$

where $\tilde{M}^*(t) = \begin{bmatrix} \tilde{M}_1^*(t) & \tilde{M}_2^*(t) & \tilde{M}_3^*(t) \end{bmatrix} \in \mathbb{R}^3$ is an auxiliary expression defined as follows

$$\begin{aligned} \tilde{M}_1^* = \ & |(m_{11} - m_{22})(\sin u_3 \cos u_3 - \sin w_3 \cos w_3)| \\ & + m_{23}|\sin u_3 - \sin w_3| \\ & + |(m_{11}(\cos^2 u_3 - \cos^2 w_3) + m_{22}(\sin^2 u_3 - \sin^2 w_3))|, \end{aligned}$$

$$\tag{B.66}$$

$$\tilde{M}_2^* = \ |(m_{11} - m_{22})(\sin u_3 \cos u_3 - \sin w_3 \cos w_3)|$$

$$+ m_{23} |\cos u_3 - \cos w_3|$$

$$+ \left| (m_{11}(\sin^2 u_3 - \sin^2 w_3) + m_{22}(\cos^2 u_3 - \cos^2 w_3)) \right|,$$

$$\text{(B.67)}$$

$$\tilde{M}_3^* = |m_{23}(\cos u_3 - \cos w_3)| + |m_{23}(\sin u_3 - \sin w_3)|, \qquad \text{(B.68)}$$

and $u_3(t)$, $w_3(t)$ are the third elements of the vectors $u(t)$, $w(t) \in \mathbb{R}^3$, respectively. After some algebraic manipulation, the expression given in (B.66) can be rewritten as follows

$$\tilde{M}_1^* = \ |(m_{11} - m_{22})((\cos u_3 - \cos w_3)\sin u_3 + (\sin u_3 - \sin w_3)\cos w_3)|$$

$$+ m_{23}|\sin u_3 - \sin w_3|$$

$$+ |(m_{11}((\cos u_3 - \cos w_3)\cos u_3 + (\cos u_3 - \cos w_3)\cos w_3)$$

$$+ \ m_{22}((\sin u_3 - \sin w_3)\sin u_3 + (\sin u_3 - \sin w_3)\sin w_3))|$$

$$\text{(B.69)}$$

which can be upper bounded as follows

$$\tilde{M}_1^* \leq \ (|m_{11} - m_{22}| + 2m_{11})|\cos u_3 - \cos w_3|$$

$$\text{(B.70)}$$

$$+ (m_{23} + 2m_{22} + |m_{11} - m_{22}|)|\sin u_3 - \sin w_3|.$$

After making use of (B.63), $\tilde{M}_1^*(t)$ can be further upper bounded by the following expression

$$\tilde{M}_1^* \leq \zeta_{m1}|\tanh(u_3 - w_3)| \leq \zeta_{m1}\|Tanh(u - w)\| \qquad \text{(B.71)}$$

where $\zeta_{m1} \in \mathbb{R}$ is a positive bounding constant. Likewise, similar bounds can be obtained for $\tilde{M}_2^*(t)$ and $\tilde{M}_3^*(t)$ of (B.67) and (B.68), respectively. These bounds can then be used in (B.65) to obtain the result given in (B.64). $\square$

Lemma B.7 *The following inequality is valid for the transformed ship dynamics of (2.95)*

$$\|V_m(u, \dot{\eta}) - V_m(w, \dot{\eta})\|_{i\infty} \leq \zeta_{v2}\|\dot{\eta}\| \ \|Tanh(u - w)\| \quad \forall u, w \in \mathbb{R}^3. \ \text{(B.72)}$$

Proof: After substituting the definition of $V_m(\cdot)$ given in (2.96) into (2.102), the following bound can be obtained

$$\|V_m(u, \dot{\eta}) - V_m(w, \dot{\eta})\|_{i\infty} \leq \max_j \left\{ \tilde{V}_{mj}(u, w, \dot{\eta}) \right\} \quad \forall j = 1, 2, 3 \quad \text{(B.73)}$$

where the elements of the vector $\tilde{V}_m(t) = \begin{bmatrix} \tilde{V}_{m1}(t) & \tilde{V}_{m2}(t) & \tilde{V}_{m3}(t) \end{bmatrix} \in \mathbb{R}^3$ are defined as

$$\tilde{V}_{m1} = \left| \dot{\psi} (m_{22} - m_{11}) (\sin u_3 \cos u_3 - \sin w_3 \cos w_3) \right|$$

$$+ \left| \dot{\psi} (m_{11} (\cos^2 u_3 - \cos^2 w_3) + m_{22} (\sin^2 u_3 - \sin^2 w_3)) \right|, \tag{B.74}$$

$$\tilde{V}_{m2} = \left| \dot{\psi} (m_{11} (\sin^2 u_3 - \sin^2 w_3) + m_{22} (\cos^2 u_3 - \cos^2 w_3)) \right| \tag{B.75}$$

$$+ \left| \dot{\psi} (m_{22} - m_{11}) (\sin u_3 \cos u_3 - \sin w_3 \cos w_3) \right|,$$

$$\tilde{V}_{m3} = \left| \dot{\psi} m_{23} (\cos u_3 - \cos w_3) \right| + \left| \dot{\psi} m_{22} (\sin u_3 - \sin w_3) \right|, \tag{B.76}$$

and $u_3(t)$, $w_3(t)$ are the third elements of the vectors $u(t)$, $w(t) \in \mathbb{R}^{\mu}$, respectively. After some algebraic manipulation, the expression given in (B.74) can be rewritten as

$$\tilde{V}_{m1} = \left| \dot{\psi} (m_{22} - m_{11}) \right.$$

$$((\cos u_3 - \cos w_3) \sin u_3 + (\sin u_3 - \sin w_3) \cos w_3)|$$

$$+ \left| \dot{\psi} (m_{11} ((\cos u_3 - \cos w_3) \cos u_3 + (\cos u_3 - \cos w_3) \cos w_3) \right.$$

$$+ m_{22} ((\sin u_3 - \sin w_3) \sin u_3 + (\sin u_3 - \sin w_3) \sin w_3))| \tag{B.77}$$

which can be upper bounded as

$$\tilde{V}_{m1} \leq \left| \dot{\psi} \right| |m_{22} - m_{11}| (|\cos u_3 - \cos w_3| + |\sin u_3 - \sin w_3|)$$

$$+ 2 \left| \dot{\psi} \right| m_{11} |\cos u_3 - \cos w_3| + 2 \left| \dot{\psi} \right| m_{22} |\sin u_3 - \sin w_3|. \tag{B.78}$$

After making use of (B.63), $\tilde{V}_{m1}(t)$ can be further upper bounded by the following expression

$$\tilde{V}_{m1} \leq \zeta_{\rho 1} \left| \dot{\psi} \right| |\tanh (u_3 - w_3)| \leq \zeta_{\rho 1} \|\dot{\eta}\| \| Tanh (u - w)\| \tag{B.79}$$

where $\zeta_{\rho 1} \in \mathbb{R}$ is a positive bounding constant. Likewise, similar bounds can be obtained for $\tilde{V}_{m2}(t)$ and $\tilde{V}_{m3}(t)$ of (B.75) and (B.76), respectively. These bounds can then be used in (B.73) to obtain the result given in (B.72). □

Lemma B.8 *The following inequality is valid for the ship dynamics of* *(2.95)*

$$\|F_1(u) - F_1(w)\| \le \zeta_{f2} \|Tanh(u - w)\| \quad \forall u, w \in \mathbb{R}^3. \tag{B.80}$$

Proof: The proof of (B.80) is straightforward from (2.92), (2.96), and the proof of Lemmas B.6 and B.7. $\square$

B.1.6   Control Signal Bounds

Lemma B.9 *The auxiliary control signal $\chi(t)$ given in (2.130) can be upper bounded by the following expression*

$$\|\chi\| \le \zeta_1 \|\varrho\| + \zeta_2 \|z\|^2 + \zeta_3 \|z\|^3 + \zeta_4 \|z\|^4 + \zeta_5 \|z\|^5 + \zeta_6 \|r\| \|z\| \tag{B.81}$$

where the composite state vector $\varrho(t)$ was defined in (2.137) and $\zeta_i \in \mathbb{R}$, $i = 1, ..., 6$ are some positive bounding constants that depend on the system parameters and the desired trajectory.

Proof: To prove Lemma B.9, the norm of (2.130) is used to obtain the following upper bound

$$\begin{aligned}
\|\chi\| \le \quad & \|M^*(\eta)\| \left\| Cosh^{-2}(e) \right\| \|(r - Tanh(e) - z)\| \\[1em]
& + \|M^*(\eta)\| \|T\| \|(z - Tanh(e))\| \\[1em]
& + \|V_m(\eta, \dot\eta_d + Tanh(e) + z)\| \|(Tanh(e) + z)\| \\[1em]
& + \|V_m(\eta, \dot\eta_d)\| \|(Tanh(e) + z)\| \\[1em]
& + \|V_m(\eta, r)\| \|(\dot\eta_d + Tanh(e) + z)\|.
\end{aligned} \tag{B.82}$$

After utilizing (2.97), (2.100), and the fact that $\left|\cosh^{-2}(\cdot)\right| \le 1$, (B.82) can be upper bounded as follows

$$\begin{aligned}
\|\chi\| \le \quad & m_2 (\|r\| + \|Tanh(e)\| + \|z\|) \\[1em]
& + m_2 \|T\| (\|z\| + \|Tanh(e)\|) \\[1em]
& + \zeta_{v1} (\|\dot\eta_d\| + \|Tanh(e)\| + \|z\|)(\|Tanh(e)\| + \|z\|) \\[1em]
& + \zeta_{v1} \|\dot\eta_d\| (\|Tanh(e)\| + \|z\|) \\[1em]
& + \zeta_{v1} \|r\| (\|\dot\eta_d\| + \|Tanh(e)\| + \|z\|).
\end{aligned} \tag{B.83}$$

To facilitate further analysis, the definition of $T(\cdot)$ given in (2.132) can be used to prove that the following inequality is valid

$$\|T\| \leq \left(1 + \|z\|^2\right)^2. \tag{B.84}$$

The result given in (B.81) is now straightforward after substituting the inequality given in (B.84) into (B.83) for $\|T(\cdot)\|$ and then using (2.137) and the facts that $|\tanh(\cdot)| \leq 1$ and $\dot{\eta}_d(t)$ is bounded. $\square$

Lemma B.10 *The auxiliary control signal $\tilde{Y}(\cdot)$ given in (2.129) can be upper bounded as follows*

$$\left\|\tilde{Y}\right\| \leq \zeta_7 \|\varrho\| \tag{B.85}$$

where ζ_7 is some positive bounding constant that depends on the system parameters and the desired trajectory.

Proof: To prove the result given in (B.85), we substitute (2.127) into (2.129) for $Y_d(\cdot)\phi$ to obtain the following expression

$$\tilde{Y} = \quad [M^*(\eta) - M^*(\eta_d)]\, \ddot{\eta}_d + [V_m(\eta, \dot{\eta}_d) - V_m(\eta_d, \dot{\eta}_d)]\, \dot{\eta}_d$$
$$+ [F_1(\eta)\dot{\eta} - F_1(\eta)\dot{\eta}_d] + [F_1(\eta)\dot{\eta}_d - F_1(\eta_d)\dot{\eta}_d] \tag{B.86}$$

where the term $F_1(\eta)\dot{\eta}_d$ has been added and subtracted. After utilizing (2.101), (2.102), (2.106), and (2.120), an upper bound can be formulated for (B.86) as follows

$$\left\|\tilde{Y}\right\| \leq \quad \zeta_m \|Tanh(e)\| \|\ddot{\eta}_d\| + \zeta_{v2} \|Tanh(e)\| \|\dot{\eta}_d\|^2$$
$$\zeta_{f1} (\|r\| + \|Tanh(e)\| + \|z\|) + \zeta_f \|Tanh(e)\| \|\dot{\eta}_d\|. \tag{B.87}$$

The result given in (B.85) follows directly from (B.87), given (2.137) and the fact that $\dot{\eta}_d(t)$ and $\ddot{\eta}_d(t)$ are bounded. $\square$

Remark B.1 *Note that if the mooring effects are included in the dynamic model of the surface ship, the expressions given in (2.138) and (B.85)could not be used to upper bound $\tilde{Y}(t)$.*

B.1.7   Matrix Property

Lemma B.11 *Given the definition for the Jacobian-type matrix $B(q)$ given in (2.177), the following property holds*

$$B^T B = I_3 \tag{B.88}$$

where I_3 is the 3×3 identity matrix.

Proof: The definition for $B(q)$ given in (2.177) can be written in the following form

$$B = \begin{bmatrix} -q_{v1} & -q_{v2} & -q_{v3} \\ q_0 & -q_{v3} & q_{v2} \\ q_{v3} & q_0 & -q_{v1} \\ -q_{v2} & q_{v1} & q_0 \end{bmatrix}. \tag{B.89}$$

Based on the structure of (B.89), we can utilize (2.170) to prove (B.88). $\square$

B.2   Chapter 3 Definitions and Lemmas

B.2.1   Supplemental Definitions

Definition B.1 *The measurable functions $\Omega_{ij}(\cdot)$ $\forall i = 1, 2, 3, 4$ introduced in (3.75) are defined as follows*

$$\Omega_{1j} = \frac{dI_{dj}}{dq}\dot{q} + \frac{B_j(q, I_j)}{L_j(q, I_j)}\dot{q} + \frac{dI_{dj}}{d\tau_d}\left(\hat{M}_m \, \dddot{q}_d + (\hat{M}_m\alpha + k_s)\ddot{q}_d \right.$$

$$\left. + k_s\alpha\dot{q}_d + \frac{\partial\tau_d}{\partial\hat{M}_m} \, \dot{\hat{M}}_m + \frac{\partial\tau_d}{\partial\hat{\theta}_m} \, \dot{\hat{\theta}}_m + \frac{\partial\tau_d}{\partial q}\dot{q}\right)$$

$$\Omega_{2j} = -\frac{dI_{dj}}{d\tau_d}\frac{\partial\tau_d}{\partial\dot{q}} \tag{B.90}$$

$$\Omega_{3j} = \frac{I_j}{L_j(q, I_j)}$$

$$\Omega_{4j} = \frac{dI_{dj}}{d\tau_d}\frac{\partial\tau_d}{\partial\dot{q}}\sum_{j=1}^{m}\tau_j(q, I_j).$$

B.2.2   Stability Analysis for Projection Cases

Lemma B.12 *Given the mechanical and electrical closed-loop error systems in (3.69) and (3.78) as well as the adaptive update laws given in (3.80-3.82), an upper bound for the expression given in (3.89) can be formulated as follows*

$$\dot{V} \leq -\lambda_3 \|x\|^2. \tag{B.91}$$

Proof: After utilizing (3.65), (3.77), and (3.80), the expression in (3.89) can be rewritten as follows

$$
\begin{aligned}
\dot{V} = \;\; & -K_s r^2 - \sum_{j=1}^{m} K_{ej} M_m \eta_j^2 \\
& + \tilde{M}_m \left(\sum_{j=1}^{m} \left(-\hat{M}_m^{-1} \left(\Omega_{4j} + \Omega_{2j} W_m \hat{\theta}_m + u_j \right) \eta_j \right) \right) \qquad \text{(B.92)} \\
& + \tilde{M}_m \left((\ddot{q}_d + \alpha \dot{e}) \, r - \Gamma_1^{-1} \, \dot{\hat{M}}_m \right).
\end{aligned}
$$

To substitute (3.81) and (3.82) into (B.126) for $\dot{\hat{M}}_m(t)$, the three following cases must be considered.

Case 1: $\hat{M}_m(t) > \underline{M}_m$

When $\hat{M}_m(t) > \underline{M}_m$, the first equation in (3.81) can be used to express $\dot{V}(t)$ as follows

$$
\dot{V} \leq -K_s r^2 - \sum_{j=1}^{m} K_{ej} M_m \eta_j^2 \leq -\lambda_3 \|x\|^2 \qquad \text{(B.93)}
$$

where (3.91) and (3.92) have been utilized. Thus, for Case 1, we can conclude that (3.89) reduces to the expression given in (B.91). In addition, the direction in which the estimate $\hat{M}_m(t)$ is updated for Case 1 is irrelevant, since the worse case scenario is that $\hat{M}_m(t)$ will move toward $\underline{M}_m$ which will be covered in Cases 2 and 3.

Case 2: $\hat{M}_m = \underline{M}_m$ and $\Omega_m \geq 0$

When $\hat{M}_m(t) = \underline{M}_m$ (i.e., $\hat{M}_m(t)$ equals the lower bound) and Ω_m is nonnegative, the second equation in (3.81) alongside (3.82) can be used to express $\dot{V}(t)$ as follows

$$
\dot{V} \leq -K_s r^2 - \sum_{j=1}^{m} K_{ej} M_m \eta_j^2 \leq -\lambda_3 \|x\|^2. \qquad \text{(B.94)}
$$

where (3.91) and (3.92) have been utilized. Thus, for Case 2, we can conclude that (3.89) reduces to the expression given in (B.91). Geometrically, $\hat{M}_m(t)$ is updated such that it moves away from the boundary case of $\hat{M}_m(t) = \underline{M}_m$ and into the region given by $\hat{M}_m(t) > \underline{M}_m$.

Case 3: $\hat{M}_m = \underline{M}_m$ and $\Omega_m < 0$

When $\hat{M}_m(t) = \underline{M}_m$ (i.e., $\hat{M}_m(t)$ equals the lower bound) and Ω_m is negative, the third equation in (3.81) alongside (3.82) can be utilized in order to express $\dot{V}(t)$ as follows

$$\dot{V} = -K_s r^2 - \sum_{j=1}^{m} K_{ej} M_m \eta_j^2$$

$$+ \tilde{M}_m \left(\sum_{j=1}^{m} \left(-\hat{M}_m^{-1} \left(\Omega_{4j} + \Omega_{2j} W_m \hat{\theta}_m + u_j \right) \eta_j \right) + (\ddot{q}_d + \alpha \dot{e}) r \right).$$
$$(B.95)$$

By definition, the following relationships must hold for this case

$$\tilde{M}_m(t) = M_m - \underline{M}_m \geq 0. \tag{B.96}$$

Moreover, the parenthesized term in (B.95) can be written as

$$\sum_{j=1}^{m} \left(-\hat{M}_m^{-1} \left(\Omega_{4j} + \Omega_{2j} W_m \hat{\theta}_m + u_j \right) \eta_j \right) + (\ddot{q}_d + \alpha \dot{e}) r = \Gamma_1^{-1} \Omega_m < 0$$
$$(B.97)$$

where we have utilized (3.82) and the fact that $\Gamma_1 > 0$ and $\Omega_m < 0$. From (B.96) and (B.97), it is clear that the second line of (B.95) is negative, and hence, (B.95) can be lower bounded as follows

$$\dot{V} \leq -K_s r^2 - \sum_{j=1}^{m} K_{ej} M_m \eta_j^2 \leq -\lambda_3 \|x\|^2 \tag{B.98}$$

where (3.91) and (3.92) have been utilized. Thus, for Case 3, we can conclude that (3.89) reduces to the expression in given in (B.91). Geometrically, this case ensures that $\hat{M}_m(t)$ stays on the boundary defined by $\hat{M}_m(t) = \underline{M}_m$ as long as it is not directed to move into the prescribed region for M_m as defined in Remark 3.4 of Chapter 3. $\square$

B.2.3   Dynamic Terms for a 6-DOF AMB System

The dynamic equation for the center of mass of the circular cylinder rotor with respect to the fixed coordinate frame (x_b, y_b, z_b) can be written as follows [1], [3]

$$M^*(q^*)\ddot{q}^* + V_m^*(q^*, \dot{q}^*)\dot{q}^* + G^* = F^* \tag{B.99}$$

where $q^* = \begin{bmatrix} x_b^o, & y_b^o, & z_b^o, & \phi, & \theta, & \psi \end{bmatrix}^T \in \mathbb{R}^6$, (x_b^o, y_b^o, z_b^o) denotes the position of o_o with respect to o_b (see Figure 3.11), (ϕ, θ, ψ) denote the Euler angles,[1] and $F^* \in \mathbb{R}^6$ is the force/torque vector. The dynamic terms

---

[1] The orientation of the rotor is given by a series of three rotations. Assuming that the rotor is initially oriented so that the moving coordinate frame (x_o, y_o, z_o) is aligned

in (B.99) are defined as follows [1], [3]

$$M^*(q^*) = \text{diag}\left\{mI_3, \quad \Gamma^{-T}H\Gamma^{-1}\right\} \in \mathbb{R}^{6\times6}$$

$$G^* = \begin{bmatrix} 0, & 0, & mg, & 0, & 0, & 0 \end{bmatrix}^T \in \mathbb{R}^6$$

(B.100)

$$V_m^*(q^*, \dot{q}^*) = \text{diag}\left\{mR\left(-R^{-1}\dot{R}R^{-1} + \Gamma^{-1}S\left(\dot{\Omega}\right)R^{-1}\right),\right.$$

$$\left.\Gamma^{-T}\left(-H\Gamma^{-1}\dot{\Gamma}\Gamma^{-1} + \Gamma^{-1}S\left(\dot{\Omega}\right)H\Gamma^{-1}\right)\right\} \in \mathbb{R}^{6\times6}$$

(B.101)

where m is the mass of the rotor, g is the gravitational constant, I_3 represents the 3×3 identity matrix, $S\left(\dot{\Omega}\right) \in \mathbb{R}^{3\times3}$ is a skew-symmetric matrix with $\dot{\Omega} = \begin{bmatrix} \dot{\phi}, & \dot{\theta}, & \dot{\psi} \end{bmatrix}^T$, $H \in \mathbb{R}^{3\times3}$ denotes the diagonal inertia matrix of the rotor with respect to the (x_o, y_o, z_o) axes, and $\Gamma, R \in \mathbb{R}^{3\times3}$ are defined as follows

$$\Gamma = \begin{bmatrix} 1 & \sin(\phi)\tan(\theta) & \cos(\phi)\tan(\theta) \\ 0 & \cos(\phi) & -\sin(\phi) \\ 0 & \sin(\phi)\sec(\theta) & \cos(\phi)\sec(\theta) \end{bmatrix}$$

(B.102)

$$R = \begin{bmatrix} c(\psi)c(\theta) & c(\psi)s(\theta)s(\phi) - s(\psi)c(\phi) & c(\psi)s(\theta)c(\phi) + s(\psi)s(\phi) \\ s(\psi)c(\theta) & s(\psi)s(\theta)s(\phi) + c(\psi)c(\phi) & s(\psi)s(\theta)c(\phi) - c(\psi)s(\phi) \\ -s(\theta) & c(\theta)s(\phi) & c(\theta)c(\phi) \end{bmatrix}$$

(B.103)

with $c(\cdot), s(\cdot)$ denoting the cosine and sine of the argument, respectively. The kinematic relationship between the position vector $q(t)$ defined in Section 3.4.1 of Chapter 3 and q^* of (B.99) is defined as follows

$$q = \begin{bmatrix} x_b^o + (d - z_b^o)\left(c(\psi)\tan(\theta) + \dfrac{s(\psi)\tan(\phi)}{c(\theta)}\right) \\ y_b^o + (d - z_b^o)\left(s(\psi)\tan(\theta) + \dfrac{c(\psi)\tan(\phi)}{c(\theta)}\right) \\ x_b^o - z_b^o\left(c(\psi)\tan(\theta) + \dfrac{s(\psi)\tan(\phi)}{c(\theta)}\right) \\ y_b^o - z_b^o\left(s(\psi)\tan(\theta) + \dfrac{c(\psi)\tan(\phi)}{c(\theta)}\right) \\ z_b^o - \dfrac{L}{2}c(\theta)c(\phi) \\ \psi \end{bmatrix} = f(q^*) \quad \text{(B.104)}$$

with the fixed coordinate frame (x_b, y_b, z_b), the rotations are defined as follows [1]: (1) a rotation ψ about the z_o axis, (2) a rotation θ about the *current* y_o axis, and (3) a rotation ϕ about the *current* x_o axis.

where $d \in \mathbb{R}$ denotes the distance between the origins o_b and o_t measured along the z_b axis (see Figure 3.11), and $L \in \mathbb{R}$ is the length of the rotor. From (B.104), the dynamic quantities defined in (3.60) can be explicitly calculated as follows

$$
\begin{aligned}
M(q) &= J^{-T}(q)M^*(q)J^{-1}(q) \\[2mm]
G(q) &= J^{-T}(q)G^* \\[2mm]
V_m(q,\dot{q}) &= J^{-T}(q)\left[-M^*(q)J^{-1}(q)\dot{J}(q) + V_m^*(q,\dot{q})\right]J^{-1}(q) \\[2mm]
\bar{F}(q) &= J^{-T}(q)F^* \\[2mm]
J(q) &= \frac{\partial f(q)}{\partial q} \\[2mm]
q^*(q) &= f^{-1}(q)
\end{aligned}
\tag{B.105}
$$

where $f(\cdot)$ was defined in (B.104).

B.2.4   Partial Derivative Definitions

Definition B.2 *The partial derivative expressions required to complete (3.143) are defined as follows*

$$
\begin{aligned}
\frac{\partial f_d}{\partial \ddot{q}_d} &= M(q) \\[2mm]
\frac{\partial f_d}{\partial \dot{q}_d} &= M(q)\alpha + V_m(q,\dot{q}) + k_s\lambda_2(q)I_6 \\[2mm]
\frac{\partial f_d}{\partial q_d} &= V_m(q,\dot{q})\alpha + k_s\lambda_2(q)\alpha I_6 \\[2mm]
\frac{\partial f_d}{\partial q} &= \frac{\partial w}{\partial q} + k_s\frac{\partial \lambda_2}{\partial q}r - k_s\lambda_2(q)\alpha \\[2mm]
\frac{\partial f_d}{\partial \dot{q}} &= \frac{\partial w}{\partial \dot{q}} - k_s\lambda_2(q)I_6
\end{aligned}
\tag{B.106}
$$

where I_6 denotes the 6×6 identity matrix. In (B.106), the partial derivative terms $\dfrac{\partial w(\cdot)}{\partial q}$ and $\dfrac{\partial w(\cdot)}{\partial \dot{q}}$ are defined as follows

$$
\frac{\partial w}{\partial q} = A_1(q,\zeta_1) + A_2(q,\zeta_2) - V_m(q,\dot{q})\alpha + \frac{\partial G(q)}{\partial q}
\tag{B.107}
$$

$$\frac{\partial w}{\partial \dot{q}} = -M(q)\alpha + A_3(q, \zeta_3) \tag{B.108}$$

where the auxiliary terms $A_1(q, \zeta_1)$ $A_2(q, \zeta_2)$, *and* $A_3(q, \zeta_3) \in \mathbb{R}$ *are defined as follows*

$$A_1(q, \zeta_1) = \left[\begin{array}{ccc} \dfrac{\partial M(q)}{\partial q_1} \zeta_1, & ..., & \dfrac{\partial M(q)}{\partial q_6} \zeta_1, \end{array} \right], \qquad \zeta_1 = \ddot{q}_d + \alpha \dot{e} \tag{B.109}$$

$$A_2(q, \zeta_2) = \left[\begin{array}{ccc} \dfrac{\partial V_m(q, \dot{q})}{\partial q_1} \zeta_2, & ..., & \dfrac{\partial V_m(q, \dot{q})}{\partial q_6} \zeta_2, \end{array} \right], \qquad \zeta_2 = \dot{q}_d + \alpha e \tag{B.110}$$

$$A_3(q, \zeta_3) = \left[\begin{array}{ccc} \dfrac{\partial V_m(q, \dot{q})}{\partial \dot{q}_1} \zeta_3, & ..., & \dfrac{\partial V_m(q, \dot{q})}{\partial \dot{q}_6} \zeta_3, \end{array} \right], \qquad \zeta_3 = \dot{q}_d + \alpha e. \tag{B.111}$$

B.3   Chapter 4 Lemmas

B.3.1   Inequality Lemma

Lemma B.13 *Based on the definition of* $sat_\beta (\cdot)$ *given in (4.13), the following inequality can be obtained*

$$(\xi_{1i} - \xi_{2i})^2 \geq (sat_{\beta i} (\xi_{1i}) - sat_{\beta i} (\xi_{2i}))^2$$
$$\forall |\xi_{1i}| \leq \beta_i, \xi_{2i} \in \mathbb{R}, i = 1, 2, ..., m . \tag{B.112}$$

Proof: To prove the inequality given in (B.112), the proof is divided into three possible cases.

Case 1: $|\xi_{1i}| \leq \beta_i, |\xi_{2i}| \leq \beta_i$

For this case, the definition of $sat_{\beta_i} (\cdot)$ given in (4.13) can be used to prove that

$$sat_{\beta_i} (\xi_{1i}) = \xi_{1i} \qquad sat_{\beta_i} (\xi_{2i}) = \xi_{2i} . \tag{B.113}$$

After substituting (B.113) into (B.112), the following expression can be obtained

$$(\xi_{1i} - \xi_{2i})^2 = \left(sat_{\beta_i} (\xi_{1i}) - sat_{\beta_i} (\xi_{2i}) \right)^2 \quad \forall |\xi_{1i}| \leq \beta_i, |\xi_{2i}| \leq \beta_i, \tag{B.114}$$

and hence, the inequality given in (B.112) is true for Case 1.

Case 2: $|\xi_{1i}| \leq \beta_i, \xi_{2i} > \beta_i$

For this case, the definition of $sat_{\beta_i} (\cdot)$ given in (4.13) can be used to prove that

$$(\xi_{2i} + \beta_i) \geq 2\xi_{1i} \qquad \forall |\xi_{1i}| \leq \beta_i, \xi_{2i} > \beta_i . \tag{B.115}$$

After multiplying (B.115) by $(\xi_{2i} - \beta_i)$ and then simplifying the left-hand side of the inequality, (B.115) can be rewritten as

$$\xi_{2i}^2 - \beta_i^2 \geq 2\left(\xi_{2i} - \beta_i\right)\xi_{1i} \tag{B.116}$$

where the fact that $\xi_{2i} - \beta_i > 0$ has been used for this case. After adding ξ_{1i}^2 to (B.116) and then rearranging the resulting expression, the following inequality can be obtained

$$\xi_{1i}^2 - 2\xi_{1i}\xi_{2i} + \xi_{2i}^2 \geq \xi_{1i}^2 - 2\beta_i\xi_{1i} + \beta_i^2 . \tag{B.117}$$

Based on the expression given in (B.117), the following facts

$$\text{sat}_{\beta_i}\left(\xi_{1i}\right) = \xi_{1i} \qquad \text{sat}_{\beta_i}(\xi_{2i}) = \beta_i \tag{B.118}$$

can be used to prove that

$$\left(\xi_{1i} - \xi_{2i}\right)^2 \geq \left(\text{sat}_{\beta_i}\left(\xi_{1i}\right) - \text{sat}_{\beta_i}(\xi_{2i})\right)^2 \qquad \forall |\xi_{1i}| \leq \beta_i, \xi_{2i} > \beta_i . \tag{B.119}$$

Case 3: $|\xi_{1i}| \leq \beta_i, \xi_{2i} < -\beta_i$

For this case, the definition of $\text{sat}_{\beta_i}(\cdot)$ given in (4.13) can be used to prove that

$$\left(\xi_{2i} - \beta_i\right) \leq 2\xi_{1i} \qquad \forall |\xi_{1i}| \leq \beta_i, \xi_{2i} < -\beta_i . \tag{B.120}$$

After multiplying both sides of (B.120) by $(\xi_{2i} + \beta_i)$ and then simplifying the left-hand side of the inequality, the following inequality can be obtained

$$\xi_{2i}^2 - \beta_i^2 \geq 2\left(\xi_{2i} + \beta_i\right)\xi_{1i} \tag{B.121}$$

where the fact that $\xi_{2i} + \beta_i < 0$ has been utilized for this case. After adding ξ_{1i}^2 to (B.121) and then rearranging the resulting expression, the following inequality can be obtained

$$\xi_{1i}^2 - 2\xi_{1i}\xi_{2i} + \xi_{2i}^2 \geq \xi_{1i}^2 + 2\beta_i\xi_{1i} + \beta_i^2 . \tag{B.122}$$

Based on the inequality given in (B.122), the following facts

$$\text{sat}_{\beta_i}\left(\xi_{1i}\right) = \xi_{1i} \qquad \text{sat}_{\beta_i}(\xi_{2i}) = -\beta_i \tag{B.123}$$

can be utilized to prove that

$$\left(\xi_{1i} - \xi_{2i}\right)^2 \geq \left(\text{sat}_{\beta_i}\left(\xi_{1i}\right) - \text{sat}_{\beta_i}(\xi_{2i})\right)^2 \qquad \forall |\xi_{1i}| \leq \beta_i, \xi_{2i} < -\beta_i; \tag{B.124}$$

hence, (B.112) is true for all possible cases. $\square$

B.3.2   Stability Analysis for Projection Cases

Lemma B.14 *Given the closed-loop dynamics in (4.102), (4.113), and (4.114), if $\hat{\theta}_m(0) \in int(\Lambda)$, then $\hat{\theta}_m(t)$ never leaves the region Λ described in Property 4.6, $\forall t \geq 0$, and an upper bound for the expression given in (4.117) can be formulated as follows*

$$\dot{V} \leq -kr^T r - ke_\lambda^T e_\lambda. \tag{B.125}$$

Proof: To prove Lemma B.14, the closed-loop dynamics given in (4.102) and (4.113) can be substituted into (4.117) for $\dot{\hat{\theta}}_1(t)$ and $\dot{\hat{\theta}}_2(t)$, respectively, to obtain the following expression

$$\dot{V} \leq -k\left(r^T r + e_\lambda^T e_\lambda\right) - e_\lambda^T \frac{h^T \tilde{\theta}_m}{h^T \hat{\theta}_m} Y_2 \hat{\theta}_2 + \tilde{\theta}_m^T \dot{\hat{\theta}}_m \,. \tag{B.126}$$

To substitute (4.114) and (4.112) into (B.126) for $\dot{\hat{\theta}}_m(t)$, the following three cases must be considered.

Case 1: $\hat{\theta}_m(t) \in int(\Lambda)$

When the parameter estimate $\hat{\theta}_m(t)$ lies in the interior of the convex region Λ, described in Property 4.6, (B.126) can be expressed as follows

$$\dot{V} \leq \ -k\left(r^T r + e_\lambda^T e_\lambda\right) - e_\lambda^T \frac{h^T \tilde{\theta}_m}{h^T \hat{\theta}_m} Y_2 \hat{\theta}_2 + \tilde{\theta}_m^T \left(\frac{h}{h^T \hat{\theta}_m} \left[Y_2 \hat{\theta}_2\right]^T e_\lambda\right). \tag{B.127}$$

Thus, for Case 1, (4.117) reduces to the expression in given in (B.125). In addition, the direction in which the estimate $\hat{\theta}_m(t)$ is updated for Case 1 is irrelevant, since the worst case scenario is that $\hat{\theta}_m(t)$ will move towards the boundary of the convex region denoted by $\partial(\Lambda)$.

Case 2: $\hat{\theta}_m(t) \in \partial(\Lambda)$ and $\mu_1^T(t)\hat{\theta}_m^\perp(t) \leq 0$

When the parameter estimate $\hat{\theta}_m(t)$ lies on the boundary of the convex region Λ described in Property 4.6 and $\mu_1^T(t)\hat{\theta}_m^\perp(t) \leq 0$, then (4.117) can be expressed as (B.127). Thus, for Case 2, we can conclude that (4.117) reduces to the expression in given in (B.125). In addition, the vector $\mu_1(t)$ has a zero or nonzero component perpendicular to the boundary $\partial(\Lambda)$ at $\hat{\theta}_m(t)$ that points in the direction towards the int(Λ). Geometrically, this means that $\hat{\theta}_m(t)$ is updated such that it either moves towards the int(Λ) or remains on the boundary. Hence, $\hat{\theta}_m(t)$ never leaves Λ.

Case 3: $\hat{\theta}_m(t) \in \partial(\Lambda)$ and $\mu_1^T(t)\hat{\theta}_m^\perp(t) > 0$

When the parameter estimate $\hat{\theta}_m(t)$ lies on the boundary of the convex region Λ described in Property 4.6 and $\mu_1^T(t)\hat{\theta}_m^\perp(t) > 0$, then (4.117) can be expressed as

$$\dot{V} \leq -k\left(r^T r + e_\lambda^T e_\lambda\right) - \tilde{\theta}_m^T \left(-\mu_1 + P_r^t(\mu_1)\right) \tag{B.128}$$

where (4.112) was utilized. Based on (B.128), Property 4.6 can be used to conclude that

$$\dot{V}_1 \leq -k\left(r^T r + e_\lambda^T e_\lambda\right)$$

$$-\tilde{\theta}_m^T\left(-\left(P_r^\perp(\mu_1) + P_r^t(\mu_1)\right) + P_r^t(\mu_1)\right) \tag{B.129}$$

$$\leq -k\left(r^T r + e_\lambda^T e_\lambda\right) + \tilde{\theta}_m^T P_r^\perp(\mu_1).$$

Because $\hat{\theta}_m(t) \in \partial(\Lambda)$, and θ_m must lie either on the boundary or in the interior of Λ, then the convexity of Λ implies that $\tilde{\theta}_m(t)$ will either point tangent to $\partial(\Lambda)$ or towards int(Λ) at $\hat{\theta}_m(t)$. That is, $\hat{\theta}_m(t)$ will have a component in the direction of $\hat{\theta}_m^\perp(t)$ that is either zero or negative. In addition, since $P_r^\perp(\mu_1)$ points away from int(Λ), the following inequality can be determined

$$\tilde{\theta}_m^T(t) P_r^\perp(\mu_1) \leq 0. \tag{B.130}$$

The inequality given in (B.130) can now be used to simplify the expression given in (B.129) to the expression given in (B.125). Furthermore, since $\dot{\hat{\theta}}_m(t) = P_r^t(\mu_1)$, the parameter estimate $\hat{\theta}_m(t)$ is ensured to be updated such that it moves tangent to $\partial(\Lambda)$. Hence, $\hat{\theta}_m(t)$ never leaves Λ. □

B.3.3   Boundedness Lemma

Lemma B.15 *The second time derivative of the force tracking error $e_\lambda(t)$ defined in (4.133) is bounded (i.e., $\ddot{e}_\lambda \in \mathcal{L}_\infty$).*

Proof: To prove Lemma B.15, the open-loop dynamics for the force tracking error $e_\lambda(t)$ of (4.133) are rewritten as follows

$$\dot{e}_\lambda = \lambda_d - \Pi^{-T}\bar{\tau}_2 + \Pi^{-T}\left(\bar{V}_{m21}\dot{u}_1 + \bar{N}_2\right)$$

$$+\Pi^{-T}\bar{M}_{21}\bar{M}_{11}^{-1}\left[\bar{\tau}_1 - \bar{V}_{m11}\dot{u}_1 - \bar{N}_1\right]. \tag{B.131}$$

:

After taking the time derivative of (B.131), the following expression for $\ddot{e}_\lambda(t)$ can be obtained

$$\ddot{e}_\lambda = \dot{\lambda}_d - \frac{d}{dt}\left(\Pi^{-T}\right)\bar{\tau}_2 - \Pi^{-T}\dot{\bar{\tau}}_2 + \frac{d}{dt}\left(\Pi^{-T}\right)\left(\bar{V}_{m21}\dot{u}_1 + \bar{N}_2\right)$$

$$+\Pi^{-T}\left(\dot{\bar{V}}_{m21}\,\dot{u}_1 + \bar{V}_{m21}\ddot{u}_1 + \dot{\bar{N}}_2\right)$$

$$+\frac{d}{dt}\left(\Pi^{-T}\right)\bar{M}_{21}\bar{M}_{11}^{-1}\left(\bar{\tau}_1 - \bar{V}_{m11}\dot{u}_1 - \bar{N}_1\right)$$

$$+\Pi^{-T}\dot{\bar{M}}_{21}\,\bar{M}_{11}^{-1}\left(\bar{\tau}_1 - \bar{V}_{m11}\dot{u}_1 - \bar{N}_1\right)$$

$$+\Pi^{-T}\bar{M}_{21}\frac{d}{dt}\left(\bar{M}_{11}^{-1}\right)\left(\bar{\tau}_1 - \bar{V}_{m11}\dot{u}_1 - \bar{N}_1\right)$$

$$+\Pi^{-T}\bar{M}_{21}\bar{M}_{11}^{-1}\left(\dot{\bar{\tau}}_1 - \dot{\bar{V}}_{m11}\,\dot{u}_1 - \bar{V}_{m11}\ddot{u}_1 - \dot{\bar{N}}_1\right).$$

$$\text{(B.132)}$$

Since $u_1(t)$, $\dot{u}_1(t)$, and $\ddot{u}_1(t) \in \mathcal{L}_\infty$ from the proof of Theorem 4.3, standard signal chasing arguments can be used to prove that all of the signals on the right-hand side of (B.132) are bounded, however, special care must be directed towards proving that $\dot{\bar{\tau}}_1(t)$, $\dot{\bar{\tau}}_2(t) \in \mathcal{L}_\infty$. To this end, the time derivative of $\bar{\tau}_1(t)$ given in (4.107) can be obtained as follows

$$\dot{\bar{\tau}}_1 = \dot{Y}_1\hat{\theta}_1 + Y_1\dot{\hat{\theta}}_1 - k\dot{r}. \qquad \text{(B.133)}$$

Since the desired position trajectory is assumed to be bounded up to its third time derivative and $e(t)$, $\dot{e}(t)$, $\ddot{e}(t) \in \mathcal{L}_\infty$, the regression matrices $Y_1(\cdot)$, $\dot{Y}_1(\cdot) \in \mathcal{L}_\infty$. Since $\dot{\hat{\theta}}_1(t)$, $\dot{r}(t) \in \mathcal{L}_\infty$, (B.133) can be used to prove that $\dot{\bar{\tau}}_1(t) \in \mathcal{L}_\infty$. After taking the time derivative of $\bar{\tau}_2(t)$ given in (4.107), the following expression can be obtained

$$\dot{\bar{\tau}}_2 = \frac{d}{dt}\left(\frac{1}{h^T\hat{\theta}_m}\right)\Pi^T\left(Y_2\hat{\theta}_2\right) + \frac{1}{h^T\hat{\theta}_m}\dot{\Pi}^T\left(Y_2\hat{\theta}_2\right)$$

$$\text{(B.134)}$$

$$+\frac{1}{h^T\hat{\theta}_m}\Pi^T\frac{d}{dt}\left(Y_2\hat{\theta}_2\right) + \dot{\Pi}^T\frac{k}{m_d}e_\lambda + \Pi^T\frac{k}{m_d}\dot{e}_\lambda.$$

From the proof of Theorem 4.3, $e_\lambda(t)$, $\dot{e}_\lambda(t)$, $\Pi^T(x)$, $\dot{\Pi}^T(x)$, $Y_2(\cdot)\hat{\theta}_2(t)$, and $\frac{d}{dt}\left(Y_2\hat{\theta}_2\right)$ are all bounded. Since the projection-based parameter update law for $\hat{\theta}_m(t)$ given in (4.111) ensures that $h^T(u)\hat{\theta}_m(t) > 0$ (see the development in Lemma B.14), $\frac{d}{dt}\left[\left(h^T\hat{\theta}_m\right)^{-1}\right]$ also remains bounded; therefore,

(B.134) can be used to prove that $\dot{\tilde{\tau}}_2(t) \in \mathcal{L}_\infty$. The fact that all of the terms on the right-hand side of (B.132) are bounded can now be used to prove that $\ddot{e}_\lambda(t) \in \mathcal{L}_\infty$. $\square$

B.3.4   State-Dependent Disturbance Bound

Lemma B.16 *Given a function $F(x) \in \mathbb{R}^{m \times n}$ and a variable $x(t) \in \mathbb{R}^k$ such that*

$$\frac{\partial F_{ij}(x)}{\partial x} \in \mathcal{L}_\infty \qquad if \qquad x(t) \in \mathcal{L}_\infty, \quad \forall i = 1, ..., m, \ j = 1, ..., n \quad (B.135)$$

where $F_{ij}(x)$ represents the ij^{th} element of $F(x)$, then

$$\left\| \tilde{F} \right\| = \| F(x_d) - F(x) \| \leq \rho_f \left(\| x_d \|, \| \tilde{x} \| \right) \| \tilde{x} \| \qquad (B.136)$$

where $\tilde{x}(t) \in \mathbb{R}^k$ represents the mismatch between $x_d(t) \in \mathbb{R}^k$ and $x(t)$ as follows

$$\tilde{x} = x_d - x, \qquad (B.137)$$

and $\rho_f(\cdot) \in \mathbb{R}$ is a positive nondecreasing bounding function.

Proof: To prove Lemma B.16, the ij^{th} element of $\tilde{F}(x)$ is first defined as follows

$$\tilde{F}_{ij} = F_{ij}(x_{d1}, x_{d2}, ..., x_{dk}) - F_{ij}(x_1, x_2, ..., x_k). \qquad (B.138)$$

After adding and subtracting the terms $F_{ij}(x_1, x_{d2}, ..., x_{dk})$, $F_{ij}(x_1, x_2, x_{d3}, ..., x_{dk})$, ..., $F_{ij}(x_1, x_2, ..., x_{k-1}, x_{dk})$ to the right-hand side of (B.138), the following expression can be obtained

$$\tilde{F}_{ij} = \ [F_{ij}(x_{d1}, x_{d2}, ..., x_{dk}) - F_{ij}(x_1, x_{d2}, ..., x_{dk})]$$

$$+ [F_{ij}(x_1, x_{d2}, ..., x_{dk}) - F_{ij}(x_1, x_2, x_{d3}, ..., x_{dk})] + \ ...$$

$$+ [F_{ij}(x_1, x_2, ..., x_{k-1}, x_{dk}) - F_{ij}(x_1, x_2, ..., x_{k-1}, x_k)]. \tag{B.139}$$

After applying the Mean Value Theorem (given in Lemma A.1 of Appendix A) to each bracketed term of (B.139), the following expression can be obtained

$$\tilde{F}_{ij} = \ \frac{\partial F_{ij}(\sigma_1, x_{d2}, ..., x_{dk})}{\partial \sigma_1} \Big|_{\sigma_1 = v_1} (x_{d1} - x_1)$$

$$+ \frac{\partial F_{ij}(x_1, \sigma_2, x_{d3}, ..., x_{dk})}{\partial \sigma_2} \Big|_{\sigma_2 = v2} (x_{d2} - x_2) \qquad (B.140)$$

$$+ \ ... \ + \frac{\partial F_{ij}(x_1, x_2, ..., x_{k-1}, \sigma_k)}{\partial \sigma_k} \Big|_{\sigma_k = v_k} (x_{dk} - x_k)$$

where $v_i(t)$ has a value in between $x_i(t)$ and $x_{di}(t)$ for $i = 1, 2, ..., k$. To simplify the notation, (B.140) is rewritten in the following matrix form

$$\tilde{F}_{ij} = \frac{\partial F_{ij}(\sigma, x, x_d)}{\partial \sigma}|_{\sigma=v} (x_d - x) \qquad (B.141)$$

where

$$x = [x_1, ..., x_k]^T, \qquad v = [v_1, ..., v_k]^T,$$
$$\sigma = [\sigma_1, ..., \sigma_k]^T, \qquad \frac{\partial F_{ij}(\cdot)}{\partial \sigma} = \left[\frac{\partial F_{ij}}{\partial \sigma_1}, ..., \frac{\partial F_{ij}}{\partial \sigma_k}\right].$$

From (B.141), the following upper bound for $\left|\tilde{F}_{ij}(x)\right|$ can be obtained

$$\left|\tilde{F}_{ij}\right| \leq \left\|\frac{\partial F_{ij}(\sigma, x, x_d)}{\partial \sigma}|_{\sigma=v}\right\| \|x_d - x\|. \qquad (B.142)$$

By noting that

$$v_i = x_{di} - c_i \tilde{x}_i \qquad \forall c_i \in (0, 1), i = 1, 2, ...k, \qquad (B.143)$$

an upper bound for $\left|\tilde{F}_{ij}(x)\right|$ can be developed from (B.135), (B.137), and (B.142) as follows

$$\left|\tilde{F}_{ij}\right| \leq \rho_{ij}(\|x_d\|, \|\tilde{x}\|)\|\tilde{x}\| \quad \text{for} \quad i = 1, ..., m, \ j = 1, ..., n \qquad (B.144)$$

where $\rho_{ij}(\cdot)$ is some positive nondecreasing scalar function. The result given in (B.136) can now be obtained. $\square$

Lemma B.17 *Given the expressions in (4.130) and (4.134), Lemma B.16 can be utilized to construct the following inequality*

$$\rho(\zeta_{dp}, \zeta_{dv}, \zeta_{da}, \|y\|)\|y\| \geq \max\left\{\|\chi\|, \left\|\tilde{Y}_v\right\|\right\} \qquad (B.145)$$

where ζ_{dp}, ζ_{dv}, and ζ_{da} were defined in (4.86), $y(t)$ is given in (4.142), and $\rho(\cdot) \in \mathbb{R}$ is a positive nondecreasing function.

Proof: To prove Lemma B.17, the result in (B.145) is first developed for $\tilde{Y}_v(\cdot)$. Based on the proof for the result in (B.145) for $\tilde{Y}_v(\cdot)$, the proof for $\chi(\cdot)$ can then be obtained. To this end, the expression given in (4.135) can be substituted into (B.146) for $\tilde{Y}_v (\cdot) \theta_v$ to obtain the following expression

$$\tilde{Y}_v = -\frac{1}{2} \frac{\partial \left(\det \left(\bar{M}_{11} \right) \right)}{\partial u_1} \dot{e} e_\lambda + \Pi^{-T} \bar{M}_{21} adj(\bar{M}_{11}) \bar{V}_{m11}(u_1, \dot{u}_1) \dot{e}$$

$$+\Pi^{-T} \bar{M}_{21} adj(\bar{M}_{11}) \left[\bar{V}_{m11}(u_1, \dot{u}_{d1}) - \bar{V}_{m11}(u_1, \dot{u}_1) \right] \dot{u}_{d1}$$

$$+\Pi^{-T} \bar{M}_{21} adj(\bar{M}_{11}) \left[\bar{N}_1(u_1, \dot{u}_{d1}) - \bar{N}_1(u_1, \dot{u}_1) \right]$$

$$- \det(\bar{M}_{11})\Pi^{-T} \bar{V}_{m21}(u_1, \dot{u}_1) \dot{e}$$

$$- \det(\bar{M}_{11})\Pi^{-T} \left[\bar{V}_{m21}(u_1, \dot{u}_{d1}) - \bar{V}_{m21}(u_1, \dot{u}_1) \right] \dot{u}_{d1}$$

$$- \det(\bar{M}_{11})\Pi^{-T} \left[\bar{N}_2(u_1, \dot{u}_{d1}) - \bar{N}_2(u_1, \dot{u}_1) \right]$$

$$\tag{B.146}$$

where common terms have been cancelled, and several terms have been added and subtracted from the resulting expression. After invoking Lemma B.16, the following upper bound can be formulated

$$\left\| \tilde{Y}_v \right\| \leq \frac{1}{2} \left\| \frac{\partial \left(\det \left(\bar{M}_{11} \right) \right)}{\partial u_1} \right\| \|\dot{e}\| \|e_\lambda\|$$

$$+ \left\| \Pi^{-T} \bar{M}_{21} adj(\bar{M}_{11}) \right\| \left\| \bar{V}_{m11}(u_1, \dot{u}_1) \right\| \|\dot{e}\|$$

$$+ \left\| \Pi^{-T} \bar{M}_{21} adj(\bar{M}_{11}) \right\|$$

$$\cdot \left(\rho_{v1}(\|u_1\|, \zeta_{dv}, \|\dot{e}\|) \|\dot{e}\| \zeta_{dv} + \rho_{n1}(\|u_1\|, \zeta_{dv}, \|\dot{e}\|) \|\dot{e}\| \right)$$

$$+ \left\| \det(\bar{M}_{11})\Pi^{-T} \right\|$$

$$\cdot \left(\left\| \bar{V}_{m21}(u_1, \dot{u}_1) \right\| \|\dot{e}\| + \rho_{v2}(\|u_1\|, \zeta_{dv}, \|\dot{e}\|) \|\dot{e}\| \zeta_{dv} \right)$$

$$+ \left\| \det(\bar{M}_{11})\Pi^{-T} \right\| \rho_{n2}(\|u_1\|, \zeta_{dv}, \|\dot{e}\|) \|\dot{e}\|$$

$$\tag{B.147}$$

where (4.86) has been utilized and $\rho_{v1}(\cdot), \rho_{v2}(\cdot), \rho_{n1}(\cdot), \rho_{n2}(\cdot) \in \mathbb{R}$ are some positive nondecreasing bounding functions. The remaining matrix norms of (B.147) can also be bounded by some positive function, hence

$$\left\| \tilde{Y}_v \right\| \leq \rho_1(\|u_1\|, \|\dot{e}\|) \|e_\lambda\| + \rho_2(\|u_1\|, \|\dot{u}_1\|, \zeta_{dv}, \|\dot{e}\|) \|\dot{e}\| \tag{B.148}$$

where $\rho_1(\cdot), \rho_2(\cdot) \in \mathbb{R}$ are some positive nondecreasing bounding functions. To express (B.148) in terms of $y(t)$ and the constant motion trajectory bounds of (4.86), the expressions given in (4.92), (4.86), and (4.128) can be used along with the following inequality

$$\|y\| \geq \|e\|, \|e_f\|, \|\eta\|, \|e_\lambda\|$$

to formulate an upper bound for $\left\|\tilde{Y}_v(\cdot)\right\|$ as follows

$$\left\|\tilde{Y}_v\right\| \leq \rho_1(\zeta_{dp}, \|y\|) \|y\| + \rho_2(\zeta_{dp}, \zeta_{dv}, \|y\|) \|y\| \leq \rho_w(\zeta_{dp}, \zeta_{dv}, \|y\|) \|y\| \tag{B.149}$$

where $\rho_w(\cdot) \in \mathbb{R}$ is some positive nondecreasing bounding function. Using a similar procedure, the following upper bound can be formulated for $\|\chi(\cdot)\|$

$$\|\chi\| \leq \rho_\chi(\zeta_{dp}, \zeta_{dv}, \zeta_{da}, \|y\|) \|y\| \tag{B.150}$$

where $\rho_\chi(\cdot) \in \mathbb{R}$ is some positive nondecreasing bounding function. The result given in (B.145) can now be directly obtained from (B.149) and (B.150). □

B.3.5   Matrix Property

Lemma B.18 *Providing the following condition is satisfied*

$$\cos(\theta) > \left| \frac{\beta_1 - \beta_2}{\beta_1 + \beta_2} \right| \tag{B.151}$$

the matrix $BR + (BR)^T$ will be positive-definite, where B and R are defined in (4.157) and (4.158), respectively.

Proof: To prove Lemma B.18, the definitions for B and R given in (4.157) and (4.158), respectively, can be used to obtain the following symmetric matrix

$$\frac{1}{2}(BR + (BR)^T) = \frac{\lambda}{z} \left[\begin{array}{cc} \beta_1 \cos\theta & -\frac{1}{2}\beta_1 \sin\theta + \frac{1}{2}\beta_2 \sin\theta \\ -\frac{1}{2}\beta_1 \sin\theta + \frac{1}{2}\beta_2 \sin\theta & \beta_2 \cos\theta \end{array} \right]. \tag{B.152}$$

Using standard linear algebra techniques, the eigenvalues of (B.152) can be determined as follows

$$\lambda_1 = \frac{\lambda}{2z} \left((\beta_1 + \beta_2) \cos\theta_0 + (\beta_1 - \beta_2) \right),$$

$$\lambda_2 = \frac{\lambda}{2z} \left((\beta_1 + \beta_2) \cos\theta_0 - (\beta_1 - \beta_2) \right) \tag{B.153}$$

where λ_i denotes the ith eigenvalue. For the symmetric matrix of (B.152) to be positive-definite, the eigenvalues given in (B.153) must be positive.

From (B.153) and the fact that λ, z, β_1, $\beta_2 > 0$, the following eigenvalues can be proven to be positive provided

$$\lambda_1 > 0 \quad \Rightarrow \quad \cos(\theta_0) > \frac{\beta_2 - \beta_1}{\beta_1 + \beta_2}$$

$$\lambda_2 > 0 \quad \Rightarrow \quad \cos(\theta_0) > \frac{\beta_1 - \beta_2}{\beta_1 + \beta_2}.$$

(B.154)

Hence, if the condition given in (B.151) is satisfied, the matrix $BR + (BR)^T$ will be positive-definite.

B.4 Chapter 5 Lemmas

B.4.1 Skew-Symmetry Property

Lemma B.19 *The transformed inertia and centripetal-Coriolis matrices introduced in the dynamic model given in (5.26) satisfy the following skew-symmetric relationship*

$$\xi^T \left(\frac{1}{2} \dot{J}^* - C^* \right) \xi = 0 \quad \forall \xi \in \mathbb{R}^3.$$

(B.155)

Proof: In order to prove (B.155), we first take the time derivative of (5.25) and then utilize (5.28) to obtain the following expression

$$\xi^T \left(\dot{J}^* - 2C^* \right) \xi \;\; = \xi^T \left(\dot{T}^{-T} J T^{-1} + T^{-T} J \dot{T}^{-1} \right) \xi$$

$$+ 2\xi^T \left(J^* \dot{T} T^{-1} + 2T^{-T} \left(J T^{-1} \dot{e}_v \right)^\times T^{-1} \right) \xi \quad \text{(B.156)}$$

$$= 2\xi^T \left(T^{-T} J \dot{T}^{-1} + T^{-T} J T^{-1} \dot{T} T^{-1} \right) \xi$$

where the fact that

$$\xi^T T^{-T} \left(J T^{-1} \dot{e}_v \right)^\times T^{-1} \xi = 0$$

(B.157)

has been utilized. Based on the expression given in (B.156), the following property can be used to prove the result given in (B.155)

$$\dot{T} = -T\dot{T}^{-1}T.$$

(B.158)

$\square$

B.4.2 Control Signal Bound

Lemma B.20 *Given the definitions in (5.23), (5.66), and (5.67), the following inequality can be developed*

$$\overline{\chi} \leq \rho \left(\|z\| \right) \|z\| \tag{B.159}$$

where $z(t)$ is defined in (5.69).

Proof: After substituting (5.64) and (5.66) into (5.67), the following expression can be obtained

$$\overline{\chi} = T^T C^* \left(e_{vf} + e_v \right) + T^T J^* \left(\eta - e_v + \frac{e_v}{\left(1 - e_v^T e_v \right)^2} \right)$$

$$-2T^T J^* e_{vf} + T^T J \dot{\omega}_d + \frac{1}{2} T^T \omega_d^\times J \omega_d - T^T N^*. \tag{B.160}$$

By taking the norm of (B.160), the following inequality can be developed

$$\|\overline{\chi}\| \leq \ \left\| T^T C^* \right\| \left(\|e_{vf}\| + \|e_v\| \right)$$

$$+ \left\| T^T J^* \right\| \left(\|\eta\| + \|e_v\| + \left\| \frac{e_v}{\left(1 - e_v^T e_v \right)^2} \right\| \right) \tag{B.161}$$

$$+2 \left\| T^T J^* \right\| \|e_{vf}\| + \|\psi\|$$

where $\psi \left(t \right) \in \mathbb{R}^3$ is defined as follows

$$\psi = T^T J \dot{\omega}_d + \frac{1}{2} T^T \omega_d^\times J \omega_d - T^T N^*. \tag{B.162}$$

After substituting (5.29) into (B.162) for $N^* \left(\cdot \right)$ and utilizing (5.60) and the fact that $a^\times b = -b^\times a \ \forall a, b \in \mathbb{R}^3$, the following expression can be obtained

$$\psi = \ \left[\left(J \tilde{R} \omega_d \right)^\times - \left(\tilde{R} \omega_d \right)^\times J - J \left(\tilde{R} \omega_d \right)^\times \right] T^{-1} \left(\eta - e_v - e_{vf} \right)$$

$$+ \left(T^T J - \tfrac{1}{2} J \tilde{R} \right) \dot{\omega}_d + \frac{1}{2} \left(T^T \omega_d^\times J \omega_d - \left(\tilde{R} \omega_d \right)^\times J \tilde{R} \omega_d \right). \tag{B.163}$$

By utilizing (5.12), (5.34), and (5.23) the following inequality can be developed

$$\|\psi\| \leq \rho_1 \left(e_v \right) \left(\|\eta\| + \|e_v\| + \|e_{vf}\| \right) + \rho_2 \left(e_v \right) \left(\left\| \frac{e_v}{\sqrt{1 - e_v^T e_v}} \right\| \right) \tag{B.164}$$

where $\rho_1(\cdot), \rho_2(\cdot)$ are some positive nondecreasing functions. Since (5.34) ensures that $\sqrt{1 - e_v^T e_v}$ is always positive and real, multiplying and dividing the last term in (B.163) by $\sqrt{1 - e_v^T e_v}$ facilitates the structure of (B.164). Based upon the definition of (5.23), (5.25), (5.28), (5.34), (5.69), and (B.164), the expression in (B.161) can now be simplified to yield the result found in (5.68). $\square$

B.5   Chapter 6 Definitions and Lemmas

B.5.1   Definitions for Dynamic Terms

Definition B.3 *The components of the dynamic model given in (6.1)* $M(q)$ $\in \mathbb{R}^{4\times 4}$, $V_m(q, \dot{q}) \in \mathbb{R}^{4\times 4}$, *and* $G(q) \in \mathbb{R}^4$ *are defined as follows*

$$
M = \begin{bmatrix} m_p + m_r + m_c & 0 \\ 0 & m_p + m_c \\ m_p L \cos\theta \sin\phi & m_p L \cos\theta \cos\phi \\ m_p L \sin\theta \cos\phi & -m_p L \sin\theta \sin\phi \end{bmatrix}
$$

$$
\begin{matrix} m_p L \cos\theta \sin\phi & m_p L \sin\theta \cos\phi \\ m_p L \cos\theta \cos\phi & -m_p L \sin\theta \sin\phi \\ m_p L^2 + I & 0 \\ 0 & m_p L^2 \sin^2\theta + I \end{matrix}
$$

(B.165)

$$
V_m = \begin{bmatrix} 0 & 0 & -m_p L \left(\sin\theta \sin\phi \dot{\theta} - \cos\theta \cos\phi \dot{\phi} \right) \\ 0 & 0 & -m_p L \left(\sin\theta \cos\phi \dot{\theta} + \cos\theta \sin\phi \dot{\phi} \right) \\ 0 & 0 & 0 \\ 0 & 0 & m_p L^2 \sin\theta \cos\theta \dot{\phi} \end{bmatrix}
$$

(B.166)

$$
\begin{matrix} m_p L \left(\cos\theta \cos\phi \dot{\theta} - \sin\theta \sin\phi \dot{\phi} \right) \\ -m_p L \left(\cos\theta \sin\phi \dot{\theta} + \sin\theta \cos\phi \dot{\phi} \right) \\ -m_p L^2 \sin\theta \cos\theta \dot{\phi} \\ m_p L^2 \sin\theta \cos\theta \dot{\theta} \end{matrix}
$$

and

$$
G = \begin{bmatrix} 0 & 0 & m_p g L \sin\theta & 0 \end{bmatrix}^T.
$$

(B.167)

B.5.2   Linear Control Law Analysis

Lemma B.21 *Given the linear controller of (6.25), the time derivative of the following function*

$$V = k_E E + \frac{1}{2} k_p e^T e \tag{B.168}$$

is given by the following expression

$$\dot{V} = 0 \tag{B.169}$$

only when

$$E(q, \dot{q}) = e(t) = 0 \tag{B.170}$$

where $E(q, \dot{q})$ and $e(t)$ are defined in (6.21) and (6.24), respectively.

Proof: To prove Lemma B.21, we define the set of all points where (B.169) is satisfied as Γ. In the set Γ, it is clear from (6.29) and (B.169) that

$$\dot{r}(t) = 0 \qquad \ddot{r}(t) = 0, \tag{B.171}$$

and hence, we can conclude from (6.16), (B.168), (B.169), and (B.171) that $x(t)$, $y(t)$, and $V_1(t)$ are constant, and that

$$\ddot{x}(t) = 0 \qquad \ddot{y}(t) = 0. \tag{B.172}$$

Furthermore, (6.22), (6.24), and (B.171) can be used to prove that

$$\dot{E}(q, \dot{q}) = \dot{e}(t) = 0. \tag{B.173}$$

Based on (B.173), $E(q, \dot{q})$ and $e(t)$ are constant, and hence, from (6.25) and (B.171), we can prove that $F(t)$ is constant. To complete the proof, the stability of the system must be analyzed for the cases when $\dot{\theta} = 0$ and when $\dot{\theta} \neq 0$.

Case 1a: $\dot{\theta}(t) = 0$ and $\dot{\phi}(t) = 0$

Based on the proposition that $\dot{\theta}(t) = 0$ and $\dot{\phi}(t) = 0$, it is straightforward to prove that

$$\ddot{\theta}(t) = 0 \qquad \ddot{\phi}(t) = 0. \tag{B.174}$$

By rearranging the first two rows of the expression given in (6.1), the following expressions can be obtained

$$\frac{F_x}{m_p L} = \frac{m_p + m_r + m_c}{m_p L} \ddot{x} + \ddot{\theta} \cos\theta \sin\phi + \ddot{\phi} \sin\theta \cos\phi \tag{B.175}$$
$$- \left(\dot{\theta}^2 + \dot{\phi}^2 \right) \sin\theta \sin\phi + 2\dot{\theta}\dot{\phi} \cos\theta \cos\phi$$

$$\frac{F_y}{m_p L} = \frac{m_p + m_c}{m_p L}\ddot{y} + \ddot{\theta}\cos\theta\cos\phi - \ddot{\phi}\sin\theta\sin\phi \qquad (\text{B.176})$$
$$- \left(\dot{\theta}^2 + \dot{\phi}^2\right)\sin\theta\cos\phi - 2\dot{\theta}\dot{\phi}\cos\theta\sin\phi.$$

Based on the expression given in (B.172), (B.174–B.176), and the proposition that $\dot{\theta}(t) = 0$ and $\dot{\phi}(t) = 0$, the following expression is obtained

$$F_x(t) = F_y(t) = 0. \qquad (\text{B.177})$$

From (6.19), (6.25), (B.171), and (B.177), it is clear that

$$e(t) = 0. \qquad (\text{B.178})$$

Furthermore, by rearranging the third row of the vector given in (6.1), the following expression can be obtained

$$\ddot{\theta} = \gamma_3 \dot{\phi}^2 \sin\theta\cos\theta - \gamma_2 \sin\theta - \frac{m_p L}{m_p L^2 + I}(\ddot{x}\cos\theta\sin\phi + \ddot{y}\cos\theta\cos\phi)$$
$$(\text{B.179})$$

where (6.2–6.3) were utilized, and $\gamma_2, \gamma_3 \in \mathbb{R}$ are positive constants defined as follows

$$\gamma_2 = \frac{m_p g L}{m_p L^2 + I} \qquad \gamma_3 = \frac{m_p L^2}{m_p L^2 + I}. \qquad (\text{B.180})$$

Based on (B.172), (B.174), and the proposition that $\dot{\phi}(t) = 0$, (B.179) can be used to prove that

$$\sin\theta = 0, \qquad (\text{B.181})$$

and hence, from (6.6), it is clear that

$$\theta(t) = 0. \qquad (\text{B.182})$$

Given (B.178) and (B.182), the expressions given in (6.16), (6.21), (6.23), and (6.24) can be used to prove Lemma (B.21) under the proposition that $\dot{\theta}(t) = 0$ and $\dot{\phi}(t) = 0$. $\square$

Case 1b: $\dot{\theta}(t) = 0$ and $\dot{\phi}(t) \neq 0$

By rearranging the fourth row of the vector given in (6.1), the following expression can be obtained

$$\gamma_1(\theta)\ddot{\phi} = -2\dot{\theta}\dot{\phi}\sin\theta\cos\theta - \left(\frac{\ddot{x}\sin\theta\cos\phi - \ddot{y}\sin\theta\sin\phi}{L}\right) \qquad (\text{B.183})$$

where $\gamma_1(\theta) \in \mathbb{R}$ is defined as follows

$$\gamma_1(\theta) = \left(\sin^2\theta + \frac{I}{m_p L^2}\right). \qquad (\text{B.184})$$

Based on (B.172), (B.183), and the proposition that $\dot{\theta}(t) = 0$, it is clear that

$$\ddot{\phi} = 0 \qquad \ddot{\theta} = 0, \tag{B.185}$$

and hence, $\theta(t)$ and $\dot{\phi}(t)$ are constant. From (B.172), (B.175), (B.176), and (B.185), the fact that $F(t)$ remains constant, and the proposition that $\dot{\phi}(t) \neq 0$, the following expression can be obtained

$$F_x = -m_p L \dot{\phi}^2 \sin\theta \sin\phi \tag{B.186}$$

$$F_y = -m_p L \dot{\phi}^2 \sin\theta \cos\phi. \tag{B.187}$$

To continue the analysis, the cases of $\sin\theta = 0$ and $\sin\theta \neq 0$ are considered. Under the additional proposition that $\sin\theta \neq 0$, (B.186) and (B.187) can be used to prove that $\sin\phi$, $\cos\phi$, and $\phi(t)$ must be constant since $F_x(t)$ and $F_y(t)$ are constant. However, the conclusion that $\phi(t)$ is constant contradicts the proposition that $\dot{\phi}(t) \neq 0$. Under the additional proposition that $\sin\theta = 0$, (B.186) and (B.187) can be used to prove that

$$F_x = F_y = \sin\theta = 0. \tag{B.188}$$

Given (6.6), (6.19), (6.25), (B.171), and (B.188), it is clear that

$$e(t) = \theta(t) = 0. \tag{B.189}$$

From (B.189), the expressions given in (6.16), (6.21), (6.23), and (6.24) can be used to prove Lemma (B.21) under the propositions that $\dot{\theta}(t) = 0$, $\dot{\phi}(t) \neq 0$, and $\sin\theta = 0$. $\square$

Case 2: $\dot{\theta}(t) \neq 0$

If either $\sin\theta = 0$ or $\cos\theta = 0$, then $\theta(t)$ would be constant. Hence, the proposition that $\sin\theta = 0$ or $\cos\theta = 0$ would lead to a contradiction with the proposition that $\dot{\theta}(t) \neq 0$. Since $\theta(t)$ is a continuous function (i.e., since $\dot{\theta}(t) \in \mathcal{L}_\infty$), it is clear that both $\sin\theta \neq 0$ and $\cos\theta \neq 0$ cannot be satisfied at the same instant in time. This fact will be used in the subsequent analysis.

To facilitate the stability analysis under the proposition that $\dot{\theta}(t) \neq 0$, each row of (6.1) can be rewritten as follows

$$P_1 \sin\phi + P_2 \cos\phi = S_1 \tag{B.190}$$

$$P_1 \cos\phi - P_2 \sin\phi = S_2 \tag{B.191}$$

$$\ddot{\theta} = -\gamma_2 \sin\theta + \gamma_3 \dot{\phi}^2 \sin\theta \cos\theta + \gamma_3 S_3 \tag{B.192}$$

$$\gamma_1 \ddot{\phi} = -2\dot{\theta}\dot{\phi}\sin\theta\cos\theta + S_4. \tag{B.193}$$

The signals $P_1(t)$, $P_2(t)$, $S_1(t)$, $S_2(t)$, $S_3(t)$, $S_4(t) \in \mathbb{R}$ introduced in (B.190–B.193) are defined as follows

$$P_1 = \sin\theta \left(\gamma_3\dot{\phi}^2\cos^2\theta - \gamma_2\cos\theta - \left(\dot{\theta}^2 + \dot{\phi}^2\right)\right) + S_3\gamma_3\cos\theta \tag{B.194}$$

$$P_2 = \ddot{\phi}\sin\theta + 2\dot{\theta}\dot{\phi}\cos\theta \tag{B.195}$$

$$S_1 = \frac{1}{m_pL}\left(F_x - (m_p + m_r + m_c)\ddot{x}\right) \tag{B.196}$$

$$S_2 = \frac{1}{m_pL}\left(F_y - (m_p + m_c)\ddot{y}\right) \tag{B.197}$$

$$S_3 = -\left[\frac{\ddot{x}\cos\theta\sin\phi + \ddot{y}\cos\theta\cos\phi}{L}\right] \tag{B.198}$$

$$S_4 = -\left[\frac{\ddot{x}\sin\theta\cos\phi - \ddot{y}\sin\theta\sin\phi}{L}\right] \tag{B.199}$$

where (B.192) has been substituted into (B.194) for $\ddot{\theta}(t)$, $\gamma_1(\theta)$ was defined in (B.184), and γ_2, γ_3 were defined in (B.180). After taking the time derivative of the expressions given in (B.190) and (B.191), the following expressions can be obtained

$$\dot{\phi}P_1\cos\phi + \dot{P}_1\sin\phi - \dot{\phi}P_2\sin\phi + \dot{P}_2\cos\phi = \dot{S}_1 \tag{B.200}$$

$$-\dot{\phi}P_1\sin\phi + \dot{P}_1\cos\phi - \dot{\phi}P_2\cos\phi - \dot{P}_2\sin\phi = \dot{S}_2. \tag{B.201}$$

In (B.200) and (B.201), the terms $\dot{P}_1(t)$, $\dot{P}_2(t)$, $\dot{S}_1(t)$, and $\dot{S}_2(t)$ can be written as follows

$$\dot{P}_1 = \dot{\theta}\left(-\gamma_2\left(1 - 2\sin^2\theta\right) + \gamma_3\dot{\phi}^2\cos^3\theta - \left(\dot{\theta}^2 + \dot{\phi}^2\right)\cos\theta\right)$$
$$-2\gamma_3\dot{\phi}\left(\gamma_1\ddot{\phi}\right)\sin\theta - 2\dot{\theta}\ddot{\theta}\sin\theta - 2\gamma_3\dot{\theta}\dot{\phi}^2\cos\theta\sin^2\theta \tag{B.202}$$
$$-S_3\gamma_3\dot{\theta}\sin\theta + \dot{S}_3\gamma_3\cos\theta$$

$$\dot{P}_2 = \dot{\theta}\ddot{\phi}\cos\theta + \phi^{(3)}\sin\theta - 2\left(\dot{\theta}^2\dot{\phi}\sin\theta - \ddot{\theta}\dot{\phi}\cos\theta - \dot{\theta}\ddot{\phi}\cos\theta\right) \tag{B.203}$$

$$\dot{S}_1 = \frac{1}{m_pL}\left(\dot{F}_x - (m_p + m_r + m_c)x^{(3)}\right) \tag{B.204}$$

$$\dot{S}_2 = \frac{1}{m_pL}\left(\dot{F}_y - (m_p + m_c)y^{(3)}\right) \tag{B.205}$$

where the expression for $\dot{S}_3(t)$ is given as follows

$$\dot{S}_3 = \frac{1}{L}\left(\dot{\theta}\ddot{x}\sin\theta\sin\phi - \dot{\phi}\ddot{x}\cos\theta\cos\phi - x^{(3)}\cos\theta\sin\phi\right)$$

(B.206)

$$+\frac{1}{L}\left(\dot{\theta}\ddot{y}\sin\theta\cos\phi + \dot{\phi}\ddot{y}\cos\theta\sin\phi - y^{(3)}\cos\theta\cos\phi\right).$$

After substituting (B.192) and (B.193) into (B.202) for $\ddot{\theta}(t)$ and $\ddot{\phi}(t)$, respectively, and then performing some algebraic manipulation, the following expression can be obtained

$$\dot{P}_1 = \dot{\theta}\left(-\gamma_2\left(1 - 4\sin^2\theta\right) + \gamma_3\dot{\phi}^2\cos^3\theta - \left(\dot{\theta}^2 + \dot{\phi}^2\right)\cos\theta\right)$$

(B.207)

$$-\gamma_3\left(\left(2\dot{\phi}S_4 + 3S_3\dot{\theta}\right)\sin\theta - \dot{S}_3\cos\theta\right).$$

After multiplying both sides of the expression given in (B.200) by $\sin\phi$, multiplying both sides of the expression given in (B.201) by $\cos\phi$, and then adding the resulting expressions, the following expression is obtained

$$\dot{P}_1 - \dot{\phi}P_2 = \dot{S}_1\sin\phi + \dot{S}_2\cos\phi.$$

(B.208)

By multiplying both sides of (B.208) by $\gamma_1(\theta)$, substituting (B.195) and (B.207) into the resulting expression for $P_2(t)$ and $\dot{P}_1(t)$, and then dividing the resulting expression by $\dot{\theta}(t)$, the following expression can be obtained

$$\gamma_1 P_3 - 2\frac{I}{m_p L^2}\dot{\phi}^2\cos\theta = \frac{S_5}{\dot{\theta}}$$

(B.209)

where (B.193) was used. In (B.209), the terms $P_3(t)$, $S_5(t) \in \mathbb{R}$ are defined as follows

$$P_3 = -\gamma_2\left(1 - 4\sin^2\theta\right) + \gamma_3\dot{\phi}^2\cos^3\theta - \left(\dot{\theta}^2 + \dot{\phi}^2\right)\cos\theta \qquad (B.210)$$

$$S_5 = \gamma_1\left(\dot{S}_1\sin\phi + \dot{S}_2\cos\phi + \gamma_3\left(\left(3S_3\dot{\theta} + 2\dot{\phi}S_4\right)\sin\theta - \dot{S}_3\cos\theta\right)\right)$$
$$+S_4\dot{\phi}\sin\theta.$$

(B.211)

The time derivative of (B.209) is given by the following expression

$$\gamma_1\dot{P}_3 + 2\dot{\theta}P_3\sin\theta\cos\theta + 2\frac{I}{m_p L^2}\left(\dot{\theta}\dot{\phi}^2\sin\theta - 2\dot{\phi}\ddot{\phi}\cos\theta\right) = \frac{\dot{S}_5\dot{\theta} - S_5\ddot{\theta}}{\dot{\theta}^2}$$

(B.212)

where the expressions for $\dot{P}_3(t)$ and $\dot{S}_5(t)$ are given as follows

$$\dot{P}_3 = \dot{\theta}P_4\sin\theta - 2\gamma_3\left(S_3\dot{\theta}\cos\theta + S_4\dot{\phi}\cos\theta\right)$$

(B.213)

$$\dot{S}_5 = 2\dot{\theta}\sin\theta\cos\theta$$

$$\cdot\left(\dot{S}_1\sin\phi + \dot{S}_2\cos\phi + \gamma_3\left(\left(3S_3\dot{\theta} + 2\dot{\phi}S_4\right)\sin\theta - \dot{S}_3\cos\theta\right)\right)$$

$$+\gamma_1\ddot{S}_1\sin\phi + \gamma_1\dot{S}_1\dot{\phi}\cos\phi + \gamma_1\ddot{S}_2\cos\phi - \gamma_1\dot{S}_2\dot{\phi}\sin\phi$$

$$+\gamma_1\gamma_3\left(\left(3\dot{S}_3\dot{\theta} + 3S_3\ddot{\theta} + 2\ddot{\phi}S_4 + 2\dot{\phi}\dot{S}_4\right)\sin\theta\right.$$

$$\left.+ \left(3S_3\dot{\theta} + 2\dot{\phi}S_4\right)\dot{\theta}\cos\theta\right)$$

$$-\gamma_1\gamma_3\ddot{S}_3\cos\theta + \gamma_1\gamma_3\dot{S}_3\dot{\theta}\sin\theta + \dot{S}_4\dot{\phi}\sin\theta$$

$$+S_4\ddot{\phi}\sin\theta + S_4\dot{\phi}\dot{\theta}\cos\theta$$

$$\text{(B.214)}$$

where (B.192) and (B.193) were used. In (B.213), the term $P_4(t) \in \mathbb{R}$ is defined as follows

$$P_4 = 10\gamma_2\cos\theta - \gamma_3\dot{\phi}^2\cos^2\theta + \left(\dot{\theta}^2 + \dot{\phi}^2\right), \qquad \text{(B.215)}$$

and in (B.213) and (B.214), the expressions for $\ddot{S}_1(t)$, $\ddot{S}_2(t)$, $\ddot{S}_3(t)$, and $\dot{S}_4(t)$ can be determined as

$$\ddot{S}_1 = \frac{1}{m_pL}\left(\ddot{F}_x - (m_p + m_r + m_c)\,x^{(4)}\right) \qquad \text{(B.216)}$$

$$\ddot{S}_2 = \frac{1}{m_pL}\left(\ddot{F}_y - (m_p + m_c)\,y^{(4)}\right) \qquad \text{(B.217)}$$

$$\ddot{S}_3 = \frac{1}{L}\left(\dot{\theta}\ddot{x} + 2\dot{\theta}x^{(3)} - 2\dot{\theta}\dot{\phi}\ddot{y}\right)\sin\theta\sin\phi$$

$$+\frac{1}{L}\left(2\dot{\theta}\dot{\phi}\ddot{x} + \dot{\theta}\ddot{y} + 2\dot{\theta}y^{(3)}\right)\sin\theta\cos\phi$$

$$+\frac{1}{L}\left(\dot{\phi}^2\ddot{x} + \dot{\theta}^2\ddot{x} + \dot{\phi}\ddot{y} + 2\dot{\phi}y^{(3)} - x^{(4)}\right)\cos\theta\sin\phi$$

$$\text{(B.218)}$$

$$+\frac{1}{L}\left(-\dot{\phi}\ddot{x} - 2\dot{\phi}x^{(3)} + \dot{\theta}^2\ddot{y} + \dot{\phi}^2\ddot{y} - y^{(4)}\right)\cos\theta\cos\phi$$

$$\dot{S}_4 = -\frac{1}{L}\left(\dot{\theta}\ddot{x}\cos\theta\cos\phi - \dot{\phi}\ddot{x}\sin\theta\sin\phi + x^{(3)}\sin\theta\cos\phi\right)$$

$$+\frac{1}{L}\left(\dot{\theta}\ddot{y}\cos\theta\sin\phi + \ddot{y}\dot{\phi}\sin\theta\cos\phi + y^{(3)}\sin\theta\sin\phi\right). \qquad \text{(B.219)}$$

By substituting (B.213) into (B.212) for $\dot{P}_3(t)$ and then multiplying the resulting expression by $\gamma_1(\theta)$, the following expression can be obtained

$$\dot{\theta}\sin\theta \left(\gamma_1^2 P_4 + 2\gamma_1 P_3 \cos\theta + \frac{2I}{m_p L^2}\dot{\phi}^2 \left(\gamma_1 + 4\cos^2\theta\right)\right) = S_6 \qquad \text{(B.220)}$$

where (B.193) was used. In (B.220), $S_6(t) \in \mathbb{R}$ is defined as

$$S_6 = \gamma_1 \left(\frac{\dot{S}_5\dot{\theta} - S_5\ddot{\theta}}{\dot{\theta}^2} + \frac{4I\dot{\phi}\cos\theta}{m_p L^2}S_4\right)$$
$$+\gamma_1^2 \left(2\gamma_3 \left(S_3\dot{\theta}\cos\theta + S_4\dot{\phi}\cos\theta\right)\right). \qquad \text{(B.221)}$$

After dividing (B.220) by $\dot{\theta}(t)\sin\theta$ and then substituting (B.209) into the resulting expression for $\gamma_1(\theta)P_3(t)$, the following expression is obtained

$$\gamma_1^2 P_4 + \frac{2I}{m_p L^2}\dot{\phi}^2 \left(\gamma_1 + 6\cos^2\theta\right) = \frac{S_6}{\dot{\theta}\sin\theta} - \frac{2S_5\cos\theta}{\dot{\theta}}. \qquad \text{(B.222)}$$

After multiplying (B.209) by $\gamma_1(\theta)$, multiplying (B.222) by $\cos\theta$, and then adding the resulting products, the following expression is obtained

$$\gamma_1^2 \left(P_3 + P_4\cos\theta\right) + \frac{12I}{m_p L^2}\dot{\phi}^2 \cos^3\theta = S_7. \qquad \text{(B.223)}$$

In (B.223), $S_7(t) \in \mathbb{R}$ is defined as

$$S_7 = \left(\frac{S_6}{\dot{\theta}\sin\theta} - \frac{2S_5\cos\theta}{\dot{\theta}}\right)\cos\theta + \frac{S_5}{\dot{\theta}}\gamma_1. \qquad \text{(B.224)}$$

The expression for (B.223) can be rewritten by using (B.210) and (B.215) as follows

$$\gamma_2\gamma_1^2 \left(9 - 6\sin^2\theta\right) + \frac{12I}{m_p L^2}\dot{\phi}^2 \cos^3\theta = S_7. \qquad \text{(B.225)}$$

To continue the analysis, the time derivative of (B.225) is determined as follows

$$4\gamma_2\dot{\theta}\gamma_1 \left(\left(9 - 6\sin^2\theta\right) - 3\gamma_1\right)\sin\theta\cos\theta$$
$$+\frac{12I}{m_p L^2}\dot{\phi}\left(-3\dot{\theta}\dot{\phi}\sin\theta + 2\ddot{\phi}\cos\theta\right)\cos^2\theta = \dot{S}_7 \qquad \text{(B.226)}$$

where $\dot{S}_7(t)$ is given by the following expression

$$\dot{S}_7 = -\left(\frac{\dot{S}_6}{\sin\theta} - 2S_5\cos\theta\right)\sin\theta$$

$$+\left(\frac{\dot{S}_6}{\dot{\theta}\sin\theta} - \frac{S_6\cos\theta}{\sin^2\theta} - \frac{S_6\ddot{\theta}}{\dot{\theta}^2\sin\theta}\right)\cos\theta$$

$$-\left(\frac{2\dot{S}_5\cos\theta}{\dot{\theta}} - 2S_5\sin\theta - \frac{2S_5\ddot{\theta}\cos\theta}{\dot{\theta}^2}\right)\cos\theta$$

$$+\frac{\dot{S}_5}{\dot{\theta}}\gamma_1 - \frac{S_5\ddot{\theta}}{\dot{\theta}^2}\gamma_1 + 2S_5\sin\theta\cos\theta.$$

(B.227)

In (B.227), the expression for $\dot{S}_6(t)$ can be determined as follows

$$\dot{S}_6 = 2\dot{\theta}\sin\theta\cos\theta\left(\frac{\dot{S}_5\dot{\theta} - S_5\ddot{\theta}}{\dot{\theta}^2} + \frac{4I\dot{\phi}\cos\theta}{m_pL^2}S_4\right)$$

$$+\gamma_1\left(\frac{\ddot{S}_5\dot{\theta} - S_5\theta^{(3)}}{\dot{\theta}^2} - 2\ddot{\theta}\left(\frac{\dot{S}_5\dot{\theta} - S_5\ddot{\theta}}{\dot{\theta}^3}\right)\right)$$

$$+\gamma_1\left(-\frac{4I\dot{\theta}\dot{\phi}\sin\theta}{m_pL^2}S_4 + \frac{4I\ddot{\phi}\cos\theta}{m_pL^2}S_4 + \frac{4I\dot{\phi}\cos\theta}{m_pL^2}\dot{S}_4\right)$$

(B.228)

$$+4\gamma_1\dot{\theta}\left(2\gamma_3\left(S_3\dot{\theta}\cos\theta + S_4\dot{\phi}\cos\theta\right)\right)\sin\theta\cos\theta$$

$$+2\gamma_1^2\gamma_3\left(\dot{S}_3\dot{\theta}\cos\theta - S_3\dot{\theta}^2\sin\theta + S_3\ddot{\theta}\cos\theta\right)$$

$$+2\gamma_1^2\gamma_3\left(\dot{S}_4\dot{\phi}\cos\theta - S_4\dot{\phi}\dot{\theta}\sin\theta + S_4\ddot{\phi}\cos\theta\right)$$

where $\ddot{S}_5(t)$ can be determined as

$$
\ddot{S}_5 = \left[2 \left(\cos^2 \theta - \sin^2 \theta \right) \dot{\theta}^2 + 2 \ddot{\theta} \sin \theta \cos \theta \right]
$$

$$
\left(\left(\dot{S}_1 \sin \phi + \dot{S}_2 \cos \phi \right) + \gamma_3 \left(\left(3 S_3 \dot{\theta} + 2 \dot{\phi} S_4 \right) \sin \theta - \dot{S}_3 \cos \theta \right) \right)
$$

$$
+ \gamma_1 \left(S_1^{(3)} \sin \phi + 2 \ddot{S}_1 \dot{\phi} \cos \phi + \dot{S}_1 \left(\ddot{\phi} \cos \phi - \dot{\phi}^2 \sin \phi \right) \right)
$$

$$
+ \gamma_1 \left(S_2^{(3)} \cos \phi - 2 \ddot{S}_2 \dot{\phi} \sin \phi + \dot{S}_2 \left(-\ddot{\phi} \sin \phi - \dot{\phi}^2 \cos \phi \right) \right)
$$

$$
+ 3 \gamma_1 \gamma_3 \left(\ddot{S}_3 \dot{\theta} \sin \theta + 2 \dot{S}_3 \left(\dot{\theta}^2 \cos \theta + \ddot{\theta} \sin \theta \right) \right.
$$

$$
+ S_3 \left(-\dot{\theta}^3 \sin \theta + 3 \dot{\theta} \ddot{\theta} \cos \theta + \theta^{(3)} \sin \theta \right) \Big)
$$

$$
+ 2 \gamma_1 \gamma_3 \left(\ddot{S}_4 \dot{\phi} \sin \theta + 2 \dot{S}_4 \left(\dot{\theta} \dot{\phi} \cos \theta + \ddot{\phi} \sin \theta \right) \right)
$$

$$
+ 2 \gamma_1 \gamma_3 \left(S_4 \left(-\dot{\theta}^2 \dot{\phi} \sin \theta + \ddot{\theta} \dot{\phi} \cos \theta + 2 \dot{\theta} \ddot{\phi} \cos \theta + \phi^{(3)} \sin \theta \right) \right)
$$

$$
- \gamma_1 \gamma_3 \left(S_3^{(3)} \cos \theta - 2 \ddot{S}_3 \dot{\theta} \sin \theta + \dot{S}_3 \left(-\dot{\theta}^2 \cos \theta - \ddot{\theta} \sin \theta \right) \right)
$$

$$
+ 4 \dot{\theta} \sin \theta \cos \theta \left(\ddot{S}_1 \sin \phi + \dot{S}_1 \dot{\phi} \cos \phi + \ddot{S}_2 \cos \phi - \dot{S}_2 \dot{\phi} \sin \phi \right)
$$

$$
+ 4 \gamma_3 \dot{\theta} \sin \theta \cos \theta \left(\left(3 \dot{S}_3 \dot{\theta} + 3 S_3 \ddot{\theta} + 2 \dot{\phi} S_4 + 2 \dot{\phi} \dot{S}_4 \right) \sin \theta \right.
$$

$$
+ \left(3 S_3 \dot{\theta} + 2 \dot{\phi} S_4 \right) \dot{\theta} \cos \theta \Big)
$$

$$
+ 4 \gamma_3 \dot{\theta} \sin \theta \cos \theta \left(-\ddot{S}_3 \cos \theta + \dot{S}_3 \dot{\theta} \sin \theta \right)
$$

$$
+ \ddot{S}_4 \dot{\phi} \sin \theta + 2 \dot{S}_4 \left(\ddot{\phi} \sin \theta + \dot{\theta} \dot{\phi} \cos \theta \right)
$$

$$
+ S_4 \left(\phi^{(3)} \sin \theta + 2 \dot{\theta} \ddot{\phi} \cos \theta - \dot{\theta}^2 \dot{\phi} \sin \theta + \ddot{\theta} \dot{\phi} \cos \theta \right).
$$

$$(B.229)$$

In (B.229), the expressions for $S_1^{(3)}(t)$, $S_2^{(3)}(t)$, $S_3^{(3)}(t)$, and $\ddot{S}_4(t)$ can be developed as follows

$$
S_1^{(3)} = \frac{1}{m_p L} \left(F_x^{(3)} - (m_p + m_r + m_c) x^{(5)} \right) \tag{B.230}
$$

$$S_2^{(3)} = \frac{1}{m_p L} \left(F_y^{(3)} - (m_p + m_c) y^{(5)} \right) \tag{B.231}$$

$$
\begin{aligned}
S_3^{(3)} = \quad & \frac{1}{L} \left(\dot{\theta} \cos\theta \sin\phi + \dot{\phi} \sin\theta \cos\phi \right) \left(\ddot{\theta} \ddot{x} + 2\dot{\theta} x^{(3)} - 2\dot{\theta}\dot{\phi}\ddot{y} \right) \\
& + \frac{1}{L} \left(\theta^{(3)} \ddot{x} + \ddot{\theta} x^{(3)} \right. \\
& + 2 \left(\dot{\theta} x^{(4)} + \ddot{\theta} x^{(3)} - \ddot{\theta}\dot{\phi}\ddot{y} - \dot{\theta}\ddot{\phi}\ddot{y} - \dot{\theta}\dot{\phi} y^{(3)} \right) \Big) \sin\theta \sin\phi \\
& + \frac{1}{L} \left(\dot{\theta} \cos\theta \cos\phi - \dot{\phi} \sin\theta \sin\phi \right) \left(2\dot{\theta}\dot{\phi}\ddot{x} + \ddot{\theta}\ddot{y} + 2\dot{\theta} y^{(3)} \right) \\
& + \frac{1}{L} \left(2\ddot{\theta}\dot{\phi}\ddot{x} + 2\dot{\theta}\ddot{\phi}\ddot{x} + 2\dot{\theta}\dot{\phi} x^{(3)} + \theta^{(3)} \ddot{y} \right. \\
& + 3\ddot{\theta} y^{(3)} + 2\dot{\theta} y^{(4)} \Big) \sin\theta \cos\phi \\
& + \frac{1}{L} \left(-\dot{\theta} \sin\theta \sin\phi + \dot{\phi} \cos\theta \cos\phi \right) \\
& \cdot \left(\dot{\phi}^2 \ddot{x} + \dot{\theta}^2 \ddot{x} + \ddot{\phi}\ddot{y} + 2\dot{\phi} y^{(3)} - x^{(4)} \right) \\
& + \frac{1}{L} \left(2\dot{\phi}\ddot{\phi}\ddot{x} + 2\dot{\theta}\ddot{\theta}\ddot{x} + \dot{\phi}^2 x^{(3)} + \dot{\theta}^2 x^{(3)} + \phi^{(3)}\ddot{y} \right. \\
& + 3\ddot{\phi} y^{(3)} + 2\dot{\phi} y^{(4)} - x^{(5)} \Big) \cos\theta \sin\phi \\
& - \frac{1}{L} \left(\dot{\theta} \sin\theta \cos\phi + \dot{\phi} \cos\theta \sin\phi \right) \\
& \cdot \left(-\ddot{\phi}\ddot{x} - 2\dot{\phi} x^{(3)} + \dot{\theta}^2 \ddot{y} + \dot{\phi}^2 \ddot{y} - y^{(4)} \right) \\
& + \frac{1}{L} \left(-\phi^{(3)} \ddot{x} - 3\ddot{\phi} x^{(3)} - 2\dot{\phi} x^{(4)} + 2\dot{\theta}\ddot{\theta}\ddot{y} + 2\dot{\phi}\ddot{\phi}\ddot{y} \right. \\
& + \dot{\theta}^2 y^{(3)} + \dot{\phi}^2 y^{(3)} - y^{(5)} \Big) \cos\theta \cos\phi
\end{aligned}
$$

$$\tag{B.232}$$

$$\ddot{S}_4 = \frac{1}{L}\left(\dot{\phi}\ddot{x} + 2\dot{\phi}x^{(3)} - \dot{\theta}^2\dot{y} - \dot{\phi}^2\ddot{y} + y^{(4)}\right)\sin\theta\sin\phi$$

$$+\frac{1}{L}\left(2\dot{\theta}\dot{\phi}\ddot{x} + \ddot{\theta}\ddot{y} + 2\dot{\theta}y^{(3)}\right)\cos\theta\sin\phi$$

$$+\frac{1}{L}\left(\dot{\theta}^2\ddot{x} + \dot{\phi}^2\ddot{x} + \dot{\phi}\ddot{y} + 2\dot{\phi}y^{(3)} - x^{(4)}\right)\sin\theta\cos\phi$$ (B.233)

$$-\frac{1}{L}\left(\ddot{\theta}\ddot{x} + 2\dot{\theta}x^{(3)} - 2\dot{\theta}\dot{\phi}\ddot{y}\right)\cos\theta\cos\phi.$$

After multiplying (B.226) by $\gamma_1(\theta)$, the following expression can be obtained

$$4\gamma_2\dot{\theta}\gamma_1^2\left((9 - 6\sin^2\theta) - 3\gamma_1\right)\sin\theta\cos\theta$$

$$-\frac{36I}{m_pL^2}\gamma_1\dot{\theta}\dot{\phi}^2\cos^2\theta\sin\theta - \frac{48I}{m_pL^2}\dot{\theta}\dot{\phi}^2\sin\theta\cos^4\theta$$ (B.234)

$$= \gamma_1\dot{S}_7 - \frac{24I}{m_pL^2}\dot{\phi}S_4\cos^3\theta$$

where (B.193) was utilized. After dividing (B.234) by $\dot{\theta}(t)\sin\theta\cos\theta$ and using (B.225), the following expression can be obtained

$$-\frac{12I\dot{\phi}^2\cos\theta}{m_pL^2}\left(8\cos^2\theta + 3\gamma_1\right) - 12\gamma_2\gamma_1^3 = S_8 - 4S_7$$ (B.235)

where $S_8(t) \in \mathbb{R}$ is defined as

$$S_8 = \left(\frac{1}{\dot{\theta}\sin\theta\cos\theta}\right)\left(\dot{S}_7\gamma_1 - \frac{24I}{m_pL^2}\dot{\phi}S_4\cos^3\theta\right).$$ (B.236)

By multiplying (B.235) by $\cos^2\theta$ and then utilizing (B.225), (B.235) can be rewritten as

$$(9 - 6\sin^2\theta)\left(8\cos^2\theta + 3\gamma_1\right) - 12\gamma_1\cos^2\theta = \left(\frac{S_9}{\gamma_2\gamma_1^2}\right)$$ (B.237)

where $S_9(t) \in \mathbb{R}$ is defined as

$$S_9 = (S_8 - 4S_7)\cos^2\theta + S_7\left(8\cos^2\theta + 3\gamma_1\right).$$ (B.238)

Given the facts that $F_x(t)$ and $F_y(t)$ are constant and $\ddot{x} = 0$ and $\ddot{y} = 0$, it is clear that

$$F_x^{(k)} = 0 \qquad F_y^{(k)} = 0 \qquad k \geq 1$$ (B.239)

$$x^{(k)} = 0 \qquad y^{(k)} = 0 \qquad k \geq 2. \tag{B.240}$$

Hence, the expressions in (B.196–B.199), (B.204–B.206), (B.211), (B.214), (B.216–B.219), (B.221), (B.224), (B.227–B.233), (B.236), and (B.238) can be used to prove that

$$\begin{aligned} S_i &= 0, \quad i = 3, 4, ...9; \\ \dot{S}_i &= 0, \quad i = 1, 2, ...9. \end{aligned} \tag{B.241}$$

After some algebraic manipulation, (B.241) can be used to rewrite (B.237) as follows

$$\begin{aligned} \left(1 + 2\cos^2\theta\right)\left(3\left(1 + \tfrac{I}{m_pL^2}\right) + 5\cos^2\theta\right) \\ -4\cos^2\theta\left(1 + \tfrac{I}{m_pL^2} - \cos^2\theta\right) = 0. \end{aligned} \tag{B.242}$$

After some further algebraic manipulation, the expression given in (B.242) can be rewritten as follows

$$\cos^4\theta + \alpha_1\cos^2\theta + \alpha_2 = 0 \tag{B.243}$$

where $\alpha_1, \alpha_2 \in \mathbb{R}$ are positive constants defined as

$$\alpha_1 = \frac{1}{2} + \frac{1}{7}\frac{I}{m_pL^2} \qquad \alpha_2 = \frac{3}{14}\left(1 + \frac{I}{m_pL^2}\right). \tag{B.244}$$

Since the expression given in (B.243) is clearly invalid, the proposition that $\dot{\theta}(t) \neq 0$ must be invalid, and hence, $\dot{\theta}(t) = 0$. The analysis given in the previous cases can now be used to prove Lemma B.21. $\square$

B.5.3   Coupling Control Law Analysis

Lemma B.22 *Given the E^2 coupling control law of (6.33), the time derivative of the following function*

$$V = \frac{1}{2}k_E E^2 + \frac{1}{2}k_p e^T e + \frac{1}{2}k_v \dot{r}^T \dot{r} \tag{B.245}$$

is given by the following expression

$$\dot{V} = 0 \tag{B.246}$$

only when

$$E(q, \dot{q}) = e(t) = \dot{r}(t) = 0 \tag{B.247}$$

where $E(q, \dot{q})$, $e(t)$, and $r(t)$ are defined in (6.21), (6.24), and (6.16), respectively.

Proof: To prove Lemma B.22, the set of all points where (B.246) is satisfied is defined as Γ. In the set Γ, it is clear from (6.38) and (B.246) that

$$\dot{r}(t) = 0 \qquad \ddot{r}(t) = 0 \tag{B.248}$$

and hence, (6.16) and (B.246) can be used to prove that $x(t)$, $y(t)$, and $V(t)$ are constant, and that

$$\ddot{x}(t) = 0 \qquad \ddot{y}(t) = 0. \tag{B.249}$$

Furthermore, from (6.22) and (B.248), it is clear that

$$\dot{E}(q, \dot{q}) = 0 \qquad \dot{e}(t) = 0 \tag{B.250}$$

and hence, $E(q, \dot{q})$ and $e(t)$ are constant.

Similar to the proof of Lemma B.21, the remainder of the analysis can be divided into two cases. For the case of $\dot{\theta} = 0$ and $\dot{\phi} = 0$, Case 1a in the proof of Lemma B.21 can be used to prove Lemma B.22. For the other cases, (6.17) can be used to rewrite (6.33) in the following equivalent form[2]

$$F = \frac{-k_d \dot{r} - k_p e - k_v \ddot{r}}{k_E E}. \tag{B.251}$$

Based on the structure of (B.251), it is clear from (B.248), (B.250), and (B.251) that $F(t)$ is constant, and similar arguments as in the proof of Lemma B.21 can be used to prove Lemma B.22. $\square$

Lemma B.23 *Given the gantry kinetic energy coupling control law (6.41), the time derivative of the following function*

$$V = k_E E + \frac{1}{2} k_v \dot{r}^T \left(\det(M) P^{-1} \right) \dot{r} + \frac{1}{2} k_p e^T e \tag{B.252}$$

is given by the following expression

$$\dot{V} = 0 \tag{B.253}$$

only when

$$E(q, \dot{q}) = e(t) = \dot{r}(t) = 0 \tag{B.254}$$

where $E(q, \dot{q})$, $e(t)$, and $r(t)$ are defined in (6.21), (6.24), and (6.16), respectively.

---

[2] Since either $\dot{\theta}(t) \neq 0$ or $\dot{\varphi}(t) \neq 0$, (6.21) can be used to show that $E(q, \dot{q}) > 0$, and hence, the denominator of (B.251) does not go to zero for the remaining cases.

Proof: To prove Lemma B.23, the set of all points where (B.253) is satisfied is defined as Γ. In the set Γ, it is clear from (6.46) and (B.253) that

$$\dot{r}(t) = 0 \qquad \ddot{r}(t) = 0 \tag{B.255}$$

and hence, (6.16) can be used to prove that $x(t), y(t),$ and $V(t)$ are constant, and that

$$\ddot{x}(t) = 0 \qquad \ddot{y}(t) = 0. \tag{B.256}$$

Furthermore, from (6.22) and (B.255), it is clear that

$$\dot{E}(q, \dot{q}) = 0 \qquad \dot{e}(t) = 0 \tag{B.257}$$

and hence, $E(q, \dot{q})$ and $e(t)$ are constant.

To complete the remaining analysis, (6.17) can be used to rewrite (6.41) in the following equivalent form

$$F = \frac{-k_d \dot{r} - k_p e - k_v \left(\det(M) P^{-1} \right) \ddot{r} - \frac{1}{2} k_v \left(\frac{d}{dt} \left(\det(M) P^{-1} \right) \right) \dot{r}}{k_E}.$$
$$\tag{B.258}$$

Based on the structure of (B.258), it is clear from (B.255) and (B.257) that $F(t)$ is constant, and similar arguments as given in the proof of Lemma B.21 can be used to prove Lemma B.23. $\square$

B.5.4   Matrix Property

Lemma B.24 *Given the definition for the matrix $P(q)$ given in (6.18), the following property holds*

$$\frac{d}{dt} P^{-1} = -P^{-1} \left(\frac{d}{dt} P \right) P^{-1}. \tag{B.259}$$

Proof: To prove Lemma B.24, the time derivative of $P(q)P^{-1}(q)$ is determined as follows

$$\frac{d}{dt} \left(PP^{-1} \right) = \left(\frac{d}{dt} P \right) P^{-1} + P \left(\frac{d}{dt} P^{-1} \right). \tag{B.260}$$

After noting that the left-side of (B.260) is equal to zero, the expression given in (B.259) can be directly obtained by premultiplying (B.260) by $P^{-1}(q)$. $\square$

References

[1] B. Etkin, *Dynamics of Flight-Stability and Control*, New York, NY: John Wiley, 1959.

[2] F. Lewis, C. Abdallah, and D. M. Dawson, *Control of Robot Manipulators*, New York, NY: MacMillan Publishing Company, 1993.

[3] J. -J. E. Slotine and L. Wi, *Applied Nonlinear Control*, Englewood Cliffs, NJ: Prentice-Hall, 1991.

[4] M. Vidyasagar, *Nonlinear Systems Analysis*, Englewood Cliffs, NJ: Prentice-Hall, 1978.

Index

Control Engineering

Series Editor

William S. Levine
Department of Electrical and Computer Engineering
University of Maryland
College Park, MD 20742-3285
USA

Aims and Scope

Control engineering is an increasingly diverse subject, whose technologies range from
simple mechanical devices to complex electro-mechanical systems. Applications are seen
in everything from biological control systems to the tracking controllers of CD players.
Some methods, H-infinity design for example, for the analysis and design of control
systems are based on sophisticated mathematics while others, such as PID control, are
understood and implemented through experimentation and empirical analysis.

The Birkhäuser series *Systems and Control: Foundations and Applications* examines
the abstract and theoretical mathematical aspects of control. *Control Engineering*
complements this effort through a study of the industrial and applied implementation of
control — from techniques for analysis and design to hardware implementation, test, and
evaluation. While recognizing the harmony between abstract theory and physical application, these publications emphasize real-world results and concerns. Problems and examples use the least amount of abstraction required, remaining committed to issues of
consequence, such as cost, tradeoffs, reliability, and power consumption.

The series includes professional expository monographs, advanced textbooks, handbooks,
and thematic compilations of applications/case studies.

Readership

The publications will appeal to a broad interdisciplinary readership of engineers at the
graduate and professional levels. Applied theorists and practitioners in industry and academia
will find the publications accessible across the varied terrain of control engineering research.

Preparation of manuscripts

We encourage the preparation of manuscripts in LATEX for delivery as camera-ready hard
copy, which leads to rapid publication, or on a diskette.

Proposals should be sent directly to the editor or to: Birkhäuser Boston,
675 Massachusetts Avenue, Cambridge, MA 02139, U.S.A.

Published Books

Lyapunov-Based Control of Mechanical Systems
M.S. de Queiroz, D.M. Dawson, S.P. Nagarkatti, and F. Zhang

Nonlinear Control and Analytic Mechanics
H.G. Kwatny and G.L. Blankenship

Qualitative Theory of Hybrid Dynamical Systems
A.S. Matveev and A.V. Savkin

Robust Kalman Filtering for Signals and Systems with
Large Uncertainties
I.R. Peterson and A.V. Savkin

Control Systems Theory with Engineering Applications
S.E. Lyshevski

Control Systems with Actuator Saturation:
Analysis and Design
T. Hu and Z. Lin

Deterministic and Stochastic Time-Delay Systems
E.K. Boukas and Z.K. Liu

Hybrid Dynamical Systems
A.V. Savkin and R.J. Evans

Stability and Control of Dynamical Systems with Applications:
A Tribute to Anthony N. Michel
D. Liu and P.J. Antsaklis, editors

Stability of Time-Delay Systems
K. Gu, V.L. Kharitonov, and J. Chen

Nonlinear Control of Engineering Systems:
A Lyapunov-Based Approach
W.E. Dixon, A. Behal, D.M. Dawson, and S.P. Nagarkatti

Forthcoming books in the Control Engineering series

PID Controllers for Time-Delay Systems
S.P. Bhattacharyya, A. Datta, and G.J. Silva

Verification and Synthesis of Hybrid Systems
E. Asarin, T. Dang, and O. Maler

Qualitative Nonlinear Dynamics of Communication Networks
V. Kulkarni